BRAIN DEATH
AND DISORDERS
OF CONSCIOUSNESS

ADVANCES IN EXPERIMENTAL MEDICINE AND BIOLOGY

Recent Volumes in this Series

Volume 540
OXYGEN TRANSPORT TO TISSUE, VOLUME XXV
Edited by Maureen Thorniley, David K. Harrison, and Philip E. James

Volume 541
FRONTIERS IN CLINICAL NEUROSCIENCE: Neurodegeneration and Neuroprotection
Edited by László Vécsei

Volume 542
QUALITY OF FRESH AND PROCESSED FOODS
Edited by Fereidoon Shahidi, Arthur M. Spanier, Chi-Tang Ho, and Terry Braggins

Volume 543
HYPOXIA: Through the Lifecycle
Edited by Robert C. Roach, Peter D. Wagner, and Peter H. Hackett

Volume 544
PEROXISOMAL DISORDERS AND REGULATION OF GENES
Edited by Frank Roels, Myriam Baes, and Sylvia De Bie

Volume 545
HYPOSPADIAS AND GENITAL DEVELOPMENT
Edited by Laurence S. Baskin

Volume 546
COMPLEMENTARY AND ALTERNATIVE APPROACHES TO BIOMEDICINE
Edited by Edwin L. Cooper and Nobuo Yamaguchi

Volume 547
ADVANCES IN SYSTEMS BIOLOGY
Edited by Lee K. Opresko, Julie M. Gephart, and Michaela B. Mann

Volume 548
RECENT ADVANCES IN EPILEPSY RESEARCH
Edited by Devin K. Binder and Helen E. Scharfman

Volume 549
HOT TOPICS IN INFECTION AND IMMUNITY IN CHILDREN
Edited by Andrew J. Pollard, George H. McCracken, Jr., and Adam Finn

Volume 550
BRAIN DEATH AND DISORDERS OF CONSCIOUSNESS
Edited by Calixto Machado and D. Alan Shewmon

BRAIN DEATH AND DISORDERS OF CONSCIOUSNESS

Edited by

Calixto Machado

Institute of Neurology and Neurosurgery
Havana, Cuba

and

D. Alan Shewmon

David Geffen School of Medicine at UCLA
Los Angeles, California

Kluwer Academic / Plenum Publishers
New York, Boston, Dordrecht, London, Moscow

Proceedings of the IV International Symposium on Coma and Death, held March 9–12, 2004, in Havana, Cuba

ISSN 0065-2598

ISBN 0-306-48482-X

233 Spring Street, New York, New York 10013

http://www.wkap.nl/

10 9 8 7 6 5 4 3 2 1

A C.I.P. record for this book is available from the Library of Congress

Printed in the United States of America

PREFACE

Calixto Machado and D. Alan Shewmon*

Since ancient times, Christians and Hebrews considered breath or *ruach* as the hallmark of life. Greeks believed that heartbeat delimited the frontier between life and death. Therefore, over the centuries people were deemed dead when they stopped breathing and their hearts stopped beating. The great pandemic disasters and the need for early burial prompted a worldwide fear of being buried alive. Winslow, in 1740, emphasized that putrefaction was the only sure sign of death. Nevertheless, the invention of the stethoscope in the middle of the 19th century enabled physicians to detect weak heartbeat and restricted respirations, raising their competence. This partially alleviated public fears of premature burial.

The remarkable technological advances of the 20th century radically changed the course of medicine, especially intensive care. The development of effective mechanical ventilators and cardiopulmonary resuscitation ("reanimation") compelled physicians in the late 1950s to grapple with a condition impossible even to imagine previously: one in which the brain is massively damaged and nonfunctional while other organs remain functioning. Is such a patient alive or dead?

This state was documented by French neurologists and neurophysiologists at the end of the 1950s. The end of the 1960s witnessed the next major milestone, with the Report of the Ad Hoc Committee of the Harvard Medical School. These years marked a turning point when brain-oriented definitions of death started to be formulated, and brain death (BD) was gradually accepted as death of the individual.

Nonetheless, there are still worldwide controversies over the very concept of human death and the putative neurological grounds for diagnosing it (whole brain, brain stem, and higher brain formulations of death). There are also disagreements over the diagnostic criteria of BD, whether clinical alone, or clinical plus ancillary tests. Moreover, a group of scholars who were strong defenders of a brain-based standard of death are now favoring a circulatory-respiratory view. Hence, the debates on human death are far from concluded.

* Calixto Machado, MD, PhD, President of the National Commission for the Determination and Certification of Death. Institute of Neurology and Neurosurgery, Havana, Cuba. Email: braind@infomed.sld.cu. D. Alan Shewmon, MD, Professor of Neurology and Pediatrics, David Geffen School of Medicine at UCLA, Los Angeles, CA 90095. Email: ashewmon@mednet.ucla.edu.

The sunny city of Havana has become the capital for discussions on human death for more than a decade. The First International Symposium on Brain Death was held in 1992, and the clear consensus among delegates was that the subject of human death was far from over. Delegates from all continents reconvened for the Second International Symposium on Brain Death in 1996 and the Third International Symposium on Coma and Death in 2000. These were remarkable conferences, attended by the most outstanding personalities in the field from multiple disciplines.

At the beginning of the new millennium, Havana will once again welcome colleagues from around the globe to hold the Fourth International Symposium on Coma and Death (March 9-12, 2004). This book is a selection of lectures from authors who are prominent scholars in this area. We hope it will serve as a valuable source of knowledge for present and future generations. The motivation to ever deepen our knowledge about human death (and life) is a matter of *human dignity*.

CONTENTS

PHILOSOPHICAL, ETHICAL, AND LEGAL ISSUES

Brain Death: Updating a Valid Concept for 2004 1
Julius Korein and Calixto Machado

The Concept of Brain Death 15
Jean-Michel Guérit

The "Critical Organ" for the Organism as a Whole: Lessons from the Lowly Spinal Cord 23
D. Alan Shewmon

"Brain Death" Is Not Death 43
Paul A. Byrne and Walt F. Weaver

The Conceptual Basis for Brain Death Revisited: Loss of Organic Integration or Loss of Consciousness? 51
John P. Lizza

Consciousness, Mind, Brain, and Death 61
Josef Seifert

The Death of Death 79
James J. Hughes

The Semiotics of Death and Its Medical Implications 89
D. Alan Shewmon and Elisabeth Seitz Shewmon

About the Continuity of Our Consciousness 115
Pim Van Lommel

Brain Death and Organ Transplantation: Concepts and Principles in Judaism 133
Z. Harry Rappaport and Isabelle T. Rappaport

Cuba Has Passed a Law for the Determination and Certification of Death 139
Calixto Machado and the National Commission for the Determination and Certification of Death

DIAGNOSIS, PATHOPHYSIOLOGY AND TREATMENT

Neuroprotection Becomes Reality: Changing Times for Cerebral Resuscitation 143
Maxwell S. Damian

Controlled Hypertension for Refractory High Intracranial Pressure 151
Philippe Hantson

On Irreversibility as a Prerequisite for Brain Death Determination 161
James L. Bernat

How Should Testing for Apnea Be Performed in Diagnosing Brain Death? 169
Christoph J. G. Lang and Josef G. Heckmann

Evoked Potentials in the Diagnosis of Brain Death 175
Enrico Facco and Calixto Machado

Recovery from Near Death following Cerebral Anoxia: A Case Report Demonstrating Superiority of Median Somatosensory Evoked Potentials over EEG in Predicting a Favorable Outcome after Cardiopulmonary Resuscitation 189
Ted L. Rothstein

Human Brain-Dead Donors and ^{31}P MRS Studies on Feline Myocardial Energy Metabolism 197
George J. Brandon Bravo Bruinsma and Cees J.A. Van Echteld

Organ Donation after Fatal Poisoning: An Update with Recent Literature Data 207
Philippe Hantson

The ABC of PVS: Problems of Definition 215
D. Alan Shewmon

Brain Function in the Vegetative State 229
Steven Laureys, Marie-Elisabeth Faymonville, Xavier De Tiège, Philippe Peigneux, Jacques Berré, Gustave Moonen, Serge Goldman, and Pierre Maquet

Global Neurodynamics and Deep Brain Stimulation: Appreciating the Perspectives of Place and Process 239
David I. Pincus

Honoring Treatment Preferences near the End of Life: The Oregon Physician Orders for Life Sustaining Treatment (POLST) Program 255
Terri A. Schmidt, Susan E. Hickman, and Susan W. Tolle

Palliative Sedation in Terminally Ill Patients 263
Paul C. Rousseau

Index 269

BRAIN DEATH AND DISORDERS OF CONSCIOUSNESS

BRAIN DEATH

Updating a valid concept for 2004

Julius Korein and Calixto Machado*

1. INTRODUCTION

Various investigators have presented criticisms about the concept of brain death.[1-8] We propose that for medical purposes the fundamental biological neurocentric definition of death of the human being is valid based on the irreversible cessation of operation of the critical system of the brain. This concept of death itself requires a paradigm shift and modification dictated by our current understanding of living systems, new observations, and further experience in applications of the diagnosis of death. However, the criterion and tests previously used to diagnose death as essentially brain death are for practical purposes unchanged. The relationships among "life," "death," "brain death," and irreversible intrinsic cessation of function of the "critical system of the brain" during the human life cycle will be detailed. The significance of the many aspects of consciousness will be stressed, and applied to the problems of persistent vegetative states and anencephaly.

2. THE BIOLOGY, CHEMISTRY, AND PHYSICS OF LIFE AND DEATH

Our fundamental concepts of life and death must be updated. The biology, chemistry and physics of living systems can now be considered in terms of biochemical evolution of organized open systems that tend to increase their complexity and stability at the expense of their environment. Such systems are far from equilibrium, and minimize the production of entropy (disorganization).[9-17] These auto-catalytic (self-sustaining) systems can only emerge spontaneously by means of phase transitions of molecular components in an environment rich in excess energy such as that produced by solar radiation or geothermal vents. Additionally these systems must have access to further molecular

---

* Julius Korein, MD, Professor Emeritus of Neurology at New York University Medical Center, New York, USA (Mailing address - 240 Central Park South, Suite 20 D, New York, NY 10019). Email: Jkorein@aol.com. Calixto Machado, MD, Ph.D. Professor of Neurology and Clinical Neurophysiology. Institute of Neurology and Neurosurgery. Apartado Postal 4268, Havana 10400, Cuba. Email: braind@infomed.sld.cu.

Brain Death and Disorders of Consciousness, Edited by Machado and Shewmon
Kluwer Academic/Plenum Publishers, New York 2004

components that are required for the manufacture of its fundamental cellular mechanisms including metabolites, boundaries, compartments, enzymes and reproductive apparatus.[18]

Given the appropriate structural and energetic thermodynamic and kinetic conditions and the required chemical and temporal environment, autocatalytic self-organizing stable open systems must evolve.[†] Subsequently the evolution of the varieties of life forms occurring are limited by three factors – (1) the rules of Darwinian evolution (variation and selection), (2) the requirements for the spontaneous emergence of self-organizing systems, and (3) the intervention of unpredictable chance circumstances.[12] The evolution of unicellular and then multicellular living organisms has developed on earth over the last 3.5 to 4 billion years. Mechanisms for survival of multicellular organisms have evolved to include hormonal, immunological, and neuronal systems among others.

The definition of life of individual unicellular examples of these diverse systems include the basic functions of the metabolic and reproductive aspects of the specific organism that allow it to develop in a direction of decreased entropy production (i.e. bacteria, amoeba, or zygote). The death of the specific organism as a whole, during a specific phase of its life cycle, can be said to occur with irreversible cessation of the functions to keep entropy production at a minimum.[9]

A virus, therefore, in its crystalline state is not considered to be alive, since prior to invading a host cell it has neither metabolic nor reproductive function. In this dormant state it only has potential for decreasing its entropy (increasing its organization). After contact and entry into the host cell, its status is open to debate.

With the advent of multicellular organisms the life of the organism as a whole can no longer be defined in terms of cellular function alone. Rather the state of functions of organ systems as they contribute to the overall drive towards decreasing entropy production must be assessed (i.e. hydra, termite, shark, chimpanzee). In many vertebrates of which man is the relevant species, and in highly developed cephalopods such as squid and octopuses, the central nervous system, more specifically the brain occupies the status of being the critical system of the organism. The irreplaceable functioning brain, specifically and especially in an individual adult member of the species Homo sapiens overwhelmingly defines the behavior of the system-as-a-whole towards decreasing entropy production. Naturally, the concept of critical system cannot be applied to all stages (or phases) of development. During the zygote or embryonic stages no functioning brain exists and a critical system, as such, cannot be identified. During the mid-fetal stage of the human organism the onset of structure and function of the incipient brain can be distinguished. In contrast plants have no nervous system at any stage of development and there may not be a single system component that can be targeted as the critical system.

An essential assumption that must be made in defining the life and death of the human organism in a specified stage is that the terms "life" and "death" represents immutable states or processes that can be studied, described and modeled.

A process can be defined as the period during which the state changes; if the period is relatively short enough the process can be considered as an event. Gradual or sigmoidal (S-shaped) processes approach a step-function that can be considered an event (Figure 1).[12] The problems of life and death are intrinsically biological in nature. Further consequence of this biological orientation of life and death is that they are independent of

---

[†] This evolutionary principle is derived from Quantum Theory and Quantum Computing as described in references 64, 65 and 66.

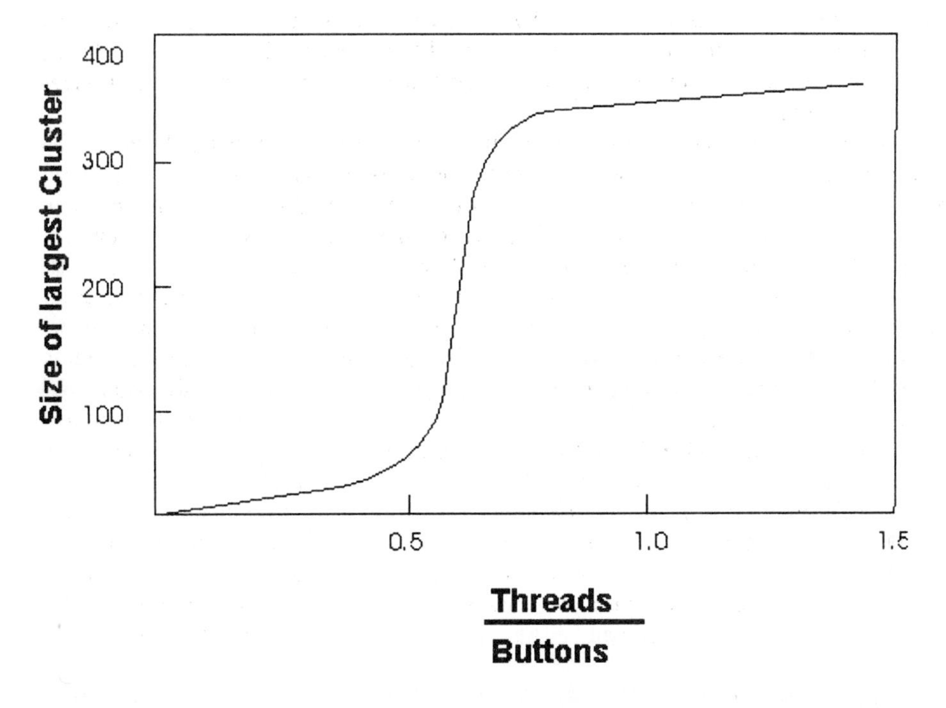

Figure 1. This S-shaped or sigmoidal curve represents a phase transition. As the ratio of threads (edges) to buttons (nodes) in a random graph passes 0.5, the size of the connected cluster slowly increases until it reaches a phase transition. (From Stuart Kauffman: At Home in the Universe, copyright-1996 by Stuart Kauffman. Used by permission of Oxford University Press, Inc.).

culture and religion. Life and death are objective states that cannot be altered or contrived within the defined parameters.[19-21]

3. THE CRITICAL SYSTEM OF THE BRAIN IN HOMO SAPIENS

The brain is the critical system of the human organism from its incipient developmental onset in the 20 to the 28-week old fetus. During its subsequent stages of normal maturation the brain continues to be the critical system with ever increasing ability to operate on and control the behavior of the organism as a whole. The functions of the brain are to receive, process, and store, information about the internal and external environment; then to select and deliver output patterns based on this information and the goal oriented mechanisms that have also been evolved and learned. These output patterns direct the system as a whole towards behavior that will increase survival of the individual and the species Thus; information is utilized to develop models of the state of the organism and the environment, along with multiple paths (plans) of goal-oriented behavior. Finally, decision-making processes occurring within the brain result in behavioral output patterns that tend to increase the organism's own organization. These processes are produced by means of parallel, distributed, neuronal feed back control networks. These networks are formed by connections among basic vertical cortical columnar units connecting to subcortical systems. Their function is ultimately to choose the most appropriate output and behavior available to the organism and its species enabling it to survive.[9,21-23]

In organisms with such brains, hierarchical mechanisms have been developed to perform unconscious and conscious modes of information processing. Different components of consciousness have been identified. We can distinguish among them, arousal, alerting, sleep-wake cycles, a variety of attentional states, and awareness. Although these states are often interdependent they can be separated. For example, in patients with blindsight we may see lack of awareness of visual input but persistence of the ability to locate (by pointing) or characterize (the color of) an object. Many clinical and experimental studies have verified that different attentional systems exist both functionally and anatomically. This is exemplified by one intentional system for processing visual information for action without awareness, and another attentional system for processing information for visual perception with awareness. Conscious awareness requires multiple, parallel, interconnections and feedback among sensory inputs, memory mechanisms, goal orientation, and output paths that function as unified entity by means of binding mechanisms. This "binding" allows focus of consciousness and is essential for awareness by means of which the brain produces such high order of information processing and selection of behavior. As a result of these characteristics the brain becomes the critical system of the phase of the organism in which it is operational.[9,21-45]

In "man" specifically, further development of the brain has resulted in self-awareness (self-consciousness), and a highly developed method of symbolic operation and transmission of information. This symbolic form of information is also used for interpersonal communication and more complex environmental modeling by means of language, logic, and mathematics. These additional functions (developed at later stages) of the critical system of the brain drives the organism as a whole still further in the direction of increased organization and is irreplaceable by any artifice.[9]

Modification of the "Critical System of the Organism" concept is required since we can further reduce the critical structure of the brain itself to the cortico-cerebral-thalamo-reticular complex. This "Critical System of the Brain" (CSB) complex includes the cerebral cortex, the basal ganglia, the thalamus, and the limbic system bilaterally, as well as the thalamic and brain stem reticular formations and components of the cerebellum. The anatomical and physiological substrate of the functions previously described requires the brain stem and thalamic reticular systems, the association nuclei of the thalamus and their connections. Binding among these functional systems occurs by means of 40 cycles per second oscillatory activity that has been intensively studied over the past decade.[30,34-48] This critical system of the brain (CSB) is the minimal irreplaceable anatomical substrate of those functions (such as the multiple aspects of consciousness) that are utilized by the organism as a whole towards behavior that will result in decreased entropy production. Thus, onset of its development and function signify the onset of the life of the organism during the stage of its earliest appearance. Irreversible cessation of intrinsic CSB function indicates the death of the organism in the stage during which the CSB exists.[9,16]

4. STAGES AND PHASES OF THE HUMAN LIFE CYCLE

In order to further clarify and utilize this concept, with all its ramifications we must consider the significance of death based on irreversible cessation of the CSB during different aspects of the human life cycle. We must simultaneously be aware that because of the virtual and practical impossibility of testing for the CSB itself we may use the entire brain including the brain stem as a first approximation of the CSB.[9,16,21]

The application of intrinsic irreversible function of the CSB to diagnose death requires us to consider the biological life cycle and the stages and phases of development of the human organism. The first classification is that of the mutually exclusive stages that have been well defined. These include the zygote formed from the gametes, through the preembryo, embryo, early, mid, and late fetal stages. Then from birth – to the newborn, neonate, infant, and stages of childhood, on through pubescence and adulthood.

In order to more clearly differentiate the different characteristics of the developmental cycle of the human organism we may additionally consider the additive qualities of development in terms of phases.[9]

The human organism falls into a natural nested set of four phases that are inclusive of one another. Each phase has its specific properties as well as nested properties derived from the previous phase. The phases of the human species and their specific stage related properties starting from the most primitive are all biological concepts (Figure 2):

4.1. Organism Phase

This phase is characterized by cellular DNA and the living state with metabolism, self-maintenance, and reproduction. (i.e. the zygote; preembryo (blastula) stage).

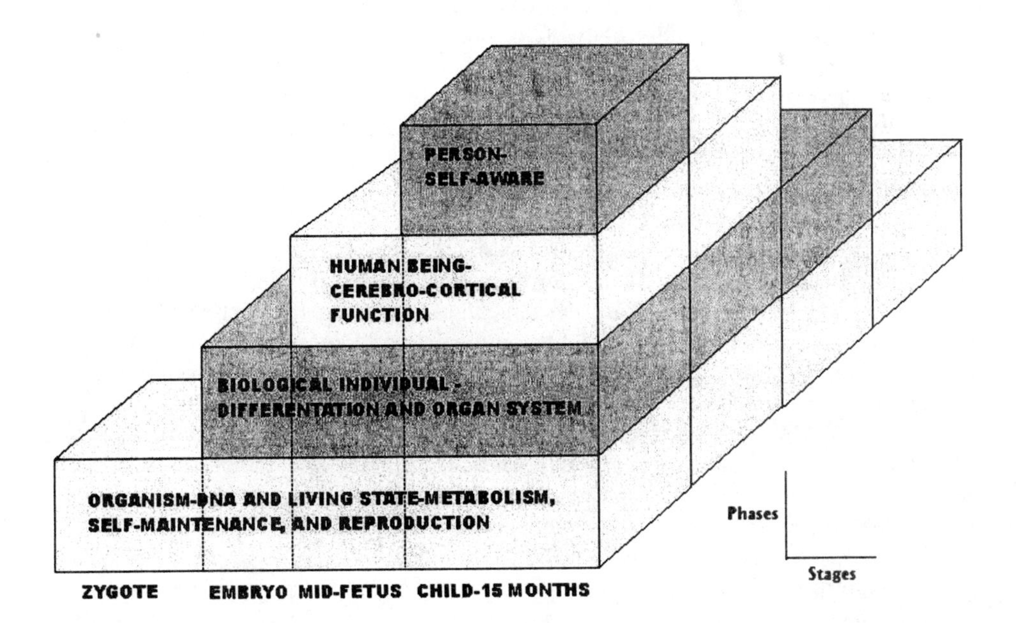

Figure 2. This figure shows the relationship of phases and stages within the human organism. (Modified from reference 9 with permission.)

4.2. Biological Individual Phase

This phase is nested within the organism phase, characterized by cell differentiation and organ development. (i.e. the six-week-old embryo; the early fetal stage).

4.3. Human Being Phase

This phase is nested within the biological individual phase, characterized by cerebro-thalamic-reticular function, i.e. the onset and incipient development of the functioning structure of the CSB in the 20 to 28 week fetus, and then further elaboration in the late fetus, the newborn, the neonate, and the infant (especially during the first year after birth) This elaboration of the CSB is reflected by the initial development and expansion of informational processing components and behavioral characteristics of consciousness. These components include arousal, alerting, sleep-wake cycles, states of attention, environmental awareness, and consciousness, as well as concomitant EEG activity. The progressive development of the incipient CSB such as early learning (which begins in utero) indicates the augmentation and versatility during the molding of the neural command and control system. (i.e. the mid fetal stage; the infant stage).

4.4. Person Phase

This phase is nested within the human being phase, characterized by continued augmentation of the CSB resulting in self-recognition and conscious self-awareness (demonstrable by identification of the self in a mirror test and self-referential communicative behavior by means of language). These behavioral characteristics indicate the brain's ability to model itself and its environment to achieve self-consciousness as well as the increasing capability of the CSB, at this phase, to be the global command system of the human organism controlling adaptive, entropy-reducing behavior. Therefore, these changes further exemplify the utility and uniqueness of the CSB illustrated by consciousness, self-consciousness, and symbol manipulation and its usage. (i.e. the 15-month-old child through the adult stages).[9]

Utilizing this framework of stages and phases of the life cycle of Homo sapiens we can consider this living system as undergoing a series of phase transformations. Each of the phases is alive. The aspects of the life during a specific phase can be described as having certain specific characteristics. For example, the most primitive **organism** phase is characterized by having cells, with a specific DNA compliment, as well as metabolism, self-maintenance, and reproduction. In the **biological individual** phase we must add cellular differentiation and the beginning of organogenesis (i.e. the circulatory system).

Of paramount interest to us is the **human being** phase. In the onset of this phase the 20 to 28-week old fetus starts to develop the structural-functional components if the brain. Developmentally it is from this time on that we may define the critical system of the human organism and apply the neurocentric definition of death. Brain death then becomes equated with the death of the human being phase of the human organism. Notice that death of the human being may occur in the mid fetal stage, prior to the development of the **person** phase. The transformation of the human being to the person does not occur until the development self-consciousness and self-referential level with the use of symbols for communication and language. This development occurs at about 15 months after birth of the human organism and can be tested for by the organism's self-recognition in a mirror, along with the ability to recognize the difference between self and environment. The human being has significant components of consciousness, but the person additionally has self-consciousness which is included in his or hers mastery of symbolic representation and communication.[9,31,46]

Note that death of human being phase must also include death of the person phase. However, death of the person phase once it has developed does not necessarily require death of the human being phase. The significance of this asymmetrical relationship will be apparent in discussing persistent vegetative states.

5. THE ROLE OF CONSCIOUSNESS

Our concept is that the brain (or more exactly the CSB) is the essential irreplaceable neurocentric mechanism that is the substrate of the conscious global organizing (entropy decreasing) behavior of the organism in the human being and the person phases of the organism. The criterion for this concept is the functioning state of consciousness in the human being phase.

Plum and Posner[46] defined consciousness as "the state of awareness of self[‡] and the environment." Two physiological components control conscious behavior: arousal and awareness.[21,46-49] Arousal represents a group of behavioral changes that occurs when a person awakens from sleep or transits to a state of alertness.[22] "Normal consciousness requires arousal, autonomic-vegetative brain function subserved by ascending stimuli from the pontine tegmentum, posterior hypothalamus and thalamus that activate wakefulness."[50]

Awareness, also known as content of consciousness, represents the sum of cognitive and affective mental functions, and denotes the knowledge of one's existence, and the recognition of the internal and external worlds.[21,46-50]

Plum[51] has recently defined not two but three components, subdividing the content of consciousness in two levels or components. According to this author, the second component or level, "which importantly regulates the sustained behavioral state function of affect, mood, attention, cognitive integration, and psychic energy (cathexis) depends on the integrity of the limbic structures including the hypothalamus, the basal forebrain, the amygdala, the hippocampal complex, the cingulun, and the septal area." The limbic system is important for the homeostasis of the internal milieu, and hence the second component of consciousness is crucial for integrating affective, cognitive and vegetative functions. Plum considers the third component as the "cerebral level, along with the thalamus and basal ganglia." This component is related to the processes of higher levels of perception, self-awareness, language, motor skill, and planning. Memory can be impaired by injury of either cerebral or limbic levels.

Arousal depends on the integrity of physiological mechanisms that take their origin in the ascending reticular activating system (ARAS): "it originates in the upper brainstem

---

[‡] As previously noted and documented in reference 9, the development of environmental awareness occurs much earlier than self awareness. Self awareness begins to manifest itself in the child at about 15 months of age. Some investigators describe "explicit consciousness" as related to the formation of self representation, self awareness, and self consciousness. These function in relation to motor activity, prediction, and goal directed planning of behavior. In contrast to "explicit consciousness" is "implicit consciousness" which is more pervasive, primitive, more involved with environmental awareness and does not attain the higher levels of self awareness (references 67 and 68).

reticular core and projects through synaptic relays in the thalamus to the cerebral cortex, where it increases excitability."[41] Moruzzi and Magoun,[52] in their pioneer studies, discovered "the presence in the brainstem of a system of ascending reticular relays, whose direct stimulation activates or desynchronizes the EEG, replacing high-voltage low waves with low voltage fast activity." Nonetheless, Steriade *et al.*[41] have recently emphasized that this desynchronization related to wakefulness "is now more apparent than real," because although large slow waves disappear during waking, the EEG shows high frequency oscillations (30-40 Hz), known as gamma oscillations, that reflect synchronized and enhanced intracortical and corticothalamic activity. Singer and Gray[53] have argued that these fast rhythms of corticothalamic neurons, known as gamma oscillations, are probably implicated in synchronizing mechanisms that respond to different features of the same perceptual object, leading to several hypotheses of high cognitive mechanisms.

Although initially the role of the mesencephalic reticular formation (MRF) and the thalamic intralaminar nuclei (ILN) was emphasized as simply mediating arousal, they are now considered as parts of integrating gating systems. Arousal is identified with the different functional states that characterize forebrain activation on the basis of brainstem modulation of corticothalamic systems, meanwhile Gating is formulated herein as a set of selective processes that may facilitate transient long-range interactions of large-scale brain networks.[44, 54-57]

Several authors have defended a present conception to identify the arousal system as interdependent on the output of cholinergic, serotoninergic, adrenergic and histaminergic nuclei located predominantly in the brainstem, basal forebrain, and posterior hypothalamus. Arousal is characterized with the different functional states related to forebrain activation on the basis of brainstem modulation of corticothalamic systems. Other researchers have tried to determine how necessary or sufficient different groups of arousal neuronal networks are, including the serotoninergic (median raphe) and noradrenergic (locus ceruleus) groups, concluding that any single group is indispensable.[55-57] Thus, "thalamo-cortical transmission may not be sufficient or even necessary to produce cortical activation."[46] Therefore, it is reasonable that arousal is due to several ascending systems stimulating the cerebral cortex and thalamus in parallel.[21,22]

By discovering that the cerebral cortex is organized in vertical columns that represent functional units was crucial for further understanding of the functional organization of the brain. "The basic functional unit of the neocortex is a vertically oriented group of cells extending across the cellular layers and heavily interconnected in the vertical direction, sparsely so horizontally."[46] At present there are arguments that the functional organization of the entire cerebral cortex is a complex of these vertical columns. Contiguous columns are interconnected by local circuits into "information-processing modules," characterized by specific afferent and efferent connections with other modular units from other cortical and subcortical areas.[22,58,59]

It has been argued the brain operates in "parallel processing," because cortical regions are linked in parallel networks with each other and with subcortical structures. Thus, a specific component of a certain cognitive function is scattered among interconnected regions, each one implicated in a distinct aspect of the cognitive ability.[22,60-62] According to Feinberg,[46] one of the most remarkable peculiarities of the brain is "the seemingly enormous redundancy, parallelism, and distributiveness" of its connections.

The cerebral cortex and thalamus make up "a unified oscillatory machine" that exhibit spontaneous rhythms and that are conditional to behavioral state and vigilance.[46] The brain uses spatiotemporally distributed systems to "capture high-order perceptual features."[46,56,60]

Plum[51] has emphasized that the ARAS substantially and inseparably activates and integrates both the arousal and the cognitive aspects of human consciousness. He recognized a brainstem-diencephalic participation not only in arousal, but also in cognitive function. In lesions affecting thalamic-mesencephalic structures that comprise the ARAS, the presence of important cognitive and affective deficits can be found. Alterations in the cerebral cortex after severe damage restricted to mesencephalic-diencephalic activating systems have been reported. They reflect transneural degeneration, and suggest that these pathways not only activate the cerebral cortex but they also trophically influence cortical neurons.[21,48,49,51] Therefore, it can be concluded that we cannot simply differentiate and locate arousal as a function of the ARAS, and awareness as a function of the cerebral cortex. Substantial interconnections among the brainstem, subcortical structures and the neocortex, are essential for subserving and integrating both components of human consciousness.[21,46-49,51,60-62] The above considerations lead one to conclude that there is no single anatomical place of the brain "necessary and sufficient for consciousness."[61]

Machado[48,49] has recently argued that consciousness provides the most significant attributes that characterize human life, and it is the most integrative function of the body. Hence, it is reasonable to state that any vestige of consciousness is inconsistent with death.

Consciousness and its components must be evaluated as prime indicators of brain function, specifically of the CSB. Thus, the CSB underlies consciousness globally and is considered as one of the key far-reaching controllers of behavior of the human being and the person phases of the human organism. When defining death within a given phase the development of the components of consciousness during that phase must be determined. The concept of death of the human being is irreversible cessation of function of the CSB. Since clear-cut isolation of the CSB is not feasible practically we will fall back on the whole brain concept (including the brain stem) as a first approximation in evaluating this neurocentric approach to death. Evaluating irreversible function of the brain as a whole (the operational brain) will automatically include irreversible cessation of the physiological substrate of the CSB. The criterion for the concept of intrinsic irreversible cessation of all brain function includes all relevant aspects of consciousness such as states of arousal, alerting, sleep-wake cycles, attention, and environmental awareness. Further tests of this criterion may include tests for respiration, ocular fixation, and EEG status. Tests of intracranial circulation are also of value.[21]

It should be made clear that somatic or systemic disintegration of the body does not necessarily immediately follow irreversible cessation of brain function! Many aspects of somatic integration are either independent of brain function or may be prolonged by appropriate replacement (i.e. hormonal) therapy.[1,9,20] This finding does not invalidate the concept of brain death. The organism (that is initially in the human being phase and subsequently in the person phase) primarily requires the functioning CSB and therefore components of consciousness (and subsequently self-consciousness and symbol manipulation) that are required to control behavior that is directed towards keeping entropy production of the system as a whole to a minimum. The other organ systems are secondarily involved and are not irreplaceable.

6. DEATH, BRAIN DEATH, PERSISTENT VEGETATIVE STATES, AND ANENCEPHALY

We may now apply the concepts previously formulated to discuss the problems of death, brain death, persistent vegetative states and anencephaly as these apply to the human organism and the human being.[9,14-16,21,46-49,63]

6.1. Death and brain death

Death, generally, can be considered only in relationship to life. The cessation of life in a previously living system refers to a specific change of state. The key aspects of this change of state relates to a unicellular living organism which had the characteristics of being a dynamic far from equilibrium open system and contained mechanisms for autocatalytic self-maintenance, metabolism, and reproduction within a boundary; the system decreased its entropy production (maintaining and increasing its own organization at the expense of environmental energy). Death occurs with the irreversible cessation of this state. In more developed unicellular organisms that have compartments for separation of functions evaluation of death is the same but may be more complex.

In multicellular organisms, the same definition may be also be used. However, if the system attains a sufficient degree of complexity, one must consider the organism as a whole, and its developmental phases in its entirety. Parts of the system may become irreversibly non-functional, but the organism as a whole may still be alive. That is, the system retains fundamental characteristics that will allow it to keep entropy production at a minimum. In such a system death of that phase of the organism can be equated with irreversible intrinsic cessation of the critical system of the organism if such an irreplaceable entropy reducing system exists. In the human being and person phases of the human organism the critical system is the brain or actually the CSB. Therefore death of the human being phase of the organism - Homo sapiens - is the advent of irreversible intrinsic cessation of the operation of the CSB – i.e. brain death. With irreversible cessation of the CSB both the human being and the person phase are dead.

An illustrative case history has been presented in which a young woman was 22 weeks pregnant and had become brain dead by appropriate criteria and tests. Her body was subsequently maintained using total resuscitation techniques including temperature maintenance and hormonal therapy, along with respiratory, nutritional, and environmental support. A normal 31-week fetus was delivered by cesarean section. Then all resuscitation related to the brain dead mother was then stopped. Clearly the brain dead mother was maintained for many weeks to allow the living human being (in the mid-fetal stage) to have a better chance at survival. The mother, as a human being only died once.[63]

6.2. Persistent Vegetative State

If only some aspects of the CSB are irreversibly non-functional and components are still functioning the person may be destroyed but the human being phase may still be alive. This is exemplified by patients in the persistent vegetative state (PVS). In persistent vegetative state (PVS) cases arousal is preserved (the PVS has periods of wakefulness), but awareness is seemingly is lost. Thus, in PVS there is an apparent dissociation of awareness from arousal.[21,48,49] It has been argued that "separate anatomic pathways mediate arousal and awareness, and that brain diseases can differentially affect each component of consciousness."[22] If the main functional finding in such patients is the preservation of arousal with an apparently loss of awareness,[21,49,62] can we deny the existence of internal awareness in PVS, because they apparently seem to be disconnected from the external world? The subjective dimension of awareness is philosophically impossible to test.[21,60,62,64-68]

6.2. Anencephaly

In anencephaly the problem is even more complicated. Variations in the type and extent of anencephaly must be considered.

First, if the CSB never developed we are not dealing with the human being phase of the organism. The maximal development could be that of a biological individual phase and death of the human being cannot be considered. The organism must be considered as analogous to an embryo. It cannot be pronounced dead as a human being.

Second, the organism has developed sufficient components of the CSB. Therefore it has reached the level of the human being phase. If it does not develop further – to the person phase it still must be considered as a living human being analogous to the PVS except it never had a prior history as a person.

Third, the organism developed a CSB and clearly reached the human being phase. Then a neurological problem arose to damage the CSB. Such a condition arises in hydranencephaly. Although this predicament may appear superficially similar to the second situation it actually is not, since further growth and development may offer a better prognosis. It is also unlike PVS since that condition usually occurs in older human organisms and the degree of recovery may be greater at a younger age.

Therefore, If there is any evidence of functioning of the CSB the patient cannot be considered dead as a human being even in the absence of the person phase of the organism.[9]

7. CONCLUSION

In conclusion one can cut through the confusion and the prior lack of knowledge by utilizing the biological neurocentric concept of brain death (irreversible intrinsic cessation of function of the CSB) as the necessary and sufficient condition for death of the human being phase of the human organism. Extension of the paradigm also resolves many ancillary problems relating to the biological concept and definition of the person phase of the organism. Additionally the concept directs us towards clearing the path in dealing with PVS and anencephaly in the future.

The classical definition of cardio-respiratory death, as used in medicine, is incomplete and cannot be applied in many situations. The mélange of partial brain death substitutes that have been proposed, involving the neocortex and brainstem only, are fundamentally defective. Utilizing the neurocentric definitions the classical diagnosis of death can be used and extended to brain death since both require irreversible cessation of the functions of the CBS.

8. REFERENCES

1. Shewmon DA. Recovery from "brain death": A neurologist's apologia. *Linacre Q* 1997;**64:**30-96.
2. Shewmon, DA. Chronic 'brain death'. Meta-analysis and conceptual consequence. *Neurology* 1998; **51:**1538-1545.
3. Singer P. Rethinking Life and Death; The Collapse of Our Traditional Ethics. New York: St. Martin's Griffen, 1994:1-256.
4. Youngner SJ. Defining death: A superficial and fragile consensus. *Arch Neurol* 1992;**49:**570-572.
5. Youngner SJ, Arnold MR, Schapiro R, eds. *The Definition of Death. Contemporary Controversies.* Baltimore: The Johns Hopkins University Press, 1999:346.
6. Zaner RM, ed. *Death: Beyond Whole Brain Criteria.* Dordrecht, Netherlands: Academic Publishers, 1988.
7. Veatch RM. Brain death and slippery slopes. *J Clin Ethics* 1992;**3:**181-187.
8. Gervais KG, ed. *Redefining Death.* New Haven, CT: Yale University Press, 1986:1-231.
9. Korein J. Ontogenesis of the brain in the human organism: definitions of life and death of the human being and person. In: Edwards RB, Bittar EE, eds. *Advances in Bioethics, Vol 2.* Greenwich Connecticut: JAI Press Inc, 1997:1-74.
10. Williams RJP, Frausto da Silva JJR, eds. *The Natural Selection of the Chemical Elements: The Environment and Life's Chemistry.* Oxford: Clarendon Press, 1996:1-646.
11. Depew DJ, Weber BH, eds. *Darwinism Evolving: Systems Dynamics and the Genealogy of Natural Selection.* Cambridge, Massachusetts: The MIT Press, a Bradford Book, 1995:429-495.
12. Kauffman SA, ed. *At Home in the Universe: the Search for the Laws of Self-Organization and Complexity.* New York: Oxford University Press,1995:1-321.
13. Collier J. The dynamics of biological order. In: Weber BH, Depew DJ, Smith JD, eds. *Entropy, Information, and Evolution: New Perspectives on Physical and Biological Evolution.* Cambridge, MA: MIT Press, 1988:227-242.
14. Korein J. Towards a general theory of living systems. In: Masturzo A, Giannino F, Figli, eds. *The 3rd International Conference of Cybernetic Medicine, Naples, Italy*, 1966:232-248.
15. Korein J, ed. Brain Death: Interrelated Medical and Social Issues. *Ann N Y Acad Sci* 1978;**Vol:**315:1-454.
16. Korein J. Ontogenesis of the fetal nervous system: The onset of brain life. *Transplant Proc* 1990; **22:**982-983.
17. Prigogine I, ed. *From Being to Becoming: Time and Complexity in the Physical Sciences.* San Francisco: Freeman Co, 1980:1-272.
18. Goodsell DS, ed. *The Machinery of Life.* New York: Springer-Verlag, 1993:1-140.
19. Bernat JL. Brain death: Occurs only with destruction of the cerebral hemispheres and the brainstem. *Arch Neurol* 1992;**49:**569-570.
20. Bernat HL. A defense of the whole-brain concept of death. *Hastings Cent Rep* 1998;**28:**14-23.
21. Machado C. Consciousness as a definition of death: Its appeal and complexity. *Clin Electroencephalogr* 1999;**30:**156-164.
22. Kinney HC, Korein J, Panigrahy A, Dikkes P, Goode R. Neuropathological findings in the brain of Karen Ann Quinlan: The role of the thalamus in the persistent vegetative state. *N Engl J Med* 1994;**330:**1469-1475.
23. Crick F, ed. *The Astonishing Hypothesis: The Scientific Search for the Soul.* New York: Charles Scribner's Sons, 1994:1-317.
24. Damasio AR, ed. *Descartes' Error: Emotion, Reason and the Human Brain.* New York: G.P. Putnam's Sons, 1994:1-312
25. Humphrey N., ed. *A History of the Mind: Evolution and the Birth of Consciousness.* New York: Simon & Schuster, 1992:1-238
26. Dennett DC, ed. *Consciousness Explained.* Boston: Little, Brown and Company, 1991:1-511.
27. Edelman GM, ed. *Neural Darwinism: The Theory of Neuronal Group Selection.* New York: Basic Books, Inc., Publisher, 1987:1-370.
28. Milner AA, Rugg MD, eds. *The Neuropsychology of Consciousness.* New York: Academic Press, 1992:1-37, 91-158 and 178-233.
29. Posner MI, Raichle ME, eds. *Images of Mind.* New York: Scientific American Library, 1994:1-257.
30. Association for Research in Nervous and Mental Disease (1998). The Conscious Brain: Normal and Abnormal. 77th Annual Program. December 5, 6, New York. New York University School of Medicine. In: The Royal Society, ed. *Philosophical Transactions: Biological Sciences.* London, 1997:1797-1942.
31. Baars BJ, ed. *In the Theater of Consciousness: The Workplace of the Mind.* New York: Oxford University Press, 1997:1-193.

32. Milner AD, Goodale MA, eds. *The Visual Brain in Action.* New York: Oxford University Press, 1995:1-248.
33. Mountcastle VB. An organizing principle for cerebral function: The unit module and the distributed system. In: Edelman GM, Mountcastle VB, eds. *The Mindful Brain: Cortical Organization and Group-selective Theory of Higher Brain Function.* Cambridge, MA: MIT Press, 1978:7-34.
34. Destexhe A, Contreras D, Steriade M. Cortically-induced coherence of a thalamic-generated oscillation. *Neuroscience* 1999;**92(2)**:427-443
35. Llinas R, Ribary U, Joliot M, Wang XJ, eds. Content and context in thalamocortical binding. In: *Temporal Coding in the Brain.* Berlin: Springer-Verlag, 1994.
36. Steriade M, Timofeev I. Neuronal plasticity in thalamocortical networks during sleep and waking oscillations. *Neuron* 2003;**37:**563-576.
37. Steriade M. Synchronized activities of coupled oscillators in the cerebral cortex and thalamus at different levels of vigilance. *Cereb Cortex* 1997;**7:**583-604.
38. Amzica F, Neckelmann D, Steriade M. Instrumental conditioning of fast (20- to 50-Hz) oscillations in corticothalamic networks. *Proc Natl Acad Sci USA* 1997;**94:**1985-1989.
39. Steriade M. The corticothalamic system in sleep. *Front Biosci* 2003;**8:**d878-d899.
40. Steriade M, Contreras D, Amzica F, Timofeev I. Synchronization of fast (30-40 Hz) spontaneous oscillations in intrathalamic and thalamocortical networks. *J Neurosci* 1996;**16:**2788-2808.
41. Steriade M. Arousal: revisiting the reticular activating system. *Science* 1996;**272:**225-226.
42. Steriade M. Corticothalamic resonance, states of vigilance and mentation. *Neurocience* 2000;**101:**243-276.
43. Steriade M, Amzica F, Contreras D. Synchronization of fast (30-40 Hz) spontaneous cortical rhythms during brain activation. *J Neurosci* 1996;**16:**392-417.
44. Steriade M. Awakening the brain. *Nature* 1996;**383:**24-25.
45. Contreras D, Steriade M. Cellular basis of EEG slow rhythms: a study of dynamic corticothalamic relationships. *J Neurosci* 1995;**15:**604-622.
46. Plum F, Posner JB, eds. *The Diagnosis of Stupor and Coma.* Philadelphia: FA Davis Company; 1980.
47. Machado C. Death on neurological grounds. *J Neurosurg Sci* 1994;**38:**209-222.
48. Machado C. A new definition of death based on the basic mechanisms of consciousness generation in human beings. In: Machado C, ed. *Brain Death (Proceedings of the Second International Symposium on Brain Death).* Amsterdam: Elsevier Science, BV; 1995:57-66.
49. Machado C. Is the concept of brain death secure? In: Zeman A, Emanuel L, eds. *Ethical Dilemmas in Neurology.* London: W. B. Saunders Company; 2000, Vol 36:193-212.
50. The Multi-Society Task Force on PVS. Medical aspects of the persistent vegetative state. *N Engl J Med* 1994;**330:**1499-1508.
51. Plum P. Coma and related global disturbances of the human conscious state. In: Peters A, ed. *Cerebral Cortex, Vol 9.* New York: Plenum Publishing Corporation, 1991:359-425
52. Moruzzi G, Magoun HW. Brain stem reticular formation and activation of the EEG. *Electroencephalogr Clin Neurophysiol* 1949;**1:**455-473.
53. Singer W, Gray CM. Visual feature integration and the temporal correlation hypothesis. *Annu Rev Neurosci* 1995;**18:**555-586.
54. Steriade M. Presynaptic dendrites of thalamic local-circuit neurons and sculpting inhibition during activated states. *J Physiol* 2003;546(Pt 1):1.
55. Kowalik ZJ, Steriade M, Skinner JE, Witte OW. Detecting oscillations in cortical and subcortical structures. *Acta Neurobiol Exp (Wars)* 2000;**60(1):**45-8
56. Steriade M. Coherent oscillations and short term plasticity in corticothalamic networks. *Trends Neurosci* 1999; **22:**337-345.
57. Amzika F, Steriade M. Integration of low frequency sleep oscillations in corticothalamic networks. *Acta Neurobiol Exp (Wars)* 2000;**60:**229-245.
58. Mountcastle, VB. An organization principle for cerebral function: the unit module and the distributed system. In: Edelman, GM, Mouncastle, VB, eds. *The Mindful Brain.* Cambridge: The MIT Press, 1978:7-50.
59. Goldman-Rakic PS. Topography of cognition: Parallel distributed networks in primate association cortex. *Annu Rev Neurosci* 1988;**11:**137-156.
60. Feinberg TE. The irreducible perspectives of consciousness. *Semin Neurol* 1997;**17:**85-93
61. Shewmon DA. "Brain Death": A valid theme with invalid variations, blurred by semantic ambiguity. In: Angstwurm H, Carrasco de Paula I, ed. *Working Group on The Determination of Brain Death and its Relationship to Human Death.* Vatican City: Pontificia Academia Scientiarum, 1992:23-51.
62. Shewmon DA. The metaphysics of brain death, persistent vegetative state, and dementia. *The Thomist* 1985;**49:**24-80.
63. Schrader D. On dying more than one death. *Hastings Cent Rep* 1986;**16:**12-17.

64. McFadden J, ed. *Quantum Evolution.* New York: W.W. Norton & Co, 2000:1-338.
65. Schwartz JM, Begley S. *The Mind and the Brain.* New York: Regan Books, Harper Collins Publishers, 2002:1-420.
66. Johnson GA, ed. *Shortcut Through Time. The path to the Quantum Computer.* New York: Alfred A Knopf, 2003:1-204.
67. Llinas RR. *I of the Vortex. From Neurons to Self.* Cambridge, Massachusetts: MIT Press, 2001:1-302.
68. LeDoux J, Debiec J, Moss H, eds. *The Self: from Soul to Brain.* Annals of the New York Academy of Sciences, 2003;**1001:**1-307.

THE CONCEPT OF BRAIN DEATH

Jean-Michel Guérit*

1. INTRODUCTION

The first description of "coma dépassé" dates from 1959.[1] Ten years later, it was followed by the introduction of the "Brain Death" ("BD") concept. BD was considered as equivalent to individual human death.[2] Since then, this concept gave rise to three partially overlapping, though qualitatively distinct, debates, which were well summarized by Bernat:[3] a *philosophical* debate on the definition of death, a *medico-philosophical* debate on the identification of a measurable criterion of death, and a *medical* debate on the definition of which tests should be used to demonstrate that this criterion has been fulfilled. Actually, these debates are closely related to each other: the choice of medical tests depends on the criterion, and the choice of the criterion depends on which definition of death has been accepted.

2. THE PHILOSOPHICAL DEBATE

Before the advent of modern resuscitation methods, death was a monolithic phenomenon. Indeed, the functional arrest of one of the major vital functions (circulation, respiration, brain function) obligatorily and almost immediately gave rise to that of all other functions. This doesn't hold true anymore, as most of these functions – except the cerebral one - can now be artificially supplanted. This raises the question whether one can define a minimal set of organic functions whose definitive loss would allow considering their owner as dead as a human being. This question subtends two corollaries: (1) Is it really possible to define such a set, and (2) of which physiological functions (and of which anatomical structures) should this set consist.

A full consensus hasn't yet been reached on the answer to the first issue. Even if most authors agree that irreversible brain destruction is necessary and sufficient for the

* Jean-Michel Guérit, M.D., Ph.D., Clinical Neurophysiology Unit, St. Luc Hospital, Catholic University of Louvain, Brussels, Belgium.

Brain Death and Disorders of Consciousness, Edited by Machado and Shewmon
Kluwer Academic/Plenum Publishers, New York 2004

irreversible loss of what constitutes the human essence of an individual – in other words, to consider that this individual is dead as a human being – other authors, like Bernat,[3] do not accept the notion of "death of the individual as a human being." Indeed, they consider that the human essence is a psychological or spiritual concept, while death is a biological one, which can only be applied to an organism. Significantly, Bernat also rejects a definition of death that would concern only the *homo sapiens* species. For that reason, Bernat considers as a prerequisite of death the irreversible loss of all 3 critical functions which allow the organism to function as a whole: *vital functions* (spontaneous breathing and control of circulation by the autonomic nervous system), *integrative functions* determining the homeostatic equilibrium, and *consciousness*. Conversely, the persistence of only one of these 3 critical functions would be sufficient to consider that "life" is still present. I do not share Bernat's position and advocate the concept of human death. Indeed, the mere social acceptance of non-vegetarian individuals, for example, justifies the definition of a specific "human-death" concept. Furthermore, even if I agree that the human essence cannot be reduced to biology, I consider there is a minimal set of critical body functions whose total and irreversible loss makes the current human specificity of an individual independent of the current integrity of all his other body functions. I consider that these critical functions are those which subtend human consciousness.

Interestingly, even if this philosophical debate leads to two different answers – one that defines death exclusively on the basis of brain criteria, and one that defines it as the loss of all critical vital functions, including consciousness – both answers agree on one point: the brain must be, at least partially, destroyed to consider an individual dead. The second, medico-philosophical, debate concerns whether brain destruction must be total or partial and, if partial, which brain structures must be destroyed to consider the individual dead as a human being.

3. THE MEDICO-PHILOSOPHICAL DEBATE

3.1. Brain Criteria of Death: Three Approaches

The issue of which parts of the brain must be destroyed before considering an individual as dead gave rise to 3 different approaches:

- The *whole-brain formulation* requires the irreversible destruction of the whole encephalon: cerebral hemispheres, brain stem, cerebellum. This approach is advocated by Bernat.[3]
- Initially proposed in 1975 by Veatch,[4] the *higher brain formulation* defines death as "the irreversible loss of what constitutes the human essence of a being" and considers the criterion that fulfils that definition to be the irreversible loss of consciousness and all cognitive functions following the destruction of the neocortex.
- The *brain-stem formulation* comes from the United Kingdom.[5,6] It considers that the total and irreversible destruction of the brain stem is a necessary and sufficient condition for death, irrespective of the possible persistence of activities in the brain hemispheres. In some way, the brain-stem formulation is the mirror image of the higher-brain formulation.

3.2. Comparison of the Three Approaches

These approaches, which belong to the medico-philosophical debate, can be regarded from both their philosophical justifications and medical implications.

3.2.1. Philosophical Justifications

As opposed to the whole-brain formulation, both the higher-brain and the brain-stem formulation result from a fundamental reflection on what constitutes the specific humanness of an individual. Veatch's approach is the most classical one and considers that the higher cortical functions provide the individual's specific humanness. Initially, the brain-stem formulation was justified by the fact that the brain stem provides the organism with what allows it to constitute more than the mere sum of its parts: midbrain integrity is a prerequisite for a correct functioning of the cortical networks mediating cognitive functions, and ponto-medullary integrity is a prerequisite for the harmony of the critical vital systems and homeostatic equilibrium.[‡] In some way, the brain-stem formulation could satisfy the advocates of the brain-death concept and those requiring combined loss of vital and integrative functions, and consciousness.

Moreover, both formulations may be reconciled with each other by the modern approaches of consciousness. Indeed, most neuroscientists agree that consciousness depends on a distributed network of neural structures, both cortical and sub-cortical. Noteworthy, Edelman[8] and Damasio[9] both consider that the brain-stem integrity is a prerequisite for the emergence of what is labelled on as "primary consciousness" by Edelman, or "core consciousness" by Damasio.

As an advocate of the human-death concept, I personally favour the Veatch's approach as that which most closely encompasses humanness. Nevertheless, as far as one accepts that both the cerebral cortex and the brain stem are a necessary (but not a sufficient) condition for consciousness, one should consider that both the loss of higher cortical functions and brain-stem death might be accepted as a sufficient criterion for the irreversible loss of what constitutes the human specificity of the individual. Moreover, and even if I consider it to rely on a too conservative approach, the whole-brain formulation actually includes the two others and should therefore be considered as *a fortiori* sufficient.

3.2.2. Medical Implications

According to the higher-brain formulation, the diagnosis of death should rely on the proof of the total destruction of the cerebral cortex, irrespective of the brain-stem status.

[‡] Another argument that was used by these who advocate the brain-stem formulation was that all patients in that situation eventually develop asystole, even with optimal resuscitation. We do not accept this argument, because it confuses a process with its outcome. Moreover, it has been challenged by a meta-analysis of Shewmon,[7] who suggests that the reasons why optimally resuscitated brain-dead patients develop asystole are more systemic than cerebral. Moreover, the "systemic survival" of these patients would decrease according to an exponential curve with two time constants: a shorter one (half-life of 2-3 months) followed 1 year later by a much longer one. One patient of Shewmon did not develop asystole even after 14 years.

The reverse holds true for the brain-stem formulation, which requires the proof that the brain stem has been destroyed, irrespective of the status of the brain hemispheres. Finally, the whole-brain approach requires proving that both the brain stem and the cerebral hemispheres have been irreversibly destroyed. In keeping with our previous suggestion that all three formulations might me considered sufficient, each of these medical demonstrations should be sufficient to diagnose individual death. How to achieve these demonstrations belongs to the third, medical, debate.

4. THE MEDICAL DEBATE

BD diagnosis relies on prerequisites (the cause of coma must be identified and be sufficient to produce BD), on the clinical examination (unreactive coma, loss of brain-stem reflexes, apnea), and on confirmatory tests, which can be subdivided into neurophysiological tests (electroencephalogram (EEG),[10] evoked potentials (EPs)[11,12] and assessments of brain circulation (angiography,[13] radioisotopes,[14] transcranial Doppler[15]). In this chapter, we shall examine how the validity of all these tests can be re-considered under the light of the medico-philosophical debate.

4.1. The Flaws of a Phenomenological[§] Approach

There is actually no test – neither clinical nor paraclinical – that is sufficient to prove the *total* destruction of the cerebral cortex or the brain stem:

- Beyond the fact that electrocerebral activity can cease reversibly in certain circumstances – in particular, drug intoxication – an isoelectric EEG doesn't prove that *all* cortical neurons have been destroyed. Of course, one could argue that the persistence of, for example, only 100 neurons is not sufficient to provide with sufficient consciousness. But what about 1,000 10,000, etc ... neurons? In other words, using the isoelectric EEG as a proof of death in a phenomenological approach exposes one to an ethically unacceptable "slippery slope" risk.
- The same argument could be used against the use of multimodality EPs: brain-stem auditory and somatosensory EPs only test limited parts of the brain stem (and not those which are likely to be implicated in consciousness), and cortical auditory, visual, and somatosensory EPs provide only a very limited sampling of cortical functions.
- The same holds true for the clinical examination of global reactivity, postural reactions or brain-stem reflexes, which only assess a very limited part of the central nervous system.
- Even those techniques that demonstrate arrest of cerebral blood flow (CBF) may be insufficient. For instance, some EEG activities may persist in BD patients in whom four-vessel angiography demonstrated CBF arrest from the four major

[§] In this paper, the adjective "phenomenological" is used to designate an approach that would pretend to exhaustively demonstrate total brain destruction, independently of the pathophysiology of that destruction.

vessels.[16,17] This phenomenon is likely due to the persistence of functional anastomoses from the external carotid artery, through perforating meningeal vessels. If this criticism holds true for four-vessel angiography, it must also hold true for intracranial Doppler and radioisotope scans, which, additionally, provide a limited evaluation of blood flow in the posterior fossa.

In summary, I actually think that none of the above-mentioned techniques is sufficient to prove the *total* destruction of brain stem or cerebral hemispheres. Even the combined use of these techniques is insufficient. Does that mean that it is impossible to prove BD? Fortunately, I believe the answer is "no," provided that one shift from a phenomenological to a pathophysiological approach.

4.2. The Pathophysiological Approach

Instead of proving the exhaustive destruction of some neural structures, one should prove the occurrence and the completion of a pathophysiological process which is known to give rise to brain destruction incompatible with any persistence of a human consciousness.

Practically speaking, this strategy is the one already routinely used in diagnosing BD. Indeed, we know that in most – not to say the quasi-totality – of cases, this diagnosis is made after head trauma or spontaneous brain hemorrhage, that is, in patients who developed transtentorial herniation following an untreated, or unsuccessfully treated, increase in intracranial pressure. We know from pathophysiology that transtentorial herniation eventually gives rise to the total destruction of the cerebral hemispheres, brain stem, and cerebellum,[18] that is, to a situation which is compatible with any of the criteria emerging from the medico-philosophical debate.

4.3. Ethical Implications of the Pathophysiological Approach

Although this pathophysiological approach should not actually affect our practical attitude toward most BD diagnoses – in particular, the choice of a confirmatory test should still be achieved on the basis of its intrinsic qualities – the pathophysiological approach could substantially modify it in two circumstances, which are traditionally considered different from BD: *anencephaly* and some extreme cases of *vegetative state.*

The issue of organ donation in anencephalic neonates is still without solution. Indeed, these babies have no cerebral cortex but their brain stem is present and functional. This means that, in keeping with the classical criteria, they are not brain dead. This implies that the prerequisite for organ harvesting is the hypothetical disappearance of brain-stem reflexes. According to a protocol developed by the Anencephalic Organ Donation Committee of Loma Linda University Medical Center, it was suggested to limit the resuscitation period to 7 days. Abbastita *et al.*[19] reported on the results of this procedure in a cohort of 11 anencephalic neonates: brain-stem reflexes never disappeared, and resuscitation was interrupted in 10 out of 11 babies, who eventually, and almost immediately, developed apnea and multisystem failure leading to the impossibility of organ harvesting. In my opinion, this strategy is prototypic of the inanity and pitfalls of the phenomenological approach. This fiasco might have been avoided with a pathophysiological approach, as anencephaly constitutes another pathophysiological process known to be incompatible with consciousness, irrespective of whether the brain stem is functioning or not.

Another domain in which the pathophysiological approach could give rise to ethical progress is that of some extreme cases of post-anoxic coma and/or vegetative state, in which it can proven that all primary components of cortical EPs have definitively disappeared.[20, 21] We know that, for metabolic reasons, the sensitivity to anoxia of the primary sensory cortices is intermediate between that of association cortex and brain stem.[18] This implies that brain anoxia sufficient to destroy the primary cortex should be sufficient to destroy all association cortex, irrespective of brain stem status. This irreversible situation is incompatible with any consciousness. It fulfils the higher-brain formulation. That is, if one accepts this formulation, this situation must allow diagnosing individual death. Significantly, this situation can be reliably identified at the patient's bedside with multimodality or, more specifically, somatosensory EPs, which show the absence of any cortical components (what we labelled as post-anoxic Grade 4 in our EP classification).[22] One major advantage of this approach is that Grade 4 somatosensory EPs constitute a pattern clearly distinct from all others, eliminating any "slippery slope" risk.

5. CONCLUSIONS

The diagnosis of individual death is subject to a triple debate: philosophical, medico-philosophical, and medical. Whatever the accepted definition of death, it must be associated with a degree of cerebral destruction that is irreversibly incompatible with consciousness. This criterion can be fulfilled by the destruction of either the whole cerebral cortex or the whole brain stem. We pointed out the flaws of a purely phenomenological approach and advocated instead a pathophysiological approach. This could give rise to an extension of the concept of "individual death," in ethically acceptable safety limits.

6. REFERENCES

1. Mollaret P, Goulon M. Le coma dépassé (mémoire préliminaire). *Rev Neurol (Paris)* 1959;**101**:3-15.
2. Ad Hoc Committee of the Harvard Medical School. A definition of irreversible coma. *JAMA* 1968;**205**: 85-90.
3. Bernat JL. A defence of the whole-brain concept of death. *Hastings Cent Rep* 1998;**28**:14-23.
4. Veatch RM. The whole-brain-oriented concept of death: an out-moded philosophical formulation. *J Thanatol* 1975;**3**:12-26.
5. Conference of Royal Colleges and Faculties of the United Kingdom. Diagnosis of brain death. *Lancet* 1976;**ii**:1069-1070.
6. Pallis C. Brain-stem death. The evolution of a concept. *Natl Med Leg J* 1987;**55**:84-107.
7. Shewmon DA. Chronic "brain death". Meta-analysis and conceptual consequences. *Neurology* 1998; **51**:1538-1545.
8. Edelman G. *Biologie de la Conscience*. Paris: Odile Jacob, 1992.
9. Damasio A. *Le sentiment-même de soi*. Paris: Odile Jacob, 1999.
10. Silverman D, Masland RL, Saunders MG, Schwab RS. Irreversible coma associated with electrocerebral silence. *Neurology* 1970;**20**:525-534.
11. Guérit JM. Evoked potentials: A safe brain death confirmatory tool? *Eur J Med* 1992; **1**: 233-243.
12. Guérit JM, Fischer C, Facco E, Tinuper P, Murri L, Ronne-Engström E, Nuwer M. Standards of clinical practice of EEG and EPs in comatose and other unresponsive states. In: Deuschl G, Eisen A, eds.

Recommendation for the practice of Clinical Neurophysiology. Guidelines of the International Federation of Clinical Neurophysiology. Amsterdam: Elsevier, 1999 (EEG Suppl.52):117-131.

13. Bradac GB, Simon RS. Angiography in brain death. *Neuroradiology* 1974;**7:**25-28.
14. Donohoe KJ, Frey KA, Gerbaudo VH, Mariani G, Nagel JS, Shulkin B. Procedural guidelines for brain death scintigraphy. *J Nucl Med* 2003;**44:**846-851.
15. Ducrocq X, Braun M, Debouverie M, Junges C, Hummer M, Vespignani H. Brain death and transcranial Doppler: experience in 130 cases of brain dead patients. *J Neurol Sci* 1998;**160:**41-46.
16. Grigg MM, Kelly MA, Celesia GC, Ghobrial MW, Ross ER. Electroencephalographic activity after brain death. *Arch Neurol* 1987;**44:**948-954.
17. Ashwal S, Schneider S. Failure of electroencephalography to diagnose brain death in comatose children. *Ann Neurol* 1979;**6:**512-517.
18. Plum F, Posner JB. *The Diagnosis of Stupor and Coma.* 3rd ed. Philadelphia: F. A. Davis Co., 1980:87-151.
19. Abbattista AD, Vigevano F, Catena G, Parisi F. Anencephalic neonates and diagnosis of death. *Transplant Proc* 1997;**29:**3634-3635.
20. Guérit JM. The interest of multimodality evoked potentials in the evaluation of chronic coma. *Acta Neurol Belg* 1994;**94:**174-182.
21. Guérit JM. Multimodality evoked potentials in the permanent vegetative state. In: C. Machado, ed. *Brain Death.* Amsterdam: Elsevier BV, 1995:171-177.
22. Guérit JM, de Tourtchaninoff M, Soveges L, Mahieu P. The prognostic value of three-modality evoked potentials (TMEPs) in anoxic and traumatic coma. *Neurophysiol Clin* 1993;**23:**209-226.

THE "CRITICAL ORGAN" FOR THE ORGANISM AS A WHOLE

Lessons from the lowly spinal cord*

D. Alan Shewmon†

1. INTRODUCTION

For over three decades now, it has been widely accepted that the death of one particular organ – the brain – constitutes human death. Much less a matter for agreement, however, has been the *reason* for that equivalence. The various proposed rationales can be subdivided into three main categories, corresponding to three fundamentally different concepts of death: (1) *sociological* (death is loss of conferred membership in society; its legal definition is culturally relative, and most modern societies happen to have chosen to recognize brain-based diagnoses); (2) *psychological* (death is loss of personhood due to loss of potential for all mental functions, and the brain is the organ of the mind); or (3) *biological* (death is loss of physiological, anti-entropic unity of an organism, and the brain is the hierarchically highest integrating organ of the body).

Historically, the standard, quasi-official, or "orthodox" rationale is the latter: the brain is regarded as the body's "central integrator" or "critical organ," destruction‡ of which entails loss of somatic integrative unity, cessation of the "organism as a whole."[3-8] It has been endorsed by a variety of influential individuals and virtually every standard-setting group that has ventured to explicitly articulate a rationale.[3-10] (This is not to say that the standard-setters have not considered consciousness important in itself and also a sign of life. It is simply that for them, lack of consciousness *per se* has not been a decisive factor in determining human death. It is not a contradiction in terms to speak of a

---

* Presented at the Third International Symposium on Coma and Death, February 22-25, 2000, Havana. Adapted, with permission from Nature Publishing Group, Specialist Journals Division (formerly Stockton-Press), from Shewmon DA. Spinal shock and 'brain death': somatic pathophysiological equivalence and implications for the integrative-unity rationale. *Spinal Cord* 1999; 37(5): 313-324.

† D. Alan Shewmon, MD, Professor of Neurology and Pediatrics, David Geffen School of Medicine at UCLA, Los Angeles, CA 90095. Email: ashewmon@mednet.ucla.edu.

‡ For all practical purposes relevant to somatic pathophysiology, "destruction"[1,2] is taken as equivalent to "total brain infarction"[3] and "irreversible nonfunction."[4]

live patient who is comatose, even if permanently so. Rather, what is conceptually pivotal is the vital status of the body as a biological organism: if the body is alive, then the person is alive even if permanently unconscious; if the body (including the brain) is dead, then so is the person.) Insofar as the "physiological kernel" of "brain death" (BD) is the brain stem,[11] the same biological rationale has been advanced also by advocates of brain-stem formulations of death.[12,13]

One difficulty that continues to plague the BD debate is the operational definition of terms such as "organism as a whole," "critical" vs. "non-critical" function, and so on. Furthermore, operational definitions of "life" or "death" in this context should avoid implicit equations of "life" with "health." "Death" is not being contrasted with "healthy," but with "sick and dying." Typically such terms are either not defined at all, or defined in an *ad hoc*, self-serving way. One way around this difficulty is the following line of reasoning.

If the brain is uniquely responsible for the organism's biological unity, so that in the absence of the brain's coordinating activity the organism becomes a mere disunited collection of organs and tissues, such somatic "dis-integration" should be just the same regardless whether the absence of brain-coordination is due to absence of a brain or merely to functional disconnection from the brain. Thus, various brain disconnection syndromes might provide important insights into the nature of BD. If the somatic physiologies are indeed similar, if not identical, as theoretically they should be, then however terms such as "organism as a whole" might be operationally defined, if they apply or do not apply to brain disconnection, they must equally apply or not apply to BD. The "central-integrator-of-the-body" rationale of BD can therefore be tested by examining the vital status of brain-disconnected bodies, so long as the somatic physiology of the two conditions is indeed equivalent. This latter assumption is what we shall examine in this chapter, focusing on the most common naturally occurring brain-body disconnection syndrome, high spinal cord transection.

This is not the first time such a comparison has been suggested, but previous allusions to it have not been developed to any significant extent. For example, one of the traditionally cited manifestations of the supposed somatic "dis-integration" in BD is hemodynamic deterioration and loss of homeostasis, intractably spiraling to imminent and irreversible asystole, [4(p.35),5(p.49),7(p.27),11(pp.35-36),13(pp.36-37),14(p.838),15(p.1368),16(p.30)] although such outcome is not universal or intrinsically necessary.[17] Still, whether BD is conceived as "total brain infarction,"[3] "irreversible cessation of all functions of the entire brain,"[4] or "permanent functional death of the brainstem,"[18] it clearly eliminates multiple neural centers involved in somatic integration.

A rarely entertained alternative explanation, however, for the somatic pathophysiology of BD is spinal shock from sudden cessation of rostral influences, as first suggested by Ibe back in 1971:

> "The clinical picture was in all cases marked by continuous loss of consciousness and persisting respiratory paralysis. The brain function was constantly extinguished. The primary spinal areflexia, on the other hand, as well as the disorders of pulse and blood pressure after overcoming the spinal shock (about 24-36 h after the accident), showed a clear tendency to recovery, possibly to complete restitution."[19]

The idea was hardly noticed, appearing in an EEG abstract, with the phrase "spinal shock" mentioned only in passing.

Later Youngner and Bartlett, critiquing the physiological rationale for "whole-brain death," alluded to certain other somatic similarities between BD and high cervical cord transection:

> "a spinal cord transection at the level of the second cervical vertebrae [sic] ... leads to the immediate and irreversible loss of spontaneous respiration, cardiovascular stability, and temperature control – all of which are integrating functions."[20(p.254)]

Again the potential significance of the comparison was largely overlooked by the neurological community.

Could it be that high cord transection and BD are somatically similar with respect not only to these functions but to others as well? The answer is highly relevant to the question of somatic integrative unity in BD, i.e., to the distinction between a sick, disabled organism and a non-organism. Bernat, in his defense of BD against recent critiques, alludes to the mentioned similarities with spinal cord dysfunction but dismisses their relevance in the following way:

> "Scholars have pointed out correctly that the spinal cord performs a number of integrative functions for the organism, and that it seems arbitrary and contrived for the whole brain proponents to emphasize the brain's role in integration to the exclusion of that of the spinal cord. While it is true that as part of the central nervous system, the spinal cord has an integrating role, it is not a critical role. For example, many patients have lived long lives with minimal support following complete destruction of their spinal cord by injury or disease. Therefore, the integrating functions served by the spinal cord clearly are not necessary for life and therefore, their absence is not necessary (and certainly not sufficient) for death. Permanent cessation of the clinical functions of the entire brain, therefore, remains the best criterion of death."[9(p.18)]

The logic behind the word "therefore" in that last sentence is unclear. The point of the critics' comparison is not that cord function is necessary for biological life (understood at the level of the "organism as a whole"), but rather that, precisely because it *isn't*, yet BD and high cord transection have similar somatic effects, then it logically follows that neither is brain function necessary for biological life (equally at the level of the "organism as a whole"). The key issue is the extent to which the two conditions are really somatically physiologically similar.

Whenever defenders of the biological rationale of BD address this comparison in the philosophical BD debate, the somatic dysfunctions in BD are typically played up while those in spinal cord transection are played down (perhaps because most of these authors' clinical experience with cord injury is with levels below the point of comparison, which is between C3 and the cervico-medullary junction). The best way to assess this comparison objectively is to consult relevant clinical literature that has neither the comparison in mind nor any philosophical axe to grind about the nature of BD, i.e., texts on the intensive care of BD and high spinal cord injury (SCI) by specialists with extensive experience in the respective conditions.

To forestall potential misunderstanding, let me re-emphasize that the purpose of this comparison is *not* to advance a claim that BD is clinically indistinguishable from SCI, which would be absurd. Nor is the issue to which the comparison is relevant the *clinical criteria* for diagnosing a dead brain (the validity of which is taken for granted for purposes of this paper), but rather one particular *conceptual rationale* (among several) for equating a dead brain with a dead individual: namely the one that claims that a dead brain equates with a dead *body*. A quite different rationale, promoted by advocates of "higher brain" or consciousness formulations, maintains that the reason why BD is death is that irreversible unconsciousness constitutes a loss of personhood, regardless of the vital status of the body as a biological organism.[20-26] Plainly, for *this* rationale the role of consciousness is critical, and any physiological comparison with SCI is irrelevant.

The interest of the comparison, therefore, has to do with its implications for the conceptual validity of the *biological rationale only* – and that interest is heightened by the fact that that rationale happens to be the mainstream or quasi-official one almost everywhere. Thus, the ignoring of consciousness in what follows is in no way intended to belittle the obviously exceeding value of intact consciousness and communicative abilities, which SCI patients possess and BD patients lack. Subjective consciousness is simply not a *sine qua non* of the "orthodox" organism-rationale, so it is equally beside the point for the present critique of that rationale.

2. METHOD

The contemporary neuroanesthesia and neuro-intensive care literatures relating to high SCI and BD were compared, focusing on somatic pathophysiology, symptomatology, treatment, and survival. To compare long-term outcomes following continued aggressive treatment, reliance was made on a compilation of some 175 BD cases with prolonged survival.[17]

3. RESULTS

We shall first examine the very extensive somatic similarities between high SCI and BD, and then the relatively few and physiologically unimportant somatic differences.

3.1. Similarities

The somatic parallels between high SCI and BD, according to the literature, are best organized into three main stages according to time from injury: (I) the induction phase of the neurologic lesion, (II) the acute phase, and (III) the chronic phase (outline structure summarized in Table 1).

I. Induction phase of neurologic lesion. In both conditions, associated multisystem damage is common and could account for somatic deterioration independently of the neurologic nonfunction.

I.A. Direct multisystem damage. The primary etiologies of SCI and BD sometimes directly damage non-neural organs as well: for example, massive trauma.[27] The early mortality rate in SCI is increased with associated injuries compared to pure

Table 1. Somatic pathophysiological *parallels* between high cervical cord transection and BD.

I. Induction phase of neurologic lesion
  A. Direct multisystem damage increases early mortality rate
  B. Indirect systemic complications from transient sympathetic storm
    1. Hypertension
    2. Subendocardial ischemia and infarcts
    3. Neurogenic pulmonary edema

II. Acute phase
  A. Irreversible apnea
  B. Quadriplegia
  C. Spinal shock
    1. "Somatic" system deactivation
      a) Flaccidity
      b) Depressed tendon reflexes and plantar responses
    2. Autonomic system deactivation
      a) Sympathetic (thoracolumbar)
        (1) Hypotension
        (2) Impaired cardiac contractility
        (3) Increased venous capacitance
        (4) Bradycardia
        (5) Cardiac arrhythmias, especially bradyarrhythmias
        (6) Cardiovascular instability can be difficult to treat and lead to early arrest
        (7) Paralytic ileus
      b) Parasympathetic (sacral) – flaccid bladder
    3. Hypothermia and poikilothermia
  D. Relatively preserved hypothalamic-anterior pituitary functions
    1. Thyroid function – euthyroid sick syndrome
    2. Other anterior pituitary functions – variably affected in BD and of little relevance to somatic unity
  E. Predisposition to infection
    1. Pneumonia, due to immobility and lack of respiratory protective reflexes
    2. Urinary tract infections due to bladder dysfunction and urinary retention
    3. Sepsis from decubitus ulcers
  F. Actuarial survival curves
    1. Biphasic: rapid drop-off, followed by relative stabilization
    2. Age effect: younger patients have greater survival potential

III. Chronic phase
  A. Resolution of spinal shock 2-6 weeks after injury – return of autonomous cord function
    1. Development of spasticity, pathological increase of tendon and cutaneous reflexes
    2. Recovery of sympathetic tone
      a) Hemodynamic stabilization
      b) Bradycardia and bradyarrhythmias resolve
      c) Gastrointestinal motility returns and enteral feedings can resume
      d) Piloerection and sweating return
    3. Recovery of sacral parasympathetic tone – automatic bladder function returns
    4. Thermoregulation tends to improve, although most patients remain poikilothermic
  B. Intercurrent infections precipitate return of spinal shock and hemodynamic decompensation
  C. Autonomic hyperreflexia (dysreflexia)

---

cord injury.[28(p.503),29(p.1167)] Similarly, the rate of early asystole in BD is increased with associated injuries compared to pure brain injury.[17]

I.B. Indirect systemic complications from transient sympathetic storm. During induction of experimental cervical cord transection, transient *hypertension* occurs, probably from massive depolarization of sympathetic preganglionic neurons.[30(p.725)] In clinical experience this is not usually observed, presumably because it has already resolved by the time medical assistance arrives.[30(p.727)] Analogously, during the process of brain herniation, *prior* to BD, transient hypertension is common, also due to massive sympathetic hyperactivity and release of circulating catecholamines.[15(p.1368),31(p.2225),32(p.818)]

This sympathetic storm can produce *subendocardial ischemia* or *neurogenic pulmonary edema* in acute SCI[29(p.1170),30(pp.724-725),33(p.1290)] as well as in severe brain injury.[15(p.1368),31(p.2225),32(p.818),34-39] The cardiac pathology associated with experimental BD can be prevented by prior sympathectomy and verapamil, confirming the causal role of sympathetic hyperactivity.[40]

II. Acute phase. The non-endocrinologic manifestations of the acute phase are nearly identical in high SCI and BD.

II.A. Irreversible apnea is a hallmark of both cord transection above C4 [30(p.722),33(p. 1287),41(p.1063)] and BD.

II.B. Quadriplegia characterizes both high cord transection[30(p.729),41(p.1063)] and BD.

II.C. Spinal shock. Immediately upon severe SCI, the intact cord below the lesion functionally shuts down through poorly understood mechanisms. This spinal shock typically lasts 2 to 6 weeks, after which autonomous cord functions gradually return.[30(p.727),33(p.1290),42(p.186)] The intensity of spinal shock increases with the height of the lesion.[30(pp.725,727),33(p.1287),43(p.1123)] Insofar as BD is, *from the cord's perspective*, a transection at the cervico-medullary junction, one should expect maximal spinal shock in BD. Indeed, in both high cervical transection and BD both "somatic"[§] and autonomic spinal functions are suppressed as follows.

II.C.1. "Somatic" system deactivation. Flaccidity and absence of tendon reflexes and of plantar responses characterize acute SCI.[43(p.1123)] Although flaccidity is not a diagnostic requirement for BD, it is so common that positive muscle tone should at least raise the question of decorticate or decerebrate rigidity, which would exclude the diagnosis. Likewise, areflexia is so common in BD that the original Harvard Committee considered it a diagnostic feature (though not absolutely required) along with absent plantar responses.[44**]

II.C.2. Autonomic system deactivation. In spinal shock, autonomic function is as depressed as segmental reflexes. With cervical lesions the entire sympathetic and sacral parasympathetic systems are affected, the sympathetic dysfunction being more deleterious.

---

§ The semantic tradition distinguishing the "somatic" from the "autonomic" nervous system is unfortunate, especially for the present purpose, implying that the autonomic system has nothing to do with the body; ironically, the autonomic system is perhaps more important for somatic unity than the "somatic" system.

** Subsequent BD criteria emphasized the diagnostic compatibility of preserved tendon reflexes without denying that areflexia is nevertheless more characteristic.[4,45,46]

II.C.2.a. Symptoms of *sympathetic paralysis* in both conditions include the following.

(1) Hypotension occurs in 68% of patients with high SCI[47] and can be of life-threatening severity.[29(p.1169),30(pp.725,727),33(pp.1287-1290),48(p.327)] Hypotension is also very common in BD.[15(pp.1368-1369),17(Table 1),31(p.2225),32(pp.818-819), 49(pp.548-549)]

(2) Impaired cardiac contractility characterizes both SCI[29(p. 1169),30(p.725)] and BD,[31(p.2225)] as does

(3) increased venous capacitance.[30(p.725),31(p.2225),33(p.1287)]

(4) In both conditions *bradycardia* is common and can help in the differential diagnosis of hypotension, suggesting sympathetic paralysis rather than hypovolemic shock (characterized by compensatory tachycardia if the sympathetic system is intact). Hypothermia or subclinical myocardial ischemia can exacerbate bradycardia in either condition.[32(p.818),50]

(5) Cardiac arrhythmias occur in 19% of high SCI, with peak incidence of bradyarrhythmias around day 4.[29(p.1170),30(p.727),33(p.1290),47,51(p.1205)] Arrhythmias, especially bradyarrhythmias, are also common in early BD.[31(pp.2225-2226),49(pp.549-550),52]

(6) The *cardiovascular instability* resulting from combined hypotension, impaired myocardial contractility and bradycardia can be difficult to treat and causes significant early mortality in high SCI.[51(p.1204),53] An autopsy study of 44 cervical cord injury victims revealed that 30 (68%) died within the first 11 days, mainly of cardiovascular complications.[43(p.1106),54] Similarly, intractable cardiovascular instability leads to asystole in about 10% of BD prospective organ donors during the maintenance phase,[15(p.1370),31(p.2226)] and "{h}ypotension and/or cardiac arrest occur in approximately two thirds of referred donors."[31(p.2225)] Hemodynamic instability renders pharmacologic and fluid management tricky in both conditions, and treatment recommendations are essentially the same.[††]

(7) Paralytic ileus often results from acute autonomic dysfunction and general systemic disturbances in both SCI[30(p.727),33(p.1290),56] and BD,[17(Table 1)] as in serious illnesses in general, requiring parenteral fluids and nutrition until motility returns.[51(p.1208)]

II.C.2.b. Sacral parasympathetic tone is abolished in both conditions, resulting in bladder flaccidity with urinary retention, requiring catheterization.[31(p.2225), 51(p.1209)]

II.C.3. Hypothermia and *poikilothermia* are common in SCI, due to cutaneous vasodilation (from sympathetic paralysis) causing increased heat loss, plus physical inactivity and inability to shiver (from skeletal muscle paralysis) causing decreased heat generation.[29(p.1170),30(p.727),33(p.1287),48(p.327),55(p.1232)] For identical

---

[††] Some standard vasopressors can have paradoxically deleterious effects in the context of sympathetic paralysis. If a pressor is necessary, dopamine is the drug of choice, but judicious volume expansion without pressors is preferable and often suffices to restore blood pressure to an acceptable range in both SCI[43(pp.1108-1110)] and BD.[31(p.2225),32,49(pp.548-549)] The high rate of early cardiac arrest in both conditions is partly iatrogenic, due to inability of the compromised heart to accommodate overly aggressive fluid resuscitation, leading to pulmonary edema and congestive heart failure in less than expert hands.[14(pp.838-839),29(p.1169),31(pp.2225-6), 43(pp.1106,1109),49(p.548),51(p.1204),55(p.1229)]

reasons, hypothermia and poikilothermia also occur in up to 86% of BD patients.[15(p. 1370),31(p.2224),32(p.819),49(p.551)] Hypothermia exacerbates cardiovascular instability in both conditions, requiring heating lamps and multiple blankets.

Hyperthermia can result from high ambient temperature or humidity, compounded by absent sympathetic control of sweating, in both SCI[29(p.1170),30(pp.725-727),33(p.1287),48(p.327),55(pp.1232-1233)] and BD.[17(Table 1)] Although in both conditions infection sometimes induces fever, usually it does not.[17(Table 1),33(p.1289)]

II.D. Hypothalamic-anterior pituitary functions, intact in SCI, tend surprisingly to be relatively preserved also in BD, presumably by virtue of the inferior hypophysial artery, which arises extradurally from the internal carotid.[15(p.1371),57-60]

II.D.1. Thyroid function is frequently altered in BD, but not as hypothalamic hypothyroidism: rather as euthyroid sick syndrome, with relatively normal TSH and T4 but decreased T3,[57-61] due to decreased peripheral deiodination of T4 to T3, perhaps from interleukin-6 [62,63] or free-radical-induced selenium deficiency.[64‡‡] It is a common epiphenomenon of critical illness in general,[75,76] including SCI.[77]

II.D.2. Other anterior pituitary functions are variably affected in BD[57-60] and less relevant for somatic unity. The only indication for even cortisol administration in organ donors is pre-existing adrenal suppression from recent steroids.[15(p.1371)]

II.E. Predisposition to infection. In both conditions immobility and lack of coughing and sighing promote pneumonia and diminish antibiotic efficacy.[17(Table 1), 33(p.1289),55(p.1231)] Bladder dysfunction promotes urinary tract infection,[17(Table 1), 30(p.727)] and immobility predisposes to decubitus ulcers with local infection and risk of sepsis. Preventive nursing care is identical for both conditions.

II.F. Actuarial survival curves

II.F.1. High cervical quadriplegia has a *biphasic* survival curve, with rapid drop-off in the first three months (especially the first). In one study three-month survival was only 43%.[78(p.82)] The subsequent chronic phase has a relatively low death rate.[78(pp.81-82),79(p.1286),80(p.1242)] Survival curves for BD treated equally aggressively also feature a rapid drop-off over the first several months (especially the first), followed by a chronic phase of relative stabilization.[17] (Unfortunately, statistics for the first week after BD are unknown.)

II.F.2. Age is a significant determinant of survival rate. Following SCI the proportion alive at a given time after injury decreases significantly with advancing age.[28(p.503),55(p.1230),78(Table 3,p.79),79(p.1286),80(p.1242)] A similar age effect characterizes survival potential in BD.[17]

III. Chronic phase

III.A. Resolution of spinal shock. Beginning 2 to 6 weeks after SCI, autonomous cord function gradually returns. In the few BD patients maintained through the acute

‡‡Thus the utility of thyroid replacement therapy for maximizing organ viability remains controversial.[15(p.1379),65] Some investigators have reported improved metabolic status and myocardial contractility following triiodothyronine administration in BD[37,66,67] or during cardiopulmonary bypass surgery,[68,69] but this probably represents a direct pharmacologic effect.[49,70-72] Others have reported no benefit from exogenous T3 in controlled placebo comparisons,[73] and in one study metabolic acidosis actually worsened.[74]

phase, spinal-shock symptomatology also gradually resolves around the same time frame, producing relative somatic stabilization.

III.A.1. Development of spasticity. Following SCI the first clinical manifestation of recovery from spinal shock is return of tendon and cutaneous reflexes, which soon become hyperactive as flaccid tone becomes spastic.[30(p.727),33(p.1290),42(p.186)] Triple flexion (hips, knees, ankles) and other withdrawal reflexes appear in response to noxious stimuli.[30(p.728),42(p.186),43(pp.1123-1124)] In prolonged BD, tone and reflexes likewise become pathologically increased (barring intrinsic damage to the cord).[17§§]

III.A.2. Sympathetic tone returns in parallel with skeletal muscle tone and is perhaps the most important systemic aspect of recovery from spinal shock.

III.A.2.a. During the weeks following SCI, *hemodynamic status stabilizes:* vasopressors can be weaned, the heart better accommodates fluctuations in intravascular volume, and fluid management simplifies.[29(p.1169),33(p.1290),41(p.1063)] Exactly the same occurs in chronic BD; prior to 4 weeks the most frequent terminal event is spontaneous asystole, whereas after 4 weeks it is treatment withdrawal.[17]

III.A.2.b. In SCI *bradycardia and bradyarrhythmias completely resolve* by 2 to 6 weeks.[47,53] Similarly, in prolonged BD cardiac rhythm also stabilizes.[17(Table 1)]

III.A.2.c. In SCI *gastrointestinal motility returns* and enteral feedings can resume.[33(p.1290),55(pp.1231,1234)] The same is true in some prolonged BD cases (through the intrinsic neurenteric plexus and intraspinal autonomic integration),[17(Table 1)] although lack of swallowing requires providing enteral feedings by tube, in contrast to SCI.

III.A.2.d. In SCI *piloerection and sweating gradually return.*[30(p.728)] Parallel data for chronic BD are lacking in the literature, although in one patient whom the author personally examined ("TK") these functions did return.[17(Table 1)]

III.A.3. Recovery of sacral parasympathetic tone. In SCI automatic bladder function gradually returns.[30(p.728)] Urinary retention and associated urinary tract infections become less problematic as bladder spasticity causes smaller capacity and frequent spontaneous voidings. Again, parallel data are lacking for prolonged BD, although patient "TK" did develop such bladder function, with spontaneous micturition about every 2 hours.[17(Table 1)]

III.A.4. In both SCI and BD *thermoregulation improves*, although most patients remain poikilothermic to some degree; hyperthermia can occur in the absence of infection, possibly from spastic hypertonia or autonomic dysreflexia (see below).[17(Table 1),55(p.1233)]

III.B. In both conditions *intercurrent infections* can precipitate *recurrence of spinal shock* with dramatic hemodynamic decompensation. The symptoms dissipate upon resolution of the infection.[17(Table 1),42(p.186)]

III.C. Autonomic hyperreflexia (dysreflexia) develops in 65-85% of SCI cases, with sudden episodes of severe hypertension triggered by noxious or visceral stimuli (e.g., defecation, full bladder, tracheal suctioning).[29(p.1169),30(p.728),33(p.1290),43(pp.

§§Unfortunately, the available information regarding the time course of these changes is too scanty for comparison with SCI.

1124-1125),48(p.327) Manifestations include profuse flushing and diaphoresis of head and neck, nasal congestion, headache and chest pain. Pathogenesis probably involves denervation hypersensitivity of peripheral adrenergic receptors.[43(p.1124)]

Reflex hypertensive responses to surgical incision in unanesthetized BD organ donors are similar (apart from heart rate – see below) yet occur much earlier than the several weeks typically required for development of autonomic dysreflexia following SCI.[15(p.1371),81-87] The timing may not represent a real point of difference, however. Little is known about the hemodynamic response to unanesthetized surgery in acute SCI, since anesthesia is typically employed. Perhaps splitting the entire thorax and abdomen is a stimulus powerful enough to overcome any spinal shock.

Less is known about autonomic dysreflexia in chronic BD, but some case reports do mention episodes of hypertension after a latency of several weeks.[17(Table 1)] Patient "TK" during the author's examination developed transient blotchy erythema of the face, neck and shoulders in response to noxious stimuli such as pinching or ice water trickling down the neck. Nurses have documented that such reaction is characteristically associated with a transient rise in blood pressure and heart rate.

3.2. Differences

The extensiveness of somatic parallels between SCI and BD contrasts markedly with the relative paucity and unimportance of the differences, as follows (outline structure summarized in Table 2).

I. Hypothalamic-posterior pituitary function is intact in SCI and variable in BD. Diabetes insipidus occurs in 38-87% of BD patients.[15(p.1370),31(p.2226),57-60, 88] The inconsistency presumably stems from vicissitudes of vascular supply from the same extradural collaterals responsible for preserving anterior pituitary functions (see above). If inadequately treated, the ensuing hypovolemia and electrolyte imbalance will exacerbate the already tenuous cardiovascular status. Diabetes insipidus can appear after a brief latency following BD, and in some chronic cases (e.g., "TK") it can also gradually resolve.[17(Table 1)]

II. Vagus nerve functions are intact in SCI and absent in BD.

II.A. Efferent functions.

II.A.1. Bradycardia. In BD loss of both sympathetic and parasympathetic functions leaves the heart's rate determined solely by its intrinsic pacemaker, which is slower than normal and lacks spontaneous variability.[89,90] The bradycardia of high SCI is even more profound due to unopposed parasympathetic tone,[29(p.1169),30(pp.725,727), 33(p.1287),53] and severe reflex bradycardia can be precipitated by vagal stimulation such as intubation, oropharyngeal suctioning, ocular pressure, or a sigh.[33(p.1287)] Such disadvantageous reactions do not occur in BD. Atropine can treat and prevent severe bradycardia in SCI [43(pp.1104,1110)] but has no chronotropic effect in BD because vagal function is already absent.[31(p.2225)]

The bradycardia associated with autonomic hyperreflexia in chronic SCI is vagally mediated.[30(p.728),33(p.1290),43(pp.1124-1125),48(p.327)] By contrast, the hypertension upon unanesthetized organ retrieval in BD (or upon somatic irritation in chronic BD) is associated with *tachycardia,* presumably mediated by unopposed spinal

Table 2. Somatic pathophysiological *differences* between high cervical cord transection and BD.

I. Hypothalamic-posterior pituitary function (diabetes insipidus in 38-87% of BD, absent in SCI)
II. Vagus nerve functions (intact in SCI, absent in BD)
  A. Efferent functions
    1. Bradycardia (more severe in SCI due to unopposed vagal tone; a hemodynamic *disadvantage* compared with BD)
    2. Recovery of gastrointestinal motility following spinal shock seems less common in BD than in SCI
  B. Afferent function (intact in SCI, absent in BD) – irrelevant if BD involves diabetes insipidus
III. Glossopharyngeal nerve function (intact in SCI, absent in BD) – irrelevant if BD involves diabetes insipidus
IV. Associated systemic complications – disseminated intravascular coagulation in 25-65% of BD, much less in SCI

---

sympathetic reflexes, including adrenally released circulating catecholamines.[15(p.1371), 81-87]

II.A.2. Recovery of *gastrointestinal motility* following spinal shock may occur less often in BD than in SCI, perhaps because of the permanent vagal suppression in BD.

II.B. Afferent function. The vagus nerve transmits signals from the baroreceptors of the cardiac atria and aortic arch to the nucleus of the tractus solitarius, which participates in regulation of blood pressure and volume through vagal efferent adjustment of heart rate, various bulbospinal pathways and hypothalamic modulation of sympathetic tone and release of antidiuretic hormone.[91(pp.315-316),92(pp.187-188),93(pp.1066-1068)] Since rostral control of sympathetic function is abolished in both SCI and BD, and since posterior pituitary function is variably affected in BD, the only meaningful comparison of vagal afferent function between the two conditions involves the subset of BD without diabetes insipidus. In SCI the atrial and aortic baroreceptors participate through vagal afferents in modulation of vagal efferent tone and antidiuretic hormone secretion,[30(p.727),43(p.1124)] whereas in BD without diabetes insipidus they do not.

III. Glossopharyngeal nerve function is intact in SCI and absent in BD. Together with vagal afferents, a branch of the ninth cranial nerve to the carotid body conveys additional baroreceptor information, all processed together and ultimately influencing homeostasis as just described.[91(pp.315-316)] The somatic differences between SCI and BD attributable to glossopharyngeal function are therefore the same as those attributable to vagal afferents, are relatively minor, and are nonexistent in the context of diabetes insipidus.

IV. Associated systemic complications. Disseminated intravascular coagulation complicates 25-65% of BD organ donors.[15(p.1370),49(p.551)] It is not described in the spinal cord literature, presumably because much less tissue thromboplastin is released by SCI than by massive brain injury.

These, then, are the somatic differences. Note that other cranial nerve functions and especially consciousness, though preserved in SCI and absent in BD, are not listed, because, important as they are for the patient, they are unrelated to the focus of this comparison, namely somatic pathophysiology insofar as it is relevant to the question whether the body in either condition is a sick organism or a mere collection of organs.

4. DISCUSSION

The similarity of effects of SCI and BD on the body is surprising at first sight and fascinating in itself, but its relevance to one of the several rationales advanced to explain why BD should be equated with death elevates the comparison from a mere physiological curiosity to a conceptually important observation.

According to the mainstream, "orthodox" rationale, the purported loss of somatic integrative unity in BD is attributable to destruction of the many brain-stem and hypothalamic integrative centers.[3-8,13] But is it *their destruction per se* or rather *the body's nonreception of their influence* that most immediately affects somatic integration? Surely the latter, because it is more proximate to the phenomenon of interest, it is the means through which the former exerts its effect, and it can also be brought about by other possible causes such as mere disconnection from cephalic structures. That the impact on somatic physiology of nonreception of rostral influence should be indifferent to the *reason* for the nonreception implies that body A with a destroyed brain and body B with a disconnected brain (e.g., due to high SCI) should have the same vital status. Logical consistency demands that if we assert that A is dead as a biological organism, we must be prepared to say the same of B; but if we insist that B is clearly alive as a biological organism (and not merely because it is conscious), then we must be willing to admit the same of A. We now consider this somatic equivalence first in theory and then in practice.

4.1. Somatic Equivalence in Theory

The anatomical pathways through which somatically integrative information is transferred between body and brain are relatively few.

In the *afferent* direction the routes are threefold:

- spinal cord,
- glossopharyngeal and vagus nerves (from the atrial, aortic and carotid baroreceptors), and
- arterial blood flow to the hypothalamus (especially to supraoptic and paraventricular nuclei).

In the *efferent* direction the routes are likewise threefold:

- spinal cord,
- vagus nerve, and
- pituitary.

The other ten cranial nerves terminate in the head or neck and are irrelevant to somatic integration. Also note that the second and third afferent pathways are relevant only in contexts where the corresponding efferent limbs are intact. Blood flow to and from the brain in general is conceivably an additional pathway, though not for all practical purposes. Certainly extra-hypothalamic brain has receptors for various hormones and circulating chemicals, but those even remotely relevant to somatic integrative unity involve the hypothalamic-pituitary axis. The brain is not known to be a secretory organ (apart from the portal system of the adenohypophysis). Similarly, the production, circulation, and venous absorption of cerebrospinal fluid could be another theoretically conceivable route for brain-body chemical interaction, but there is no evidence that such exchange serves any somatically integrating role.

Plainly, disruption of any one of the significant pathways of encephalo-somatic communication does not destroy the unity of the "organism as a whole." Every endocrinology clinic has patients with diabetes insipidus or even panhypopituitarism who live perfectly normal lives on replacement therapy. Every major rehabilitation center cares for ventilator-dependent patients with high cervical quadriplegia. Sometimes for therapeutic reasons the vagus nerve is ablated surgically or pharmacologically.

What about elimination of two of these three routes? For example, suppose that a high cervical quadriplegic were given atropine to treat bradycardia. The somatic physiology of such a patient would be virtually identical to that of a BD patient without diabetes insipidus (the only difference being preserved carotid-body modulation of antidiuretic hormone in the SCI patient).

To complete the analogy, suppose that the atropinized SCI victim was an endocrinology patient with chronic panhypopituitarism, stable on replacement therapy. The somatic physiology relevant to integrative unity is now absolutely identical to that of total brain infarction. The *only* difference lies in consciousness and those cranial nerves restricted to the head and neck. Is such a body an implacably disintegrating "collection of organs," or a live "organism as a whole" that happens to be severely disabled and dependent on medical technology? If the former, then we would have the bizarre anomaly of a "conscious corpse"; if the latter, then the BD body must equally be an "organism as a whole" despite *its* severe disability and technological dependence (its unconsciousness being an additional disability, but not one that *per se* settles the question whether this is an unconscious organism or a non-organism).

From the *body's perspective*, BD and atropinized high cord transection are virtually indistinguishable (comparing SCI with the subset of BD *without* diabetes insipidus, or "total" BD with the subset of SCI *with* pharmacologically controlled diabetes insipidus), because the caudal margin of total brain infarction *is* in fact a cervico-medullary junction infarction. Thus, *regardless how one might choose to define operationally terms such as "integrative unity" or "organism as a whole," if they are defined carefully enough to apply properly to any ventilator-dependent quadriplegic with diabetes insipidus, then* ipso facto *they will apply as well to any BD patient.*[94]

Not only is this conclusion inescapable on theoretical grounds; it is fully ratified in clinical experience.

4.2. Somatic Equivalence in Practice

The literatures on intensive care of acute SCI and BD are so similar that they can almost be mutually transformed one into the other merely by interchanging the terms "SCI" and "BD." Yet this curious fact seems to have passed largely unnoticed, even by authors of respective chapters in the same book (e.g., Kofke et al.[33] and MacKenzie and Geisler[43] on the one hand, and Lew and Grenvik[15] on the other).

Induction phase. During the process of brain herniation or spinal cord infarction, but prior to BD or complete cord transection, secondary cardiopulmonary pathology (subendocardial microinfarcts, neurogenic pulmonary edema) due to massive sympathetic discharge is common and contributes substantially to the acute hemodynamic instability and early mortality in both BD and SCI. In some cases multiple non-neural organs are also damaged directly by whatever primary etiology damaged the brain or cord.

The time course of subsequent relative stabilization, although attributable largely to recovery from spinal shock, also corresponds to resolution and healing of such multisystem complications. That young patients have much greater potential for survival in both conditions also reinforces the idea that the tendency to early demise is attributable more to somatic than to neurologic factors.

It is therefore fallacious to attribute the increased incidence of early arrhythmias and cardiovascular collapse in BD to the absence of brain function *per se*. (It is also inconsistent, if a parallel attribution is not made for SCI.) Rather, the early systemic instabilities are more likely attributable to *systemic pathology antedating the BD*, as suggested also by Novitzky.[37] If the brain or cord become infarcted in a way that does not overactivate the sympathetic system, cardiopulmonary complications should be much less of a problem (cf.[40]).

Spinal shock. The comparison of SCI and BD literatures reveals that every described manifestation of spinal shock also occurs in BD, and conversely, every non-endocrinologic systemic dysfunction characteristic of BD (excluding associated multisystem injury) is explainable in terms of spinal shock. Nevertheless, despite the similarities of somatic symptomatology and the theoretical grounds for equating the somatic pathophysiology of BD with that of cervico-medullary junction transection, a marked explanatory asymmetry has prevailed.

The SCI literature attributes a particular set of signs and symptoms to *functional suppression* of the structurally intact spinal cord *below* the lesion, *below* the foramen magnum, and calls it "spinal shock." By contrast, the BD literature attributes the same set of signs and symptoms directly to *destruction* of brain-stem vegetative control centers *above* the foramen magnum and calls it "loss of somatic integrative unity."

For example, depressed sympathetic tone in SCI is usually attributed to deactivation of second-order sympathetic neurons in the intermediolateral cell column, but in BD it is typically attributed to destruction of first-order sympathetic neurons in the hypothalamus and their axons in passage through the medulla (e.g.,[95]). But since BD includes cervico-medullary junction infarction, in the context of which the rostral integrity of the sympathetic system is physiologically irrelevant, the most parsimonious explanation of impaired sympathetic tone in BD is spinal shock, just as in SCI. As another example, it is illogical to attribute hypothermia in BD to destruction of hypothalamic thermoregulatory centers (as does Gert[10]), when the same hypothermia would result from the caudal end of the brain pathology alone (i.e., cervico-medullary junction infarction) even if the hypothalamus and medulla were intact, as in SCI.

The same could be said about every other non-endocrinologic somatic dysfunction in BD. This explanatory asymmetry is probably motivated by an *a priori* conviction that BD *ought* to be equated with organismal death and SCI not; regardless, it is a logical double standard without physiological basis.

Moreover, because unopposed parasympathetic tone is more disadvantageous hemodynamically than absence of both sympathetic and parasympathetic tone, one could argue that, *precisely because vagal function is intact in SCI* and often needs to be pharmacologically suppressed, *SCI bodies are ironically even less integrated than BD bodies* (at least those without diabetes insipidus). Therefore, if despite this disadvantage SCI patients still possess enough integrative unity to qualify as living "organisms as a whole," all the more do BD patients.

The similarity of survival curves of the two conditions also argues that high SCI is every bit as somatically "dis-integrating" as BD during the acute phase, yet no one concludes therefrom that the bodies of SCI patients are *ipso facto* "dead" or are mere collections of organs without unity at the level of the "organism as a whole." Conversely, the somatic instability in BD is potentially every bit as transient as in SCI. If it looks like spinal shock, acts like spinal shock, resolves like spinal shock – why not call it spinal shock?

Whether the somatic symptomatology of BD is best understood as manifesting "loss of integrative unity" or "spinal shock" is answerable through considering the constellation and temporal evolution of clinical signs. The comparison is greatly obscured by the fact that BD patients are typically not supported for the 2 to 6 weeks necessary to manifest the resolution characteristic of spinal shock. But the case reports of BD patients who *have* been so maintained[17] indicate that the respective somatic symptomatologies of BD and SCI do in fact closely parallel one another. The parallel is not only qualitative (signs, symptoms) and quantitative (severity, high rate of early asystole) but also temporal (gradual return of autonomous cord function and somatic stabilization) – unless, of course, the BD patient also sustained diffuse spinal cord infarction or injury, a common yet under-appreciated association, especially if the etiology is anoxic-ischemic.

Endocrinologic differences. Apart from consciousness and the strictly cephalic cranial nerves, the only significant difference between high SCI and BD is hypothalamic-pituitary function. But hypothalamic dysfunction or nonfunction has *never* been a diagnostic requirement for BD: neither according to the original Harvard Committee,[44] nor the President's Commission,[4] nor the British Conference of Medical Royal Colleges,[12] nor the Swedish Committee,[3] nor the Task Force for the Determination of Brain Death in Children,[96] nor the myriad other proposers of diagnostic criteria.[97] Diabetes insipidus is absent in 13-62% of BD potential organ donors.[15(p.1370), 31(p.2226)] Significantly, preservation of such somatically integrative hypothalamic function is declared explicitly *compatible* with the diagnosis of BD according to the most recent practice parameter of the American Academy of Neurology,[45] even though it contradicts the very definition of "whole-brain death."[98-100] For *this* subgroup of BD, the somatic physiologic distinction from atropinized high spinal cord transection is virtually nonexistent.

5. CONCLUSION

In summary, if the loss of brain-regulation of the body in the one context (SCI) is insufficient to constitute cessation of the "organism as a whole," then the same loss of brain-regulation must be equally insufficient in the other context (BD). To be sure, total brain destruction is a fatal lesion, but "fatal" in the sense of a strong tendency to bodily death (which can be opposed at least for a while by medical intervention) rather than *per se* equaling bodily death. The brain cannot be construed with physiological rigor as the body's "central integrator," in the sense of *conferring* unity top-down on what would otherwise be a mere collectivity of organs. Neither is any other organ "the central integrator." A living body possesses not an integr*ator* but integr*ation*, a holistic property deriving from the mutual interaction among all the parts.[94]

If BD is to be coherently equated with death, "death" must therefore be understood in a non-biological sense, as proponents of "higher-brain" or consciousness formulations have advocated.[20-26] Whether society will want to adopt such a concept of death – which amounts to a notion of the human *person* as not only conceptually distinct, but *actually dissociable*, from a living human body – remains to be seen. That debate is purely philosophical in nature and exceeds the present scope. What is clear from a somatic pathophysiological comparison with SCI is that the mainstream assertion that BD represents biological death of the human "organism as a whole" is physiologically untenable.

6. REFERENCES

1. Byrne PA, O'Reilly S, Quay PM. Brain death - an opposing viewpoint. *JAMA* 1979;**242:**1985-1990.
2. Byrne PA, O'Reilly S, Quay PM, Salsich PW, Jr. Brain death - the patient, the physician, and society. *Gonzaga Law Rev* 1982/83;**18:**429-516.
3. Swedish Committee on Defining Death. *The concept of death. Summary.* Stockholm: Swedish Ministry of Health and Social Affairs, 1984.
4. President's Commission for the Study of Ethical Problems in Medicine and Biomedical and Behavioral Research. *Defining Death: Medical, Legal, and Ethical Issues in the Determination of Death.* Washington, DC: U.S. Government Printing Office, 1981.
5. Bernat JL. The definition, criterion, and statute of death. *Semin Neurol* 1984;**4:**45-51.
6. Bernat JL. *Ethical Issues in Neurology.* Boston: Butterworth-Heinemann, 1994.
7. Korein J. The problem of brain death: development and history. *Ann NY Acad Sci* 1978;**315:**19-38.
8. White RJ, Angstwurm H, Carrasco de Paula I. Final considerations formulated by the scientific participants. In: White RJ, Angstwurm H, Carrasco de Paula I, eds. *Working Group on the Determination of Brain Death and its Relationship to Human Death. 10-14 December, 1989.* (Scripta Varia 83). Vatican City: Pontifical Academy of Sciences, 1992:81-82.
9. Bernat JL. A defense of the whole-brain concept of death. *Hastings Cent Rep* 1998;**28:**14-23.
10. Gert B. A complete definition of death. In: Machado C, ed. *Brain Death. Proceedings of the Second International Conference on Brain Death. Havana, Cuba, February 27-March 1, 1996.* Amsterdam: Elsevier, 1995:23-30.
11. Pallis C. Whole-brain death reconsidered - physiological facts and philosophy. *J Med Ethics* 1983;**9:**32-37.
12. Conference of Medical Royal Colleges and their Faculties in the United Kingdom. Diagnosis of death. *Br Med J* 1979;**1:**3320.
13. Lamb D. *Death, Brain Death and Ethics.* Albany, NY: State University of New York Press, 1985.
14. Guerriero WG. Organ transplantation. In: Narayan RK, Wilberger JEJ, Povlishock JT, eds. *Neurotrauma.* New York: McGraw-Hill, 1996:835-840.

15. Lew TWK, Grenvik A. Brain death, vegetative state, donor management, and cessation of therapy. In: Albin MS, ed. *Textbook of Neuroanesthesia with Neurosurgical and Neuroscience Perspectives*. New York: McGraw-Hill, 1997:1361-1381.
16. Pallis C, Harley DH. *ABC of Brainstem Death*. London: BMJ Publishing Group, 1996.
17. Shewmon DA. Chronic "brain death": meta-analysis and conceptual consequences. *Neurology* 1998;**51:** 1538-1545.
18. Conference of Medical Royal Colleges and their Faculties in the United Kingdom. Diagnosis of brain death. *Br Med J* 1976;**2:**1187-1188.
19. Ibe K. Clinical and pathophysiological aspects of the intravital brain death [abstract]. *Electroencephalogr Clin Neurophysiol* 1971;**30:**272.
20. Youngner SJ, Bartlett ET. Human death and high technology: the failure of the whole-brain formulations. *Ann Intern Med* 1983;**99:**252-258.
21. Gervais KG. *Redefining Death*. New Haven: Yale University Press, 1986.
22. Green MB, Wikler D. Brain death and personal identity. In: Cohen M, Nagel T, Scanlon T, eds. *Medicine and Moral Philosophy. A Philosophy and Public Affairs Reader*. Princeton, NJ: Princeton University Press, 1982:49-77.
23. Lizza JP. Persons and death: What's metaphysically wrong with our current statutory definition of death? *J Med Philos* 1993;**18:**351-374.
24. Machado C. A new definition of death based on the basic mechanisms of consciousness generation in human beings. In: Machado C, ed. *Brain Death. Proceedings of the Second International Conference on Brain Death. Havana, Cuba, February 27-March 1, 1996*. Amsterdam: Elsevier, 1995:57-66.
25. Veatch RM. The impending collapse of the whole-brain definition of death [erratum in Hastings Cent Rep 1993;23(6):4]. *Hastings Cent Rep* 1993;**23(4):**18-24.
26. Zaner RM. Death: Beyond Whole-Brain Criteria. In: Engelhardt HTJ, Spicker SF, eds. *Philosophy and Medicine*. Dordrecht/Boston: Kluwer Academic Publishers, 1988.
27. Matjasko MJ. Multisystem sequelae of severe head injury. In: Cottrell JE, Smith DS, eds. *Anesthesia and Neurosurgery*. 3rd ed. St. Louis: Mosby, 1994:685-712.
28. Geisler WO, Jousse AT. Life expectancy following traumatic spinal cord injury. In: Frankel HL, ed. *Spinal Cord Trauma*. Amsterdam: Elsevier Science Publishers, 1992:499-513.
29. Teeple E, Heres EK. Anesthesia management of spinal trauma. In: Narayan RK, Wilberger JE, Jr., Povlishock JT, eds. *Neurotrauma*. New York: McGraw-Hill, 1996:1167-1177.
30. Albin MS. Spinal cord injury. In: Cottrell JE, Smith DS, eds. *Anesthesia and Neurosurgery*. 3rd ed. St. Louis: Mosby, 1994:713-743.
31. Darby JM, Stein K, Grenvik A, Stuart SA. Approach to management of the heartbeating 'brain dead' organ donor. *JAMA* 1989;**261:**2222-2228.
32. Field DR *et al.* Maternal brain death during pregnancy: medical and ethical issues. *JAMA* 1988;**260:**816-822.
33. Kofke WA, Yonas H, Wechsler L. Neurologic intensive care. In: Albin MS, ed. *Textbook of Neuroanesthesia with Neurosurgical and Neuroscience Perspectives*. New York: McGraw-Hill, 1997:1247-1347.
34. Antonini C *et al.* Morte cerebrale e sopravvivenza fetale prolungata. [Brain death and prolonged fetal survival]. *Minerva Anestesiol* 1992;**58:**1247-1252.
35. Novitzky D *et al.* Pathophysiology of pulmonary edema following experimental brain death in the chacma baboon. *Ann Thorac Surg* 1987;**43:**288-294.
36. Novitzky D, Rose AG, Cooper DK. Injury of myocardial conduction tissue and coronary artery smooth muscle following brain death in the baboon. *Transplantation* 1988;**45:**964-966.
37. Novitzky D. Heart transplantation, euthyroid sick syndrome, and triiodothyronine replacement. *J Heart Lung Transplant* 1992;**11(4 Pt 2):**S196-198.
38. Samuels MA. Cardiopulmonary aspects of acute neurologic diseases. In: Ropper AH, ed. *Neurological and Neurosurgical Intensive Care*. 3rd ed. New York: Raven Press, 1993:103-119.
39. Yoshida K-I, Ogura Y, Wakasugi C. Myocardial lesions induced after trauma and treatment. *Forensic Sci Int* 1992;**54:**181-189.
40. Novitzky D *et al.* Prevention of myocardial injury during brain death by total cardiac sympathectomy in the Chacma baboon. *Ann Thorac Surg* 1986;**41:**520-524.
41. Tator CH. Classification of spinal cord injury based on neurological presentation. In: Narayan RK, Wilberger JE, Jr., Povlishock JT, eds. *Neurotrauma*. New York: McGraw-Hill, 1996:1059-1073.
42. Eidelberg E. Consequences of spinal cord lesions upon motor function, with special reference to locomotor activity. *Prog Neurobiol* 1981;**17:**185-202.

43. MacKenzie CF, Geisler FH. Management of acute cervical spinal cord injury. In: Albin MS, ed. *Textbook of Neuroanesthesia with Neurosurgical and Neuroscience Perspectives*. New York: McGraw-Hill, 1997:1083-1136.
44. Beecher HK *et al.* A definition of irreversible coma. Report of the Ad Hoc Committee of the Harvard Medical School to Examine the Definition of Brain Death. *JAMA* 1968;**205:**337-340.
45. American Academy of Neurology - Quality Standards Subcommittee. Practice parameters for determining brain death in adults (Summary statement). *Neurology* 1995;**45:**1012-1014.
46. Wijdicks EF. Determining brain death in adults. *Neurology* 1995;**45:**1003-1011.
47. Lehman KG, Lane JG, Piepmeier JM, Batsford WP. Cardiovascular abnormalities accompanying acute spinal cord injury in humans: incidence, time course and severity. *J Am Coll Cardiol* 1987;**10:**46-52.
48. Stoelting RK, Dierdorf SF, McCammon RL. *Anesthesia and Co-Existing Disease*. New York: Churchill Livingstone, 1988.
49. Robertson KM, Cook DR. Perioperative management of the multiorgan donor. *Anesth Analg* 1990; **70:**546-556.
50. Drory Y, Ouaknine G, Kosary IZ, Kellermann JJ. Electrocardiographic findings in brain death; description and presumed mechanism. *Chest* 1975;**67:**425-432.
51. Rodts GE, Jr., Haid RW, Jr. Intensive care management of spinal cord injury. In: Narayan RK, Wilberger JE, Jr., Povlishock JT, eds. *Neurotrauma*. New York: McGraw-Hill, 1996:1201-1212.
52. Logigian EL, Ropper AH. Terminal electrocardiographic changes in brain-dead patients. *Neurology* 1985; **35:**915-918.
53. Winslow EBJ, Lesch M, Talano JV, Meyer PR, Jr. Spinal cord injuries associated with cardiopulmonary complications. *Spine* 1986;**11:**809-812.
54. Wolman L. The disturbance of circulation in traumatic paraplegia in acute and late stages: a pathological study. *Paraplegia* 1965;**2:**213-226.
55. Cahill DW, Rechtine GR. The acute complications of spinal cord injury. In: Narayan RK, Wilberger JE, Jr., Povlishock JT, eds. *Neurotrauma*. New York: McGraw-Hill, 1996:1229-1236.
56. Frost FS. Gastrointestinal dysfunction in spinal cord injury. In: Yarkony GM, ed. *Spinal Cord Injury. Medical Management and Rehabilitation*. Gaithersburg, MD: Aspen Publishers, 1994:27-39.
57. Arita K *et al.* The function of the hypothalamo-pituitary axis in brain dead patients. *Acta Neurochir (Wien)* 1993;**123:**64-75.
58. Gramm H-J *et al.* Acute endocrine failure after brain death? *Transplantation* 1992;**54:**851-857.
59. Howlett TA *et al.* Anterior and posterior pituitary function in brain-stem-dead donors. A possible role for hormonal replacement therapy. *Transplantation* 1989;**47:**828-834.
60. Keogh AM, Howlett TA, Perry L, Rees LH. Pituitary function in brain-stem dead organ donors: a prospective survey. *Transplant Proc* 1988;**20:**729-730.
61. Powner DJ *et al.* Hormonal changes in brain dead patients. *Crit Care Med* 1990;**18:**702-708.
62. Boelen A *et al.* Induced illness in interleukin-6 (IL-6) knock-out mice: a causal role of IL-6 in the development of the low 3,5,3'-triiodothyronine syndrome. *Endocrinology* 1996;**137:**5250-5254.
63. Hashimoto H *et al.* The relationship between serum levels of interleukin-6 and thyroid hormone in children with acute respiratory infection. *J Clin Endocrinol Metab* 1994;**78:**288-291.
64. Berger MM, Lemarchand-Béraud T, Cavadini C, Chioléro R. Relations between the selenium status and the low T3 syndrome after major trauma. *Intensive Care Med* 1996;**22:**575-581.
65. Karayalçin K *et al.* Donor thyroid function does not affect outcome in orthotopic liver transplantation. *Transplantation* 1994;**57:**669-672.
66. Novitzky D, Cooper DK, Morrell D, Isaacs S. Change from aerobic to anaerobic metabolism after brain death, and reversal following triiodothyronine therapy. *Transplantation* 1988;**45:**32-36.
67. Washida M *et al.* Beneficial effect of combined 3,5,3'-triiodothyronine and vasopressin administration on hepatic energy status and systemic hemodynamics after brain death. *Transplantation* 1992;**54:**44-49.
68. Clark RE. Cardiopulmonary bypass and thyroid hormone metabolism. *Ann Thorac Surg* 1993;**56(1 Suppl):**S35-41; discussion S41-42.
69. Klemperer JD, Klein I, Gomez M, et al. Thyroid hormone treatment after coronary-artery bypass surgery. *N Engl J Med* 1995;**333:**1522-1527.
70. Hsu R-B, Huang T-S, Chen Y-S, Chu S-H. Effect of triiodothyronine administration in experimental myocardial injury. *J Endocrinol Invest* 1995;**18:**702-709.
71. Sypniewski E. Comparative pharmacology of the thyroid hormones. *Ann Thorac Surg* 1993;**56(1 Suppl):** S2-6; discussion S6-8.
72. Utiger RD. Altered thyroid function in nonthyroidal illness and surgery. *N Engl J Med* 1995;**333:**1562-1563.
73. Goarin J-P *et al.* The effects of triiodothyronine on hemodynamic status and cardiac function in potential heart donors. *Anesth Analg* 1996;**83:**41-47.

74. Randell TT, Höckerstedt KAV. Triiodothyronine treatment in brain-dead multiorgan donors -- a controlled study. *Transplantation* 1992;**54:**736-738.
75. Chopra IJ. Clinical review 86: Euthyroid sick syndrome: Is it a misnomer? *J Clin Endocrinol Metab* 1997; **82:**329-334.
76. Rolih CA, Ober KP. The endocrine response to critical illness. *Med Clin North Am* 1995;**79:**211-224.
77. Cheville AL, Kirshblum SC. Thyroid hormone changes in chronic spinal cord injury. *J Spinal Cord Med* 1995;**18:**227-232.
78. Mesard L, Carmody A, Mannarino E, Ruge D. Survival after spinal cord trauma. A life table analysis. *Arch Neurol* 1978;**35:**78-83.
79. Carter RE, Graves DE. Spinal cord injury occurring in older individuals. In: Narayan RK, Wilberger JE, Jr., Povlishock JT, eds. *Neurotrauma*. New York: McGraw-Hill, 1996:1281-1287.
80. Piepmeier JM. Late sequelae of spinal cord injury. In: Narayan RK, Wilberger JE, Jr., Povlishock JT, eds. *Neurotrauma*. New York: McGraw-Hill, 1996:1237-1244.
81. Fitzgerald RD *et al.* Cardiovascular and catecholamine response to surgery in brain-dead organ donors. *Anaesthesia* 1995;**50:**388-392.
82. Fitzgerald RD *et al.* Endocrine stress reaction to surgery in brain-dead organ donors. *Transpl Int* 1996;**9:** 102-108.
83. Gramm H-J *et al.* Hemodynamic responses to noxious stimuli in brain-dead organ donors. *Intensive Care Med* 1992;**18:**493-495.
84. Gramm H-J, Schäfer M, Link J, Zimmermann J. Authors' reply and report on another manifestation of a possible rise in sympathoadrenal activity during retrieval surgery. *Intensive Care Med* 1994;**20:**165-166.
85. Hill DJ, Munglani R, Sapsford D. Haemodynamic responses to surgery in brain-dead organ donors [letter; comment]. *Anaesthesia* 1994;**49:**835-836.
86. Pennefather SH, Dark JH, Bullock RE. Haemodynamic responses to surgery in brain-dead organ donors. *Anaesthesia* 1993;**48:**1034-1038.
87. Pennefather SH. Hemodynamic responses to noxious stimuli in brain-dead organ donors [letter; comment]. *Intensive Care Med* 1994;**20:**165.
88. Fiser DH, Jimenez JF, Wrape V, Woody R. Diabetes insipidus in children with brain death. *Crit Care Med* 1987;**15:**551-553.
89. García OD *et al.* Heart rate variability in coma and brain death. In: Machado C, ed. *Brain Death. Proceedings of the Second International Conference on Brain Death. Havana, Cuba, February 27-March 1, 1996.* Amsterdam: Elsevier, 1995:191-197.
90. Goldstein B *et al.* Autonomic control of heart rate after brain injury in children. *Crit Care Med* 1996;**24:** 234-240.
91. Blaustein AS, Walsh RA. Regulation of the cardiovascular system. In: Sperelakis N, Banks RO, eds. *Essentials of Physiology*. 2nd ed. Boston: Little, Brown and Company, 1996:309-321.
92. Kopp UC, DiBona GF. Neural control of volume homeostasis. In: Brenner BM, Stein JH, eds. *Body Fluid Homeostasis*. New York: Churchill Livingstone, 1987:185-220.
93. Moss NG, Colindres RE, Gottschalk CW. Neural control of renal function. In: Windhager EE, ed. *Handbook of Physiology. A critical, comprehensive presentation of physiological knowledge and concepts. Section 8: Renal Physiology*. New York: Oxford University Press, 1992:1061-1128.
94. Shewmon DA. The brain and somatic integration: insights into the standard biological rationale for equating 'brain death' with death. *J Med Philos* 2001;**26(5)**:457-478.
95. Litvinoff JS. Maternal brain death during pregnancy [letter]. *JAMA* 1989;**261:**1729.
96. Task Force for the Determination of Brain Death in Children. Guidelines for the determination of brain death in children. *Ann Neurol* 1987;**21:**616-617.
97. Black PM. Brain death (Second of two parts). *N Engl J Med* 1978;**299:**393-401.
98. Halevy A, Brody B. Brain death: reconciling definitions, criteria, and tests. *Ann Intern Med* 1993;**119:** 519-525.
99. Taylor RM. Reexamining the definition and criteria of death. *Semin Neurol* 1997;**17:**265-270.
100. Truog RD. Is it time to abandon brain death? *Hastings Cent Rep* 1997;**27:**29-37.

"BRAIN DEATH" IS NOT DEATH

Paul A. Byrne and Walt F. Weaver*

1. INTRODUCTION

We draw attention to differences and difficulties in language and in concepts between "brain death" and true death that was published 24 years ago.[1] We also focus on failure to utilize the scientific method, sound reasoning, and available medical technology in the determination of *one* of the two most important states known to man: *death*. The other condition, *life*, is obviously related because of the interdependence of the two conditions. Life and true death cannot and do not exist at the same time in the same person.

2. THE COMMITTEE

"Brain death" was not propagated via a medical scientific method. A committee of experts was convened to deal with issues that could affect disposition and/or utilization of these patients. The first words of the "Report of the Ad Hoc Committee of the Harvard Medical School to Examine the Definition of Brain Death"[2] are as follows: "Our primary purpose is to define irreversible coma as a new criterion for death." Was this the hubris of a few academicians or was it simply a surrender to fear of legal chastisement regarding perceived economic and utilitarian needs in 1968, especially the desire to get healthy living vital organs for transplantation?

The primary purpose of the Committee was not to determine ***IF*** irreversible coma was an appropriate criterion for death but *to see to it* that ***IT WAS*** established as a "new criterion for death." With an agenda like that at the outset, the data could be made to fit the already arrived at conclusion. It seems that there was a serious lack of scientific method in this process.

* Paul A. Byrne, MD FAAP, Director of Dept. of Pediatrics and Neonatology, St. Charles Mercy Hospital and Clinical Professor of Pediatrics, Medical College of Ohio. Address: 2600 Navarre Ave., Oregon, OH 43616. Walt F. Weaver, MD FACC, Clinical Associate Professor of Medicine, University of Nebraska School of Medicine, Lincoln, NE. In the past: Member International Society for Heart Transplantation; Chairman Board of Governors, American College of Cardiology and Member Task Force on Clinical Research, American College of Cardiology.

Brain Death and Disorders of Consciousness, Edited by Machado and Shewmon
Kluwer Academic/Plenum Publishers, New York 2004

3. THE BRAIN

The brain consists not of a single part but of several closely interrelated ones (cortex. cerebellum, midbrain, medulla, etc.). Though composed of superficially similar tissues that are closely linked together both anatomically and physiologically, these parts can continue to have activity independently of one another, even when one or more of them have been destroyed. As one might then expect, the brain as a whole has no physiologically identifiable function or functions that could rightly be called the "life-giving function or functions." Rather, there exists a large multiplicity of different functions that are characteristic of the different parts. Although the characteristic functions of the brain-parts normally are closely coordinated, the parts have different functions that often cannot be carried out without the other parts. Further, none of these parts is in complete control of the others.

The brain is an organ whose varied functions serve to integrate physiologically (i.e., by biophysical, biochemical, or other neuronal mechanisms) the different parts of the body. Such physiological operations of integration are, in fact, the ordinary conditions for the continuance of the organismic unity of the body. The brain's ceasing to function does not imply, *a priori,* its destruction but only absence of physiological activity at the time of the evaluation. If the persistence of absence of physiologic activity is accompanied by asystole, hypotension, and other detrimental responses, then this tends quickly, if not instantaneously, to destruction of the brain and disintegration of the body that we call death. However, with immediate institution of life support measures, the brain tissue may end up being only stunned. Often at the time of the initial absence of physiological functioning, this will have caused the patient to be declared "brain dead." Even if another examiner consults (as required in some situations), an apnea test or variant of it[3] will likely further compromise recovery of brain tissue. By this time the *treatment will have shifted* from attempting to reduce further neurological damage to the donor *to preservation of his healthy vital organs for the benefit of the recipient.*

4. MULTIPLE CRITERIA—"BRAIN DEAD" BY ONE BUT NOT THE OTHERS

Chaos has occurred surrounding the label "brain death." The multiple criteria that have evolved worldwide testify to the presence of multiple subsets of this all-inclusive term "brain death." However, the very diagnosis of "brain death" militates against any further attempt to evaluate outcome of these different critical subsets of legally deceased patients, since their true physiologic death comes when they are utilized for vital organ donation, subjects for teaching or research, (permitted under the Uniform Anatomical Gift Act), or when life-support efforts are discontinued. All subsets utilized life-support measures since their value as a live human is maintained in this manner and justified by the perceived "good." There has been little interest or effort to study these patients in terms of classification trials to evaluate long-term response/recovery outcomes with present day life-support efforts in subsets of "brain death" patients.

5. NOT BASED ON VALID SCIENCE

Brain-related criteria are not based on valid scientific data. The Harvard Criteria were published without any patient data and there were no references to basic science reports. The Minnesota Criteria[4] evolved from a study of 25 patients. Only 9 had an EEG done and of these, 2 had "biologic" activity in their EEG after they had been declared "brain dead." Their conclusion: No longer is it necessary to do an EEG.

It seems scientifically invalid not to use an EEG in the diagnosis of "brain death" if any degree of certainty is to be obtained. The British Criteria do not include the EEG.[5] This was apparently due to the influence of the Minnesota Criteria, which do not require an EEG. The National Institutes of Health Criteria were based on a very limited study and, "Accordingly, these criteria are recommended for a larger clinical trial."[6] This has never been done.

By 1978, more than 30 sets of criteria had been published.[7] Many more have appeared subsequently for various reasons and in different countries. In most cases, physicians are free to choose any one of these. Thus, a patient could be determined to be dead by one set, but not by another.

6. FLEXIBLE CRITERIA

No matter how seemingly rigid the criteria are, the ease with which they can be bent is manifested in the report by the President's Commission, where it is written: "An individual with irreversible cessation of all functions of the entire brain, including the brain stem, is dead. The 'functions of the entire brain' that are relevant to the diagnosis are those that are clinically ascertainable"[8] (page 162). In one sentence, whatever stringency there was has been reduced to no more than what can be "clinically ascertainable." Thankfully, there is more physiology taking place in all of us than what is "clinically ascertainable."

If one uses the Minnesota Criteria, the British Criteria, or the published Guidelines of the President's Commission, it is not necessary to include EEG evaluation in determining "brain death." In which case, if the cortex is still functioning, but is wholly cut off from manifesting its activity clinically by damage elsewhere in the brain–something that does occur and which an EEG can clearly show–then this functioning (which could involve memory, feelings, emotion, language skills, etc.) is suddenly considered irrelevant to the person's life or death. According to the NIH Study, 8% of those declared dead on the basis of criteria that omit the EEG, still have cortical activity when evaluated by non-clinical means (EEG). Thus, action such as excision of a donor's beating heart causes death in *at least* one out of twelve cases under such circumstances. As Dr. Walker (Clinical Neurosciences, 1975)[9] wrote, this represents "... an anomalous and undesirable situation." The general public might use much stronger words!

7. "BRAIN DEAD" PREGNANT MOTHERS

It is worrisome to note that "brain dead" pregnant mothers given *modern* life support efforts have survived for as long as 107 days until delivery of a normal child.[10] Yet, in the usual prospective donors there often seems to be a utilitarian based urgency to declare "brain death" and move ahead with vital organ transplantation. Transplant cardiologists know it is important to protect and preserve the vital organs by this urgency, but one must wonder: could it be that it is also urgent to move ahead *before* any signs of *recovery of brain function* would appear and embarrass the physician who had declared death? It is of interest that in "brain dead" victims of homicidal assault, lawyers rarely file charges until the victim is truly and certainly dead. In similar manner, to our knowledge undertakers never embalm until "brain dead" patients are truly and unequivocally dead. Sometimes common sense overrules utilitarian reason!

8. CESSATION OF FUNCTIONING, FUNCTION, FUNCTIONS OR DESTRUCTION

If there is an irreversible loss of all the characteristic functions of the brain, must we say that the brain has been wholly destroyed?[1] "Destroy" is used in its primary sense: "to break down or disintegrate the basic structure of," "to disrupt or obliterate the constitutive and ordered unity of." "Destruction" does not imply abruptness or physical violence. For the brain, "destruction" implies such damage to the neurons that they disintegrate physically, both individually and collectively. The converse, of course, is obvious: the total destruction of the entire brain does imply irreversible cessation of every kind of brain function and functions, but not loss of life. (T.K. in Shewmon's meta-analysis.)[11]

There are evidently many varieties of reversible cessation of brain-functioning known. Most of these are nondestructive. But we know of no medical principle that requires that a nondestructive cessation of function or functions must always be reversible. There is no evident contradiction in supposing the existence of permanent synaptic barriers, permanent analogs of botulinus toxin, or yet other mechanisms that would block all brain functioning while leaving the brain's neuronal structure intact and ready for action (at least until such time as the effects of this nonfunctioning on the rest of the body might react back on the brain in a destructive manner). Therefore, there is no reason to think that cessation of function, whether reversible or irreversible, necessarily implies total or even partial destruction of the brain; still less, death of the person.

Thus, the statutes that have sought to turn a loss of brain function into a general criterion of death are all vitiated by a fundamental category mistake: they take *that which functions* to be simply identical with the act of functioning. Yet, if something irreversibly ceases to function, its existence is not necessarily extinguished thereby; it merely becomes permanently idle. Nonfunction, no matter what qualifiers are used with it, is not the same thing as destruction. The few existing pathological studies of brains in "brain dead" patients do not confirm diffuse damage; in fact some specimens have been reported as showing only minor changes.[12]

In any case in which all functioning of the brain has irreversibly ceased, destruction of the brain and death will follow fairly quickly unless therapeutic action is taken. But if proper supportive action is administered, such an irreversible lack of brain function might last for a long time before the patient would begin to suffer destruction of brain tissue.

In such circumstances one would certainly not be free to treat a patient as dead. So long as we are dealing solely with cessation of function, we are dealing with a living patient. If, further, he happens to be dying, by this very fact he is alive and not dead. Whatever room there may be for discussion, pro and con, concerning obligations to maintain the supportive action that prevents the situation from deteriorating, at least as long as destruction of the brain has not occurred, the patient is alive.[13] As far as we can now know, there would even remain some possibility that in some cases a successful therapy might be found, but at present there are no markers or studies by which these patients can be selected. This is the primary reason we call for studies to evaluate these patients *scientifically*, not just for their body parts.

9. IRREVERSIBILITY: NON-EMPIRICAL CONFUSION

In addition to confounding what functions with its functioning, the criteria for "brain death" introduce further obfuscation through the use of the term "irreversibility" and its cognates. Now, irreversibility as such is not an empirical concept and cannot be empirically determined. Both destruction of the brain and the cessation of its functions are, in principle, directly observable; such observations can serve as evidence. Irreversibility, however, of any kind, is a property about which we can learn only by inference from prior experience. It is not an observable condition. Hence, it cannot serve as evidence, nor can it rightly be made part of an empirical criterion of death.

In brief, to regard the irreversibility of cessation of brain function (at best, a deduction from a set of symptoms) as synonymous or interchangeable with destruction of the entire brain (one but not the only possible cause of these symptoms) is to commit a compound fallacy: identifying the symptoms with their cause and assuming a single cause when several are possible.

Perhaps the strongest argument against the identification of irreversible cessation of all brain functions with death is this: those who initially accepted "brain death" did not really accept the identification themselves. The Harvard Committee was well aware of their intent and actions by clearly stating that they recommended that the patient be declared dead before any effort is made to take the patient off a respirator. Their reasoning for this recommendation was to provide legal protection to those involved, "Otherwise, the physicians would be turning off the respirator on a person who is under strict, technical application of law, still alive."

For, if "irreversible cessation of all brain functions" were merely other words for saying "complete destruction of the entire brain," why would there be the least hesitation on the part of the proponents to drop all reference to "brain function" and to ease their opponents' fears by substituting "complete destruction of the entire brain?" But, in fact, the proponents have vigorously resisted efforts to make this replacement. Yet surely, no function of a brain could survive that brain's complete destruction. Unfortunately, valuable evidence to settle these questions could have been obtained if the brains were studied at the time of organ harvesting over the years since 1968.

We may be permitted to wonder what lies behind this resistance to the identification they themselves have so constantly used and without which their basic arguments collapse. If the only brain functions remaining were firings of a few isolated neurons or the like, perhaps all this would not matter much. But since death is to be constituted by irreversible cessation of all brain functions as determined in accordance with one of the more than 30 disparate sets of criteria within acceptable standards of medical practice, and since one or more of the other sets might not be fulfilled, there is nothing to prevent *any* of the characteristic functions of the component brain parts from being declared "peripheral." For it is certain that no one of them can be declared to be that function that alone makes the whole person live. Cortical activity was evidently regarded as peripheral by the Minnesota criteria when reticular formation function has ceased, and by the British criteria when the brainstem's functions are gone due to structural damage. Many today argue that midbrain activity or brainstem activity is peripheral once the cortex has ceased to function. There is no limit to what real functions may be declared peripheral when the only non-peripheral function is imaginary. There continues to be no global consensus,[14] and unresolved issues remain worldwide.[15]

Further, if complete destruction of the brain were what really is intended, then why is so much written concerning indefinite ventilation of "cadavers" and the like? If a patient whose whole brain has been destroyed is on a ventilator, then, even by the older criteria, with only rare exception would he survive more than a week. If, however, his brain is not destroyed but merely nonfunctioning, then ventilatory support *should* be continued, at least as long as there is any chance of effecting a recovery or even of seeking an as yet unknown way to reverse his presently irreversible lack of function.[13]

10. RECENT INVESTIGATION

Elegant and innovative research by Dr. Cicero Coimbra[16] in brain-injured animals using criteria similar to "brain death" confirms that with modern day technology, varying degrees of recovery can occur. We do know that the major vital organs have the ability to regenerate cells and/or shift performance to other normal or less compromised areas of the same organ. A good example is the myocardium, which we know can occasionally either regenerate or recover "stunned" or "hibernating" myocardium to improved levels of functioning.[17,18] We know that the brain can be "stunned" in many ways. A prospective study of "brain dead" patients could have a high cost, but markers for those with potential for partial or even total recovery might be identified. Obviously, the potential would be very worthwhile for those who would otherwise die following removal of their vital organs or utilized as subjects for research/teaching or simply to have life-support measures discontinued.

Many theological and religious aspects were not presented here, but we have presented our commentaries on these recently.[19]

11. CONCLUSIONS

Brain related criteria for death, from initially using the term, "brain death," right up to the present time, was not and has not been based on studies that would be considered valid for any other medical purposes. The Harvard Committee had an agenda. The criteria were published without any patient data. And things have only gone downhill from there. "Brain death" is not true death. Further studies are indicated, but can anything be done to change something false to be the truth?

12. REFERENCES

1. Byrne PA, O'Reilly S, Quay PM. Brain DeathAn Opposing Viewpoint. *JAMA* 1979;**2**:242(18):1985-1990.

2. Report of the Ad Hoc Committee of the Harvard Medical School to Examine the Definition of Brain Death. Special Communication. *JAMA* 1968;**205(6):**85-88.
3. Goudreau JL, Wijdicks EFM, Emery SF. Complications during apnea testing in the determination of brain death: Predisposing factors. *Neurology* 2000;**55:**1045-1048
4. Mohandas A and Chou SN. Brain death. A clinical and pathological study. *J Neurosurg* 1971;**35(2):**211-218.
5. Diagnosis of brain death. Conference of Royal Colleges and Faculties of the United Kingdom. *Lancet* 1976;**2(7994):**1069-1070.
6. An Appraisal of the Criteria of Cerebral Death: A Summary Statement. A Collaborative Study. *JAMA* 1977;**237(10):**982-986.
7. Black PM. Brain Death. Part 1. N Engl J Med 1978 Aug 17; 299(7): 338-344; part 2, *N Engl J Med* 1978; **299(8):**393-401.
8. Guidelines for the determination of death. Report of the medical consultants on the diagnosis of death to the President's Commission for the Study of Ethical Problems in Medicine and Biomedical and Behavioral Research. *JAMA* 1981;**246(19):**2184-2186.
9. Walker AE. Cerebral Death. In: Chase TN, ed. *The Clinical Neurosciences, Vol 2*. New York: Raven Press 1975:75-89.
10. Bernstein IM, Watson M, Simmons GM, Catalano PM, Davis G, Collins R. Maternal Brain Death and Prolonged Fetal Survival. *Obstet Gynecol* 1989;**74(3 part 2):**434-7.
11. Shewmon DA. Chronic "brain death." *Neurology* 1998;**51:**1538-1545.
12. The NINCDS Collaborative Study of Brain Death, U.S. Department Of Health And Human Services, Public Health Service, National Institutes of Health. NIH Publication No. 81-2286, December 1980. NINCDS Monograph No.24.
13. Byrne PA, O'Reilly S, Quay PM, and Salsich P. The Patient, The Physician and Society. *Gonzaga Law Rev* 1982/83;**18(3):**429-526.
14. Wijdicks EFM. Brain death worldwide: Accepted fact but no global consensus in diagnostic criteria. *Neurology* 2002;**58(1):**20-25.
15. Swash M, Beresford R. Editorial: Brain death: Still-unresolved issues worldwide. *Neurology* 2002;**58(1):**9-10.
16. Coimbra CG. Implications of ischemic penumbra for the diagnosis of brain death. *Braz J Med Biol Res* 1999;**32(12):**1479-1487.
17. Kim SJ, Peppas A, Hong SK, Yang G, Huang Y, Diaz G, Sadoshima J, Vatner DE, Vatner SF. Persistent stunning induces myocardial hibernation and protection - flow/function and metabolic mechanisms. *Circ Res* 2003;**92:**1233.
18. Parmar G, Lalani AV.The Sleeping Heart: Hibernating Myocardium. *Perspectives in Cardiology* 2001; **April:**44-53.
19. Bruskewitz FW, Vasa RV, Weaver, WF, Byrne PA, Nilges, RG, and Seifert. Are Organ Transplants Ever Morally Licit? *Catholic World Report* 2001;**11(3):**50-56.

THE CONCEPTUAL BASIS FOR BRAIN DEATH REVISITED

Loss of organic integration or loss of consciousness?

John P. Lizza*

When a neurological criterion for determining death was formally introduced in the recommendation of the 1968 Ad Hoc Committee of the Harvard Medical School, the conceptual basis for accepting the criterion was unclear.[1] As Martin Pernick notes, the Committee shifted back and forth between endorsing the loss of consciousness, as opposed to the loss of bodily integration, as the conceptual foundation for the new criterion that the Committee eventually proposed.[2] Indeed, the Committee's characterization of the criterion as "irreversible coma" reflected this ambiguity, as the term had been used in the past to describe the condition of individuals in deep coma or persistent vegetative state.[3] In fact, although the Committee proposed a new criterion for determining death, it had little to say about the definition or concept of death for which the criterion was proposed.

One of the first attempts to articulate clearly the conceptual foundation for the neurological criterion appeared in the work of James Bernat, Charles Culver, and Bernard Gert, when they proposed "the permanent cessation of functioning of the organism as a whole" as a definition of death.[4] One problem they had with accepting the irreversible loss of consciousness as the conceptual basis for the neurological criterion was that it would entail classifying individuals who had irreversibly lost consciousness, but who had retained brain stem function, e.g., individuals in permanent vegetative state, as dead. Since Bernat, Culver, and Gert rejected the idea that such individuals could be classified as dead, they invoked the loss of organic integration as the basis for the neurological criterion. They then could maintain that, because individuals in permanent vegetative state retain the integrative functions of the brain stem, they were still integrated organisms and therefore still alive. The President's Commission for the Study of Ethical Problems in Medicine and Biomedical and Behavioral Research took a similar position in 1981.[5]

* John P. Lizza, Ph.D., Department of Philosophy, Kutztown University of Pennsylvania, Kutztown, PA 19530, USA.

Brain Death and Disorders of Consciousness, Edited by Machado and Shewmon
Kluwer Academic/Plenum Publishers, New York 2004

However, in recent literature, there has been significant criticism of the whole-brain neurological criterion of death by those who argue that the human organism as a whole may remain alive despite the loss of all brain function. Cases of post mortem pregnancy and the extraordinary case reported by D. Alan Shewmon in which a male with no brain function has been sustained for over 13 years challenge the claim that brain function is necessary for organic integration.[6,7,8,9,10] Also, the view that individuals who have lost all brain function but receive artificial life support are not integrated organisms but merely collections of organic parts has been effectively critiqued by many scholars, including Becker,[11] Troug,[12] Halevy and Brody,[13,14] Grant,[15] Seifert,[16] Byrne *et al*,[17] Taylor,[18] Veatch,[19,20] Gervais,[21] Wikler,[22] and Shewmon.[23,24] Indeed, at the last International Conference on Coma and Death, Dr. Shewmon delivered on his claim to "drive the nails into the coffin" of the idea that organic integration requires brain function.[25] At bottom, individuals who have lost all brain function but continue to function in such biologically integrated ways for such lengthy timeframes are integrated organisms of some sort and cannot be classified as corpses or dead organisms.

If these challenges to the neurological criterion have merit, then we are faced with a choice: we must either retain the definition of death as the loss of organic integration and give up the current whole-brain, neurological criterion or come up with an alternative definition of death that is consistent with accepting a neurological criterion. I propose that we do the latter and shall argue that the irreversible loss of organic integration is an incorrect conceptual basis for accepting a neurological criterion of death. The correct basis appeals to the essential significance of consciousness in the lives of people. I think that, all along, the real reason that we have been willing to accept "brain death" as death is not because we were sure that it entailed the loss of integration of the organism as a whole, but because we were sure that it constituted the irreversible loss of consciousness and every other mental capacity and function. Because so many of us believe that the potential for consciousness is essential (necessary) to the kind of being that we are, we are willing to accept its loss as the end of our lives.

If an organism is alive in cases such as "post mortem" pregnancy, the question arises: What is alive? It does not automatically follow that the person is alive, as claimed by those who wish to maintain that death is the loss of organic integration but reject the neurological criterion for death. Advocates of a consciousness-related formulation of death do not consider such a being to be a living person. In their view, a person cannot persist through the loss of all brain function or even the loss of just those brain functions required for consciousness and other mental functions. Thus, if an advocate of the consciousness-related formulation of death wishes to maintain that the person dies even though some being is alive in these cases, what remains alive must be a different sort of being. It must either be a human being, as distinct from a person, or a being of another sort, e.g., a "humanoid" or "biological artifact." By "humanoid" or "biological artifact," I mean a living being that has human characteristics but falls short of being human, a form of life created by medical technology. Indeed, this may be the most sensible thing to say about such a living being. Whereas a person is normally transformed into a corpse at his or her death, technology has intervened in this natural process and has made it possible for a person to die in new ways. Instead of a person's death being followed by remains in the form of an inanimate corpse, it is now possible for a person's remains to take the form of an artificially sustained, living organism devoid of the capacity for consciousness and any other mental function.

This distinction between the person and the organism may cause some consternation. Aren't persons organisms? The answer to this question is not so simple. However, how one answers this question ultimately affects what one is willing to accept as the definition and criteria of death. It is beyond the scope of this paper to give a detailed and sustained argument for a particular answer to this question. However, what I hope to do is (1) outline three ways that philosophers have tried to answer this question, (2) show how some of the participants in the debate over the definition of death have adopted (wittingly or unwittingly) these alternative philosophical views about the relation between the person and organism, and (3) show how the various answers affect the definition and criteria of death. I will also briefly point out some difficulties for two of the three views and argue by elimination for the third view.

The conceptual basis for a consciousness-related formulation of death has been clouded, in part because some of the main proponents of this formulation, such as Veatch, Green, and Wikler, have relied on one of the two problematic ways of understanding the relation between the person and organism. I thus will outline what I think is the most coherent conceptual basis for a consciousness-related formulation of death and for accepting a neurological criterion for determining death. In my view, Karen Gervais comes closest to correctly formulating the conceptual basis for accepting a neurological criterion of death, since she invokes a substantive concept of the person, rather than the more problematic, functionalist concept that Veatch, Green, and Wikler have assumed.[21] Also, by emphasizing how a neurological criterion of death is related to personal life and distinguishing human biological life from human personal life, Tristam Engelhardt elucidates some important aspects of the concept of death underlying acceptance of a neurological criterion.[26]

There are three main views of the person that entail different ways of understanding the relation of the person to the organism.[†] Simplifying somewhat, the first view identifies the person with the human organism. To be a person is to be a physical specimen of the human species. While some proponents of this view hold that the organism must be alive, at least one proponent, Fred Feldman, holds that the specimen need not be alive.[27] Just as dead butterflies that are carefully preserved and mounted for display are still members of their respective biological species, Feldman argues that the embalmed Aunt Ethel is still a member of her biological species. The person, Aunt Ethel, is literally the corpse at the wake and the body that is interred. Feldman admits that it is not a very satisfying way in which people continue to exist after their death. In philosophical circles, this view is referred to as "animalism," and is associated with the work of Fred Feldman,[27] Eric Olson,[28] and P. F. Snowdon.[29] Critics of the whole-brain neurological criterion of death, such as Hans Jonas, appear to accept this "species" meaning of person.[30]

The second concept of person identifies the person with certain abilities and qualities of awareness. It is a "qualitative" or "functionalist" view of the person. It has its roots in the work of John Locke[31] and David Hume,[32] and finds it contemporary expression most clearly in the work of Derek Parfit.[33] These philosophers treat the person as a set of mental qualities, including consciousness, memory, intentions, and character traits, rather than as a substantive entity or subject. Locke, for example, distinguishes the person from the human animal (organism) and questions of personal identity from those of the identity

---

[†]Two other views that I do not have time to discuss in this short paper are (1) persons are identical to immaterial substances (souls) and (2) persons are identical to brains.

of the human animal (organism). In his discussion of the hypothetical case of a prince and cobbler swapping bodies, Locke argues that personal identity travels with one's psychological states and memories.[31] Thus, if the body of the cobbler woke up one day with the psychological states and memories of the prince and the body of the prince, with the psychological states and memories of the cobbler, Locke concludes that the prince and cobber would have swapped bodies. For Locke, personal identity over time consists of the connectedness between psychological states evident in memories, regardless of whatever substance, material or immaterial, may underlie those psychological states over time.

Because the substantive matter underlying the psychological states is irrelevant to personal identity, Locke's view is a precursor to contemporary functionalist theories of the mind and personal identity. Functionalism as a philosophy of mind rejects the idea that the mind is a substantive entity, whether material or immaterial. Instead, the mind is conceived as a function that can be described abstractly by a machine table of inputs, internal states, and outputs. While the function needs to be embodied in some medium, e.g., the neurophysiological processes of the brain, the function can be described independently of whatever underlies or instantiates the function. As K. T. Maslin points out,

> *"the Lockean/Parfit proposal is analogous to treating you as a function or program run on the hardware of the brain, the material embodiment being strictly irrelevant to your identity and survival. You could go from body and brain to body and brain, just as information on a floppy disc can be transferred intact to another disc if the original becomes damaged."*[34]

Bernat, Culver, and Gert's definition of "person" in terms of "certain abilities and qualities of awareness" is another example of a use of this qualitative or functionalist view of the person. In their view, "death" cannot be applied literally to persons, because death is a biological concept appropriate to biological organisms, not to roles, functions, abilities, or qualities of awareness. Since Bernat, Culver, and Gert believe that individuals in permanent vegetative state have lost all the qualities that they believe define persons, e.g., consciousness, memory, and personality, they hold that a person who suffers an irreversible loss of consciousness has ceased to exist. However, since they claim that a death had not occurred with the irreversible loss of consciousness, they maintain that persons do not literally die and that expressions like "people die every day" involve a metaphorical use of "dying" or "death." Culver and Gert write:

> *"Person is not a biological concept but rather a concept defined in terms of certain kinds of abilities and qualities of awareness. It is inherently vague. Death is a biological concept. Thus, in a literal sense, it can be applied directly only to biological organisms and not to persons."*[35]

Some of the main proponents of a consciousness-related or "higher-brain" formulation of death, such as Robert Veatch, Michael Green, and Daniel Wikler, have also accepted a qualitative or functionalist view of the person. However, Veatch believes that it is a mistake to try to conceptually ground the definition of death on such a concept, as Green and Wikler have tried to do.[36] Veatch's concern is that the functionalist theory of personhood may entail that a person no longer exists in cases of dementia in which all traces of rationality and many other cognitive abilities are lost. Veatch believes, however, that a death has not occurred in such cases. In addition, Veatch is specifically critical of Green and Wikler's argument, because their view defines death in terms of the loss of personal identity. Veatch believes that we can conceive of cases in which personal identity in the Lockean sense may be lost, but a death has not occurred. According to Veatch, Green and Wikler's theory would commit them to drawing the absurd conclusion that someone who has lost psychological continuity, for example, by suffering complete amnesia, has necessarily died. However, Veatch asks us to suppose that this human being who has suffered complete amnesia subsequently develops a new set of beliefs, memories, and other psychological characteristics that we associate with personhood. According to Veatch, even though we might regard such a being as a new person, it is counterintuitive to say, as he believes Green and Wikler must say, that a death occurred in such a case. Thus, like Bernat, Culver, and Gert, Veatch assumes that *persons* are functional or qualitative specifications of human beings, rather than substantive entities, and concludes that such a view about persons and personal identity cannot provide the proper grounding for a definition of death. Like Bernat, Culver and Gert, Veatch believes that human beings, not persons, are the kind of thing that dies. However, Veatch accepts what he says is the traditional Judeo-Christian concept of a *human being* as an essential union of mind and body.[36] Since such a union may still exist even though someone suffers complete amnesia, the human being would still exist. A death has not occurred. However, since an irreversible loss of consciousness, as in permanent vegetative state, would entail the destruction of the essential union of mind and body, Veatch believes that the traditional Judeo-Christian concept of the human being warrants accepting a consciousness-related criterion of death. He regards individuals in permanent vegetative state as dead. The essential union of the mind and body in those individuals has been destroyed.

Karen Gervais has also criticized Green and Wikler's view on grounds similar to Veatch. She argues that as long as the biological substrate for consciousness remains intact, despite the complete loss of memories, a death has not occurred. However, Gervais believes that the same *person* continues to exist.[37] At this point, she is invoking a substantive concept of personhood and rejecting the functionalist view that she attributes to Green and Wikler and assumed by Veatch. In contrast to Veatch, Gervais treats persons as the kind of thing that can literally die. I believe that she is correct in this, and that Veatch erred in failing to recognize a substantive concept of person. She also reinterprets Veatch's argument in a way that essentially equates her substantive concept of person with Veatch's substantive concept of human being. What is common to Gervais and Veatch is that they both accept the idea that what dies is a substantive entity that is essentially mind and body. Gervais calls such a being a "person," whereas Veatch refers to it as a "human being." Gervais, like Veatch, also accepts a consciousness-related formulation of death that would treat individuals in permanent vegetative state as dead. Veatch's error thus lies in his assumption that there is no alternative to the qualitative or functionalist view of the person and personal identity. Insofar as Gervais invokes a substantive concept of person, she puts the argument for a consciousness-related formulation of death on more coherent, conceptual grounds.

The third concept of person is the one that has been suggested in Gervais's writings. It is a "substantive" concept that treats the person not as some qualitative or functional specification of some more basic kind of thing, e.g., a human organism, but as a primitive substance that necessarily has psychological and corporeal characteristics. P. F. Strawson's definition of a person as an individual to which we can necessarily apply both predicates that ascribe psychological characteristics (P-predicates) and predicates that ascribe corporeal characteristics (M-predicates) is an example of the use of "person" in this substantive sense.[38, 39] In Strawson's view, and even assuming with Bernat, Culver, and Gert that death is a biological or corporeal concept, it is neither a category mistake nor a metaphor to predicate death to persons. This view of the person is reflected in common expressions such as "people die every day," and differs from the "species meaning" in that it entails that the person must have the capacity or realistic potential for psychological functions. This cannot be said about a corpse or about some living members of the biological species *Homo sapiens*, e.g., anencephalics and individuals in permanent vegetative state.

With these distinctions drawn, I shall now give reasons in the brief space allowed for rejecting the species and qualitative views of persons and accepting the third, non-reductive, substantive view of persons. The main difficulty for the species view of persons is that it entails beliefs that most of us would reject. For example, in Feldman's version, all members of the species *Homo sapiens*, regardless of whether they are alive or not, are persons. However, as David Wiggins has pointed out, this view seems to commit us to counting "among present persons Jeremy Bentham, and even the Pharaohs, presumably more competently preserved."[40] Also, Wiggins notes:

> *"the discomfort that almost everyone feels about any straightforward equation between himself and his body; the idea that the lifeless corpse is not the person; and the fact that there is something absurd – so unnatural that the upshot is simply falsity – in the proposition that people's bodies play chess, talk sense, know arithmetic, or even run or jump or sit down."*[41]

Stipulating that the member of the species must be alive is no help, since many of us would not identify ourselves with brain-dead bodies that are artificially sustained or, hypothetically, with decapitated human bodies that could be artificially sustained. Again, such bodies may be integrated organisms (perhaps still "human" organisms), but most of us would not view them as a continuation of ourselves. In such cases, we would view ourselves as having died. Thus, the species view should be rejected, because it does not square with our intuitions about who gets counted among the living "we."

There are two main difficulties with the functionalist view of person, particularly with Bernat, Culver, and Gert's claim that we commit a category mistake by predicating death literally to persons. First, as I have pointed out elsewhere, it conflicts with what many ordinary people, as well as philosophers, have always said about persons and death.[42] Ordinarily, we think of people as the kind of thing that can literally die. We also think of them as the kind of thing that can literally live, eat, sleep, jump, breathe, etc. However, according to Bernat, Culver, and Gert, these predicates apply only metaphorically to persons. To say that these common expressions are metaphorical departs from ordinary usage and understanding of the terms. Whereas the general public and many philosophers have treated persons as substantive entities and predicated death to persons, Bernat, Culver, and Gert assume that such predication is conceptually impossible and that the public and these philosophers have been speaking metaphorically all along. While their departure from common usage, of course, does not mean error, it shows what is at stake in terms of conceptual and linguistic revision, if we accept their view. Moreover, since they have claimed that their view accords with the ordinary, non-technical understanding of death among the general public, and have argued that this agreement provides support for their view, it is important to show how their view is at odds with the ordinary understanding of "person" as a kind of thing that can literally die.

The second major problem with the functionalist concept of person is that it fails to do justice to the moral dimension of persons. Locke's proposal to treat the person as a mode or set of psychological qualities of a substance was challenged by Molyneux[43] and Reid,[44] precisely on such grounds. For example, Molyneux pointed out that Locke's memory criterion of personal identity entailed the unacceptable implication that there would be no reason to hold someone responsible for an immoral act that he or she committed while drunk, if the person had no memory of having done the act.[43] Locke, himself, believed that the concept of person was a "forensic" term and ultimately recognized the force of Molyneux's criticism.[45] Christine Korsgaard has similarly challenged Parfit's qualitative view of the person on grounds that it fails to adequately account for moral responsibility.[46] The upshot of these criticisms is that only a view that treats the person as a substantive entity, not as a mode or quality of some substantive entity, can provide adequate grounding for the moral dimension of persons.

David Wiggins has suggested that our theory of personhood tries to hold together in a single focus three aspects of persons:

> *(1) an object of biological, anatomical, and neurophysiological inquiry;*
> *(2) a subject of consciousness; and*
> *(3) a locus of all sorts of moral attributes and the source or conceptual origin of all value.*[47,48]

Commenting on Wiggins's insight, Christopher Gill writes:

> *"These ideas of the person are intelligible to us because, and in so far as, we deploy (or at least presuppose) them in the process of interacting with, and making sense of other human beings. Similarly, it is worthwhile for us as thinkers to try to hold these different ideas 'in a single focus', because, in our ordinary relationships with other human beings, we are capable of treating them as 'persons' in the three relevant senses, and of doing so in an interconnected way.*
>
> *A crucial point in Wiggins's argument is that this type of interpretation is mutual and reciprocal. Human beings, considered as persons, 'are subjects of fine-grained interpretations by us and are the would-be exponents of fine-grained interpretations of us'. We interpret persons, that is to say, as being like ourselves (and as capable of interpreting us as beings like themselves); and this fact has implications for our understanding of human beings as persons in all three of the senses Wiggins identifies as well as in their interconnection. In particular, it helps to explain the moral scruple with which we treat those we regard as persons: we think of them as being (like ourselves) consciousness-bearing, embodied individuals,*

> *capable of originating action and of interacting with us as persons. This explains our proper sense of moral constraint at the thought of treating a human being as a thing, or non-person, whose presence can be ignored or (at worst) deleted by willful killing."*[47]

In Wiggins's view, the questions "What is a person?" and "Who are 'we'?" are biological, metaphysical, and moral. I have suggested that the species and qualitative views of personhood are mistaken, because they fail to do justice to one or more of these dimensions of persons. If persons are living, substantive entities with the capacity or realistic potential for some mental functions, then the irreversible loss of that capacity or potential would mean the ceasing to exist or death of the person. Persons whose brains have been destroyed to the point where they have lost the potential for consciousness and every other mental function are therefore dead. Even though technology may intervene to sustain human or humanoid bodies that have irreversibly lost the potential for consciousness, this should not obscure the reality that the person, understood as a substantive being, has died. We are not just our bodies or qualities of them.[‡]

REFERENCES

1. Ad Hoc Committee of the Harvard Medical School to Examine the Definition of Brain Death. A definition of irreversible coma. *JAMA* 1968;**205:**337-340.
2. Pernick MS. Brain death in a cultural context. In: Youngner SJ, Arnold RM, Schapiro R, eds. *The definition of death: contemporary controversies.* Baltimore: Johns Hopkins University Press, 1999: 12.
3. Joynt RJ. A new look at death. *JAMA* 1984;**252:**682.
4. Bernat J, Culver C, Gert B. On the criterion and definition of death. *Ann Intern Med* 198;**94:**389-94.
5. President's Commission for the Study of Ethical Problems in Medicine and Biomedical and Behavioral Research. *Defining death: medical, ethical, and legal issues in the determination of death.* Washington, DC: US Government Printing Office, 1981.
6. Field DR, Gates EA, Creasy RK, Jonsen AR, Laros, RK. Maternal brain death during pregnancy. *JAMA* 1988;**260(6):**816-822.
7. Bernstein IM, Watson M, Simmons GM, Catalano M, Davis G, Collins R. Maternal brain death and prolonged fetal survival. *Obstet Gynecol* 1989; **74(3)**: 434-437.
8. Anstötz A. Should a brain-dead pregnant woman carry her child to full term? The case of the 'Erlanger baby'. *Bioeth* 1993; **7(4)**: 340-350.
9. Shewmon DA. Chronic 'brain death': meta-analysis and conceptual consequences. *Neurology* 1998; **51**: 1538-1545.
10. Shewmon DA. Letters and replies. *Neurology* 1999; **53**: 1369-1372.
11. Becker LC. Human being: the boundaries of the concept. *Philos Public Aff* 1975;**4:**335-59.
12. Troug RD. Is it time to abandon brain death? *Hastings Cent Rep* 1997;**27(1):**29-37.
13. Halevy A, Brody B. Brain death: reconciling definitions, criteria, and tests. *Ann Intern Med* 1993;**119:**519-525.
14. Brody B. How much of the brain must be dead? In: Youngner SJ, Arnold RM, Schapiro R, eds. *The definition of death: contemporary controversies.* Baltimore: Johns Hopkins University Press, 1999: 71-82.
15. Grant AC. *Human brain death in perspective: comments on the spinal dog and decapitated frog.* Presented at The Third International Symposium on Coma and Death, Havana, February 22-25, 2000.
16. Seifert J. Is 'brain death' actually death? *The Monist* 1993;**76(2):**175-202.

[‡] In this paper, I have tried to show how a substantive concept of personhood provides the conceptual basis for accepting a consciousness-related neurological criterion for death. I have said little about which specific neurological criterion, e.g., whole-brain or "higher-brain" we should adopt in law and public policy. This is a different issue that requires consideration of many other factors that are beyond the scope of this short paper, e.g., what means are currently available for reliably determining when the loss of consciousness is irreversible.

17. Byrne PA, O'Reilly S, Quay P, Salsich Jr PW. Brain death – the patient, the physician, and society. In: Potts M, Byrne PA, Nilges RG, eds. *Beyond brain death: the case against brain based criteria for human death.* Dordrecht: Kluwer, 2000:21-89.
18. Taylor RM. Re-examining the definition and criterion of death. *Semin Neurol* 1997;**17**:265-279.
19. Veatch RM. Brain death and slippery slopes. *J Clin Ethics* 1992;**3(3)**:181-187.
20. Veatch RM. Maternal brain death: an ethicist's thoughts. *JAMA* 1982;**248**:1102-1103.
21. Gervais KG. *Redefining death.* New Haven: Yale University Press, 1986:146-147.
22. Wikler D. Who defines death? medical, legal and philosophical perspectives. In: Machado C, ed. *Brain death: Proceedings of the Second International Symposium on Brain Death.* Amsterdam: Elsevier, 1995:13-22.
23. Shewmon DA. 'Brainstem death,' 'brain death,' and death: a critical reevaluation of the purported evidence. *Issues Law Med* 1998;**14**:125-145.
24. Shewmon DA. Recovery from 'brain death': a neurologist's apologia. *Linacre Q* 1997;**64(1)**:30-96.
25. Shewmon DA. The 'critical organ' for the 'organism as a whole': lessons from the lowly spinal cord. *Presented at The Third International Symposium on Coma and Death,* Havana, February 22-25, 2000.
26. Engelhardt Jr HT. Medicine and the concept of person. In: Beauchamp TL, Perlin S, eds. *Ethical issues in death and dying.* Engelwood Cliffs, NJ:Prentice-Hall, 1978:271-284.
27. Feldman F. *Confrontations with the reaper: a philosophical study of the nature and value of death.* New York: Oxford University Press, 1992.
28. Olson ET. *The human animal: personal identity without psychology.* New York: Oxford University Press, 1997.
29. Snowdon PF. Person, animals and ourselves. In: Gill C, ed. *The person and the human mind: issues in ancient and modern philosophy.* Oxford: Clarendon Press, 1990:83-107.
30. Jonas, H. Philosophical Essays: *From ancient creed to technological man.* Engelwood Cliffs, NJ: Prentice-Hall, 1974.
31. Locke J. *An essay concerning human understanding.* 2nd ed., London, 1694.
32. Hume D. *A treatise of human nature.* London, 1739.
33. Parfit D. *Reasons and persons.* Oxford: Oxford University Press, 1986.
34. Maslin KT. *An introduction to the philosophy of mind.* Malden, MA: Blackwell, 2001:275.
35. Culver C, Gert B. *Philosophy in medicine: conceptual and ethical issues in medicine and psychiatry.* New York: Oxford University Press, 1982:182-183.
36. Veatch RM. The impending collapse of the whole-brain definition of death. *Hastings Cent Rep* 1993; **23(4)**:18-24.
37. Gervais KG. *Redefining death.* New Haven: Yale University Press, 1986:126.
38. Strawson PF. *Individuals.* London: Methuen, 1959:87-116.
39. Strawson PF. Persons. In: Feigel H, Scriven M, Mazwell G, eds. *Minnesota studies in the philosophy of science II.* Minneapolis: University of Minnesota Press, 1958:330-353. Reprinted in Rosenthal DM, ed. *The nature of mind.* New York: Oxford University Press, 1991:104-115.
40. Wiggins D. *Sameness and substance.* Cambridge, MA: Harvard University Press, 1980: 162.
41. Wiggins D. *Sameness and substance.* Cambridge, MA: Harvard University Press, 1980: 163-4.
42. Lizza JP. Defining death for persons and human organisms. *Theor Med Bioeth* 1999; **20**:439-453.
43. Molyneux W. *Letter of Molyneux to Locke, 23 Dec. 1693.* In: The works of John Locke VIII. London, 1794:329.
44. Reid T. *Essays on the intellectual powers of man.* London, 1785.
45. Allison, HE. Locke's theory of personal identity: a re-examination. *J Hist Ideas 1966*; **27**:41-58.
46. Korsgaard CM. Personal identity and the unity of agency: a Kantian response to Parfit. *Philos Public Aff* 1989;**18(2)**:101-132.
47. Gill C. The human being as an ethical norm. In: Gill C, ed. *The person and the human mind.* Oxford: Clarendon Press, 1990:156.
48. Wiggins D. The person as object of science, as subject of experience, and as locus of value. In: Peacocke A, Gillett G, eds. *Persons and personality: a contemporary inquiry.* Oxford: Oxford University Press, 1987:56.

CONSCIOUSNESS, MIND, BRAIN, AND DEATH

Josef Seifert[*]

The arguments in favor of equating brain death with actual human death have shifted from medicine and biology to philosophy. For the most common argument in favor of brain death from the alleged loss of integration has in the meantime received its final blow, particularly through the results of Prof. Shewmon's studies expounded in his two Havana papers 1996 and 2000, namely: (1) a careful comparison of the disintegration occurring in BD with other similar forms of even farther-reaching disintegration in clearly living organisms, and (2) his demonstrating the somatic pathophysiological equivalence between the state of "brain death" with the state of patients in "spinal shock," who have conscious life, and the implications of this equivalence for the integrative-unity rationale used in the defense of equating death with brain death. In the 2000 Symposium no single objection to his results has been raised in discussions, nor during the last four years has a refutation of his results been offered.

Not repeating Shewmon's criticisms of BD, and not even addressing their philosophical aspects, this paper seeks to face the new challenges because, after confirming that they are not dead, the question how to treat the so-called "brain-dead" individuals is posed again to philosophers and ethicists. It is alleged that brain-dead individuals, even if clearly alive biologically speaking, do not live as persons, do not possess rational consciousness, and therefore have no mind or soul endowed with a dignity that would forbid us to kill them biologically speaking by extracting their vital organs.

Confronting these arguments, I will seek to show that: a) consciousness and the mind of persons cannot be reduced to the brain or to supervenient properties of brain processes; and b) the dignity of the human person cannot be reduced to the value of persons in an awakened state because human dignity has other roots besides consciousness. Therefore, not only the argument for brain death from the alleged loss of integration but also the argument from loss of rational consciousness, quality of life, and dignity of the "brain dead person" is untenable.

In view of the decline of the traditional medical, biological, and bio-philosophical arguments for identifying brain death with the actual death of human persons, the arguments in favor of maintaining the legal and ethical use of the brain death criteria has shifted to the philosophical level. With reference to this fact, in the discussions of the

---

[*] Josef Seifert, Ph.D., Rector, International Academy for Philosophy, Liechtenstein.

Brain Death and Disorders of Consciousness, Edited by Machado and Shewmon
Kluwer Academic/Plenum Publishers, New York 2004

present and the third *International Symposium on Coma and Death* 2000 two themes have turned out to be of paramount importance:

(1) the question of whether the human person has a mind or soul irreducible to the brain or whether conscious life and the conscious subject can be explained sufficiently by matter and specifically by the brain. From the latter position, if it were true, many consequences would follow both for the theoretical understanding of man and for medical theory and praxis: man would ultimately be a machine, freedom would be an illusion, knowledge would be impossible; "psychic" or "mental" diseases would actually be just brain diseases and could one day sufficiently be treated by psychopharmaca or brain surgery; so-called "brain death" would be quite certainly the death or even the destruction of the person, etc.

(2) The second decisive question has been whether the dignity of the person is exclusively rooted in higher consciousness and whether therefore brain dead individuals, even if they are alive, may be treated as if they were dead *qua* persons, and whether even patients in PVS might be killed by starvation and dehydration in an ethical way, as long as we do not have to assume that they are conscious, or possess a significant degree of consciousness and quality of life beyond that of animals.

To these two questions this chapter seeks to answer in a purely philosophical way that (a) the human person (mind) cannot be reduced to matter (the brain) and that (b) the dignity of the person is also but not only rooted in awakened rational conscious life. Let us then turn to our first theme first:

1. CONSCIOUSNESS AND ARGUMENTS FOR THE EXISTENCE OF A MIND (SOUL) THAT IS IRREDUCIBLE TO THE BRAIN

We can identify four groups of phenomenological and purely philosophical arguments for the existence of the mind or soul.

1.1. Arguments from Consciousness and Its Subject as Such

(1) Let us start with a simple experience of which the Nobel Laureate and great neuroscientist Sir John Eccles said that it is far more evident than all evidences of neuroscience:[†] In all conscious experiences of the human person we find the arch-datum of the absolutely indivisible "I," of the subject of conscious experience. Now, this immediately given subject possesses dimensions of unity which are inexplicable in principle through matter: I am, as one and the same subject that is not composed of non-identical parts, present in myriads of experiences; I, the identical Self, have to remain present, even for the smallest experience (hearing a melody, for example) to be possible, in innumerable temporally separated mental operations which would not be experienced and related to each other without the indivisible and experienced identity of myself as subject of consciousness. We also find the same subject present in an enormous variety of kinds of conscious experiences, such as perceiving, thinking, judging, questioning, willing, loving, etc.

[†] This did not prevent David Hume and other non-ownership theoreticians of the mind to deny it with sophistical and contradictory arguments.[1]

Neither the subject of consciousness nor conscious experiences themselves are just objects *of which we have intentional consciousness;* they do not just stand over against conscious acts. Rather we experience them from within—as being identical with ourselves; we experience them in a "lateral consciousness," not just as objects of *frontal consciousness.*[2–11] And all of our conscious experiences are conscious actuations of one and the same identical and indivisible "I" which is irreducible to their multitude but present in all and in each of them; moreover, this individual and unique subject is likewise *given* and *experienced* from within in a new mode: not just as my perceptions, anguishes or hopes but as "I myself," the subject *who* experiences them.

No material entity including our brain could explain such a unity and identity of the conscious subject, because (a) the matter of our brain has astronomically many non-identical and divisible neurons and their dendrites and other parts, and (b) there is not even an analogy to the simple human subject in the brain: any single biological focal point or center to which all information would flow is missing there.[12,13] Even less could an epiphenomenon or supervenient product of brain activity account for the personal "I," because in virtue of its nature of an accident, any such dependent quality or product of something else (brain events) would even lack the most elementary character of a person, the ontological self-standing in being, i.e., the character of a subject or substance. But this point leads us to a new theme: the ontology of the person, to which we will return.

(2) Each subject of consciousness possesses an unrepeatable individuality and uniqueness. This absolutely unrepeatable individuality of the subject of consciousness cannot ever be explained by the brain, among whose billions of neurons and parts only the elementary particles possess a comparable unrepeatable individuality of an indivisible element. These elementary particles of matter, however, one of which no lesser philosopher than Roderick M. Chisholm wanted to identify with the Self whose evident absolute indivisibility he recognized, do not live, whereas the conscious subject lives, nor are they as such connected with consciousness.[14,15] Therefore, they can never explain the conscious subject. Also in that they occupy space and have a certain size, in their spatial "exteriority," they can never possess the wholly different simplicity of the conscious Self which necessarily lacks size and parts outside each other in space. Moreover, elementary particles possess only an anonymous individuality of a "building block," a character opposite to the profound individuality of the person, as we will see better when turning later to an ontology of the person. The unique unrepeatable individuality of the simple personal Self can also not be explained genetically through matter (for example through a unique genetic code) because each genetic code can be reduplicated in identical twins, clones, in each cell, and, however slim the chances of a repetition of a genetic code in independent organisms are, it could in principle be repeated in the course of history, whereas the personal subject is absolutely unique and unrepeatable in any possible world.

1.2. Arguments from Specific Conscious Experiences and Acts

(1) Each act of knowledge depends on, and is determined in its content by, the nature of the known object. This grasp of the nature of anything would be absolutely impossible, and the rationality of cognition would be destroyed, if chemical or physical forces would not only be conditions but causes by which knowledge and its contents would be determined. In that case the content of knowledge would not depend on the nature of things but on blind chains of physical and chemical forces. It would be different not in accordance with the differences in things but in dependence on different physical events in the

brain. But knowledge and its rationality require far more than a strict dependence in its content on the nature of the object known in an intentional act of cognition: cognition entails not only an objective being determined by the object but involves a unique transcendence of the subject without which knowledge, i.e., an actual reaching of the object in the intentional cognitive act, would not be possible. Knowledge, and most of all the clearly and indubitably given phenomena of evidence and of knowing that we know something, would be entirely impossible if our knowledge were determined "from the back" through blind physical causes and if it were to change in its content not in accordance with changes in things but with mere changes in our physical make-up and material forces operating on us. But we *do actually possess at least some indubitable knowledge.* And the evidence of this knowledge, for example of logical laws, or of our own existence, proves that our knowing and we as conscious subjects cannot be the side-effects, epiphenomena, or supervenient properties of brain events which would cause our knowledge. Therefore, the brain can neither be the subject nor the *cause* of knowledge but only play a subordinate though crucial role as *condition* of knowledge: as the *id sine qua non,* but not as that *through which* knowledge arises.

(2) Especially rational knowledge of universals, the formation of concepts and judgments, the perception of beauty, and most of all the knowledge of eternal, invisible, spiritually intelligible, necessary and immaterial truths and essences as well as of objects such as ideal numerical relations could never be accounted for as effects of brain events;[16, 17] for neither are the essences and intelligible objects material things nor could these account for their knowledge: matter can only enter into a purely participatory and objective relation to universals, eternal truths, and immaterial essences, never into an experienced conscious contact with an immaterial world.

(3) Freedom as such and self-determination of the subject through its own will is another indubitable phenomenon: even if we could be deceived about our very being (which is absolutely impossible), says Augustine, it would still be indubitably clear that we would not want to be thus deceived; and in this evidently known free resistance to error our freedom is clearly manifested, as already Plato stated.[18–23] But the self-determination of a subject in freedom would be absolutely impossible if its subject and act were a material object or even a causal effect (or epiphenomenon) of material processes; thus the existence and evidence of freedom refutes any materialism.

In general, we must distinguish identity, condition, cause, and effect. The brain and brain processes can very well be conditions and effects of truly rational acts and experiences of the described sort, but never be identical with them or their cause.

1.3. Arguments from an Ontology of the Substantial Nature and Subject of Human Consciousness: The Notion of Person

(1) The personal I, the subject of consciousness, is not only concluded to but itself clearly given. And it is absolutely evident that this I and free Self can never be an accident, a property of something else. As center of consciousness and subject of free acts, it stands in itself in being as substance, as subject (hypostasis). This is what Thomas Aquinas or Boethius, who defined the person as "(Persona est) rationabilis naturae individua substantia,"[24] or as "naturae rationalis individua substantia."[25] And even Kant, though he fiercely denied, in his section of the *Paralogisms of Pure Reason,* the possibility of knowing the substantial being of the soul, later returns to this evidence:[26, 27, 27a]

> *"But I know of the soul:*
>
> *That it is a substance; or: I am a substance. The "I" means the subject, inasmuch as it is no predicate of another thing. What is no predicate of another thing is a substance.*
>
> *The "I" is the . . . subject of all predicates, thoughts, actions . . . it is absolutely impossible that the "I" be the predicate of something else. The "I" cannot be the predicate of another being. In fact, predicates belong to me. The "I" itself, however, cannot be predicated of another thing, I cannot say: Another being is the "I." Consequently, the "I" or the soul which is being expressed through the "I" is a substance."*[‡]

(2) As our experience of sleep or loss of consciousness but also a deeper ontological reflection show: while the conscious subject, and not something "behind it," is the person, the person still cannot be reduced to its conscious state and awakening as conscious subject. What awakens here precedes temporally and ontologically the conscious state. Besides, this ultimate subject of consciousness possesses the single marks of substantiality more perfectly than it is even thinkable in a material thing:

(A) The person, as indivisible, unique Self and as living, conscious and freely acting subject, stands more clearly in itself than any material thing could ever do; it does not inhere in another substance, possessing *inseitas* (subject-character) in an exemplary fashion inaccessible to matter. Therefore it cannot only not be the brain, as we have seen. It can even less be an effect of the brain which could never be a substantial entity/subject standing on its own feet in being.

(B) Moreover, the person possesses an absolutely irreplaceable individuality as the simple and unrepeatable subject of conscious and free acts in a way in which no material thing can ever be an individual *tóde ti* (this there), as Aristotle calls substance. Only here, this unrepeatable thisness is not a mere objective individuality as also each stone and tree possesses it, but possesses an absolutely unrepeatable thisness and individualness experienced from within. Only a spiritual and immaterial subject can possess these characteristics, a personal soul which neither is a composite whole, as all complex matter, nor a simple part thereof (elementary particle or *atomon*), which, while no longer being really divisible, is still extended in space and mathematically divisible.[28] A being characterized by size and the "exteriority" of material parts or energy fields never can experience from within its own individuality. The personal soul, being absolutely delineated from its "surroundings" and never a mere part or building block of composite atomic, molecular and more complex structures, is also the opposite of ontological anonymity and inauthentic individuality of elementary particles of wholes.

[‡] My translation of Kant,[26] pp. 201–202:

> Ich erkenne aber von der Seele:
>
> 1) daß sie eine Substanz sey; oder: Ich bin eine Substanz. Das Ich bedeutet das Subject, sofern es //PM202// kein Prädicat von einem andern Dinge ist. Was kein Prädicat von einem andern Dinge ist, ist eine Substanz.
>
> Das Ich ist das . . . Subject aller Prädicate, alles Denkens, aller Handlungen . . . . Es geht also gar nicht an, daß das Ich ein Prädicat von etwas anderm wäre. Ich kann kein Prädicat von einem andern Wesen seyn; mir kommen zwar Prädicate zu; allein das Ich kann ich nicht von einem andern prädiciren, ich kann nicht sagen: ein anderes Wesen ist das Ich. Folglich ist das Ich, oder die Seele, die durch das Ich ausgedrückt wird, eine Substanz.

(C) The person possesses also more perfectly a third characteristic of substance mentioned by Aristotle, a sameness enduring throughout change and time: while body cells and all other body parts can and do change, without the material subject, which requires no *strict self-identity,* disappearing, the personal Self remains the same from beginning to the end of life and neither has nor changes parts of which it would be composed; any such loss of *strict self-identity* which in atoms and cells appears normal, is incompatible with the self-identity of the person. The personal Self thus possesses a far deeper self-identity than any material thing could ever possess. Moreover, the person lives consciously and remembers its enduring being; it is an individual subject *awakened to himself or herself,* an evident fact which likewise accounts for a far more significant sense of self-identity than the brain could ever possess.

The person is not only also a substantial being but substance in a more proper sense than matter ever can be. From this self-identity, which even the most complex and admirable structure of the brain can in no way account for, the spiritual and immaterial nature of the mind (soul) becomes obvious.

Besides possessing each mark of substance more perfectly, the person above all possesses them in a mutual union more perfect than conceivable in matter: for example, the individual thisness is here not found on another level than the character of the substantial subject, as in a material object whose essence and form does not reside on the level of the ultimate "substantial" micro-parts. In the person, the simple subject himself is individual, enduring, and delineated from all other beings.

1.4. Arguments from the "Lived Body" and Union of Body and Soul

One might object that man is a person who has a body, an incarnate spirit, a soul-in-a-body, and that the body-soul unity is opposed to any Cartesian conception of the human person as a pure spirit to whom the body is merely external. But precisely the phenomenon of the *Leib* (the "lived body") reveals the distinctness between body and soul and cannot be accounted for without admitting it. Because the conscious inner awareness of our own body through ourselves as the single and indivisible subject of conscious life provides both evidence of the distinctness of body and mind and holds the clue for the unified human body which would never be a human and unified body without the presence of the soul (mind) in it.

2. DIFFERENT TYPES OF MIND/BRAIN (BODY-SOUL) RELATIONSHIPS

Obviously, the human mind is not that of an angel. It is closely and intimately related to the human body in many ways.

2.1. Static Relationships

The contingent[§] condition without which consciousness or concrete experiences cannot take place is evidently distinct from an efficient cause through the power of which for

[§] In contradistinction to *absolute* conditions of conscious life (such as a subject), these contingent conditions do not apply absolutely and can therefore cease to be conditions of conscious life in an immortal existence.[29]

example a state of feeling nervous may be caused, just as the motor and the processes of combustion in it that drive a car, are distinct from the conditions of its motion inside the motor and in the surroundings.

Another static relation of the body to the soul completely distinct from a condition is found in the body as a visible and audible expression of the person. If we look at the tone of the human voice, at the shape of the human face, the upright posture, the forehead, the eyes, the whole aesthetic quality of the human body, we find an amazing fact: that which belongs to the order of the physical expresses and manifests in an immediately accessible way the dignity and nature of the human soul, of the human person. In this sense, Ludwig Wittgenstein rightly said that the best image of a person is the human body. Therefore the appearance of a pig, quite natural for a certain animal, would be a horrible disharmony of the human form to the essence of the person, while the human body is a fantastic medium of manifesting the nature of personhood in the sensible world.

2.2. Dynamic Body/Mind Relationships

This broad category includes again many types:

2.2.1. From Body to Mind

a) Causal relationships: Non-rational experiences as physical pain and non-intentional emotional states which are not meaningfully and intentionally related to objects, or also disturbances of the rational conscious contact with objects, allow for being caused by brain events. Moreover, we have many empirical evidences of causal relations through which somatic factors lead to psychic states. Alcohol, drugs, wounds, headaches, poisons, etc., obviously exercise effects on the mind. Even the concrete possession of mental faculties can be causally destroyed by brain lesions or certain diseases.

b) Very different is the role of the body and the brain as *medium* of receptive acts which have their origin in physical events in the external world that exert some causal influence on the body but where the cognitive and meaningful relations to the world are merely *mediated, not caused* by the body and specifically by the brain: In sense perceptions the body plays such a medial role because the causal chain of physical and physiological events stands in the service of an entirely different reality—the intentional consciousness of an object which becomes the main source of the content of these conscious acts: for example, we listen to a Mozart Opera; here it is Mozart's music that we get to know and on which the content of our knowledge of it, and of its good or bad performance depends.

The body and the brain can exert this mediating influence on intentional acts unconsciously (for example, by the brain). The body can also be, however, a consciously lived medium of intentional relationships to objects in sense perceptions and relations to other persons such as in touch.

The medial function of the body is much more complicated and indirect when it operates in relationship to the recognition of objects as belonging to a certain species or genus.

2.2.2. From Mind to Body and through the Body to the World

a) Inner free acts of concentration on an object can certainly have effects on, and conditions in, the brain, but they can never be caused by the brain nor can the brain or other parts of the body be their source (except via motivation, as when we make decisions because of our knowledge of having a brain tumor). The evidence of the necessary essence and of the actual existence of inner free acts that cannot be created by any cause outside the mind (the soul), and of their conscious subject, proves that their origin can only lie in the free mind and never in brain events; but not only can unfree mental events have *causes* in the brain but also free acts can have *conditions* as well as *effects in the brain*. The evidence of freedom and of free influence of the mind on the body cannot be sufficiently explained by quantum-physical theories of the brain. Certainly, quantum physics "opens up" the physical universe (and therefore also the brain) to a sphere of causation which is not subject to strict natural laws, the prime example of which experienced by us lies in the free human dominion over material events, and provides thereby a sort of proof of the condition of the possibility of freedom. Nonetheless, the evident free dominion of the mind over the body (and hence also over the brain) occurs in virtue of *conscious and intentional acts exerting power over the body* (and unconsciously over the brain), and hence differs in this and many respects entirely from microphysical "chance events" quite apart from the need to distinguish many meanings of "chance," some of which are absurd and in no way proven through quantum physics.[30] Moreover, the type of influence the free will and ingenuity of human minds exert over matter when a person is writing a book or building a great palace is so immense, ordered, regular, etc., that it could never be reduced to, or sufficiently explained by, the type of mere statistical and chaotic irregularities modern quantum physics detected.[31] Besides all this, it clearly and in an ordered fashion belongs to the macro-physical order that is subject in a large measure to our wills, whereas quantum physics refers to the micro-physical order.

b) When emotions and volitional attitudes find a bodily expression, what happens in the body is not only an effect of concrete mental acts but renders them visible (expresses them).

The problem how exactly the spiritual soul, *not being a body,* can nonetheless *really* influence the body and be influenced by it, how exactly it does so, remains a mystery regarding which we lack immediate experience and can only develop speculative theories.[32]

No one should expect a total solution of the mystery of this body-mind relationship regarding which both highly intelligible, evident relationships and impenetrable mysteries exist which make a person-*in-carne* a puzzle which no scientist and no philosopher will ever be able to solve entirely.

3. THE UNION OF BODY AND SOUL AND THE HUMAN SOUL AS *"FORMA CORPORIS"*

The soul is the *forma corporis*. What does "form" mean here?

First, one can mean by "form" of the body the external impression and shape of a material thing. This is, of course, not the human soul.

Secondly, "form" of matter can refer to the "form" of a material thing which makes it from within into a determinate kind of matter (silver, lead, etc.). This "form" could not explain consciousness any better than, according to a crass materialistic interpretation, the formed body of the brain can produce consciousness. Moreover, identifying the human soul with the inner formedness of the body would contradict fragmentation and lack of simplicity of this form as well as the experience of the partial identity between the body of the living human person and the corpse.

In a third sense the life-principle of plants and animals can be characterized as "form" of the organic body. "Form" in this sense is not a body but something in and for a body.[10, 33]

In a fourth sense one can speak of "form" when one characterizes the human spiritual soul as the substantial form of the body. The spiritual human soul not only does not "exhaust itself" in its animating and informing influence on the body; it is not merely "more" than form of the body (matter), but in regard to its spiritual and most meaningful acts it is not at all primarily to be understood or to be defined as "form of the body." Moral goodness, love or religious acts represent a spiritual world in themselves which does not allow that the essence of the human soul be conceived of primarily in terms of its being "form" of the body.

The mind, or more precisely, the spiritual soul of man, while being infinitely more than a *forma corporis*, still is *forma* of the body. The deepest reason for the human body becoming a *Leib* (a lived body in the specifically human sense) lies in its being essentially destined to be united with, and an expression of, the mental and spiritual life of the soul. In that the human body stands in the service of the "Logos" (spirit, mind, meaning), as Gregory of Nyssa especially emphasizes, and in that it serves a spiritual life which transcends it, the human body receives from this spiritual life its "form" in the deepest sense. The body is not only assumed into the life of the spirit but is also, according to its innermost structure, ordained to union with the spiritual soul. The body visibly expresses the mental and spiritual life which unfolds the wealth of its meaning in the human soul; and the human body cannot at all be understood in its significance without reference to the spirit-soul. The *logos*-bearing functions of the human eye, mouth, hands, voice, etc., provide as it were the sole key for understanding the essence and nature of the human body, even for comprehending its biological properties and exterior shape and form and the functions of the brain, all of which receive their ultimate and specifically human character through their reference to the spiritual soul (mind, self) of man. In this sense the human spirit is indeed the form (the principle of unity and intelligibility and meaning) of the human body.

The human soul is precisely form of the body by not primarily being form of the body. It informs the body, both raising it up and "incarnating," concretizing and "expressing the life of the spirit in the body" most deeply in those dimensions of its life which are not "for the sake of the body."

The unity of man as consisting of body and soul is not threatened or rendered impossible by the mind, but the difference of the mind from the body constitutes the only conceivable basis for doing justice to the unity of man, in whom body and soul are joined to form a much more unified being than two sorts of matter could ever constitute, as Bergson said:

> *"By pushing dualism to an extreme, we appear to have divided body and soul by an unbridgeable abyss. In truth, however, we were indicating the only possible means of bringing them together."*[34]

4. THE FOURFOLD ROOT OF THE DIGNITY OF THE HUMAN PERSON**

4.1. The Nature of Human Dignity and Its Standing under Attack

We have come across the question of human dignity repeatedly in our discussions. Are human soul and human life just neutral facts and can they be disposed of according to the whims and wishes of individuals or political powers? Is the destruction of millions of human lives in Auschwitz or in the Gulag Archipelago just a neutral fact? In this case our proofs for the existence of the soul would be of little existential impact and interest. Or is this human life, in virtue of its very essence, endowed with a high value and preciousness, in a word, with a dignity that constitutes a decisive standard for medical ethics?

Peter Singer's book, *Should the Baby live?* which was also published in Germany,[35] raises the question of human dignity and its source and shows its dramatic importance for medical ethics. According to Singer, higher developed and healthy animals—such as pigs, dolphins, or dogs—deserve more legal protection than those human beings which, for one reason or another, are unable to exercise fully their specifically human capacities of thinking, speaking, etc. To believe that each living being of the species "homo sapiens" is endowed with an incomparable dignity Singer considers a mere expression of pride, a new sort of pride to which Singer gives a jaw-breaker of a name derived from Latin words: "speciecism."

Singer's explicit program of "unsanctifying human life" denies the essential dignity of the human person, implying that human life deserves legal protection only at a certain level of consciousness and "quality" of life, a thesis which was also suggested by some speakers during the *Third International Symposium on Coma and Death*. This view depends on the notion that human dignity is only derived from consciousness and a certain quality of conscious life. In order to examine this view, let us first investigate the meaning of "dignity."

The term "dignity" designates first a unique value which endows each person not only with an intrinsic and objective preciousness—which likewise belongs to animals, all living beings, as well as to works of art—but which constitutes in addition a higher level of value incomparable with any other value.

Dignity signifies an excellence of value which is so closely linked to the nature of a person that it cannot be understood independently of grasping the essence of the person in whose nature it is rooted. Hence the definition of the person through Alexander of Hales in terms of this dignity: "The person is a substance which is distinguished through a property related to dignity."[36] The essence and real existence of persons give rise to an ontological dignity, of which we understand that it belongs inalienably to a being endowed with understanding, freedom, moral conscience, the capacity of religious acts, an absolutely irreplaceable and indivisible self (in contradistinction to the *Verzweigbarkeit* of plant life),[33] etc., in a word, to a person.

"Dignity" not only means an excellence of value as such but also bears an intrinsic relationship to being the object of morality and of moral imperatives; it signifies a preciousness of the person which is both so essential for personhood and so imposing that it issues moral imperatives to respect it; even more: Dignity imposes a strict obligation to

** For a deeper discussion of the whole problem of body and soul, and the root of dignity, cf. Seifert.[28]

respect a being endowed with it both legally and morally in an essentially higher and more absolute manner than those beings which also possess morally relevant value but lack this dignity, as for example animals.

Certainly, also cruelty to animals is morally evil, but it cannot be compared to the violation of persons endowed with this lofty and morally imposing value: dignity. The violation of dignity constitutes not merely an immoral act but a special moral outrage. "Dignity" signifies such a morally relevant value that actions which are essentially directed against this dignity are also essentially directed against morality, that is, they are essentially and intrinsically evil and cannot become good and permissible under certain circumstances or when they are performed for certain good purposes. Dignity indicates a certain "sacrosanctness" of the person which makes the person's value inviolable for any reason whatsoever. Kant says that dignity is a value which must not be "verrechnet" (calculated), that is, which has an absolute character in that it does not permit its violation for any pragmatic reasons, even for a quantitatively speaking higher good. Kant goes on saying that the dignity of the person excels over anything that has a price, it is the irreplacable value of a being which can never be negotiated (das "über allen Preis erhaben ist, mithin kein Äquivalent verstattet.")[37] Therefore respect (*Achtung*) is owed to man absolutely and in a way which does not allow ever to use a human being just as a means.[37, 38] The dignity of the person forbids for example absolutely to rape a woman, even if such an action can save ten other women from a similar fate: because such an attack against freedom and sexual integrity is essentially an attack against the inviolable human dignity and is therefore essentially and intrinsically wrong. Any form of purely consequentialist foundation of moral norms fails to recognize this aspect of dignity and must therefore be rejected.[39]

Inviolable dignity of course cannot have the sense that that which possesses it cannot be violated in a moral sense, but it has the double sense that (a) a person endowed with dignity *ontologically speaking possesses a value that cannot be destroyed by our behavior*, and (b) the human person *morally speaking never ought to be violated* with respect to that which constitutes this dignity; the person, in virtue of his dignity, is *morally speaking inviolable precisely because this dignity never ought to, but can be, violated,* i.e., treated in a fashion unworthy and offensive of the person.

4.2. The First Source of Human Dignity: The Individual Living Person Himself

Authors who restrict human dignity to certain exemplars of the human kind seek to assign dignity to accidental features of the person and remove it from his essence. But we can clearly see that the being of the person, and hence also his dignity of a person *qua* person, does not reside only on the level of acts and accidents but resides on the level of his essence and substantial nature, and hence is given with his very existence and life.

This becomes immediately evident when we consider that to be a person is always to be—in Aristotle's terms—a substance, that is, to be a subject that stands in itself in being and is not just a quality or attribute or function of something else, as we have seen above.

Moreover, the being of the person, which is the first source of human dignity, requires both the rational and intellectual essence as well as the concrete individual existence and life of the subject which we designate as person. Persons are never abstract essences but always existing and incommunicable individuals. This was clearly recognized by Richard of St. Viktor who said that the person is "the incommunicable existence of an intellectual nature" (persona est intellectualis naturae incommunicabilis existentia), and that "it exists in itself alone according to a singular mode of rational existence" (existens per se solum juxta singularem quendam rationalis existentiae modum).[40]

The reality of individual existence, and even the absolute uniqueness and non-substitutability of the person as a subject endowed with a rational nature, is a condition of human dignity and of any personal dignity. Thus the essence of a rational being and the real existence and life of an unsubstitutable individual of such a nature interpenetrate each other in the origin of personal dignity.

We may express this point by saying: a human being possesses inalienable dignity not only "when he functions as a person," but he possesses this dignity in virtue of "being a person."[41] The root of the dignity of the person thus lies in his substantial reality which in turn includes his life—and this excludes that the person possesses his dignity only in terms of functioning as a person.

In Aristotelian terms, it is the substantial being of a human person and his inalienable potencies which ground this dignity, and not only their actualization. Also when humans are sleeping, they possess this dignity. Also the embryo who cannot use his intellect yet—but possesses it as a condition of the possibility of ever using it—and even the irreversibly comatose patient who retains those potencies that belong to his essence and substantial being even if he loses forever in this life the concrete ability to use them—is endowed with this dignity of the person.

These facts—the rootedness of personal dignity in the substantial being of the human person and its independence from its actualizations—are overlooked in the logical consequence of two philosophical positions of which we can recognize, however, that they do not do justice to the person: I mean an actualism which reduces the person to conscious acts or at least to a consciously lived center of acts. Such a position overlooks the great discovery of Aristotle: that acts cannot simply arise from nowhere but presuppose a subject that has being in itself, stands in being in himself and not only as inhering in another thing. Only the substantial being of the person and the insight of Boethius that the person is an individual substance of rational nature[††] can ground personal dignity.

The other great error which obscures this insight quite necessarily is any form of materialism which identifies the brain and its functions with the Self of the person. Of course, under this assumption we must take it for granted that human beings which cannot function as persons are not persons because the organ of their brain is exactly what gives rise to their personhood and dignity. But we have seen that the Self of the person who thinks and wills must be simple and cannot consist in the ten billions of brain cells nor in the billions times billions of their parts and distinct and partly unrelated functions, nor in the mere organic life in the brain and in the body which lacks the indivisible unsubstitutable center of the I. Thus the existence of a spiritual mind, of a soul as source of cognition and of free acts is indeed necessary to constitute this first and most basic level of human dignity: the dignity of the human substance, a dignity which inalienably belongs to man *qua* man and *qua* person and which does not possess degrees. Every human being has it in an equal degree.

[††] For this reason, it is unfortunate that Scheler thinks that he can save the category of person only by denying the substantial being of the person as standing in himself in being:[42]

> Das Zentrum des Geistes, die "Person," ist also weder gegenständliches noch dingliches Sein, sondern nur ein stetig selbst sich vollziehendes (*wesenhaft* bestimmtes) *Ordnungsgefüge von Akten*.

4.3. The Second Source of Human Dignity: Consciousness and Actualization of Personhood

Singer[43] makes the "proposal that we reject the sanctity of human life."[43] He thinks that the value of human beings can only lie in their actual capacity to think, to will, etc. He goes as far as to defend the view that retarded and otherwise handicapped children are less endowed with dignity than pigs or chimpanzees.[‡‡]

While we totally reject any negation of the first and most foundational level of human dignity, we do not deny that actual consciousness and lived conscious life gives rise to a second and new dimension of the dignity of persons, and is inseparable from what the person is called to be.

This second source of the dignity of the human person lies in the conscious actualization of the person, in the awakened personal being. This can be absent in seriously retarded persons or in human beings in the permanent vegetative state. Fichte and Engelhardt assume that the state of actualized personhood is reached by normal children only after the second year of life; one could also put it at a much later or earlier date. For the consciously awakened being of the person undergoes an infinity of shades and degrees, from the first prenatal experiences in the embryonic state to the early childhood until adulthood. Also in this consciously awakened being of the human person we find indeed the root of a new dimension of dignity which expresses itself in the acquisition of human rights which are not grounded—as the right to life and not to be sexually abused—in the very being of a person, in the substantial character of personhood, but in the different degrees of consciousness and maturity. For example the human right to freedom of speech and of movement, or to education cannot be attributed to a small child, as little as the human right to marry, to educate children, etc. These rights which are grounded only in awakened and consciously lived personhood of a certain level of maturity are unlike the right not to be subjected to murder, to mutilation, or to undignified treatment, etc., which are rooted in the first source of human dignity, the living human person as such.

This second source of human dignity and of human rights can indeed be lost through so-called brain death, even if brain death is not actual death of the person, irreversible coma, etc.[§§] Thus this second dimension of the dignity of a human person and the rights grounded in it are not as inalienable as the dignity and rights which are grounded simply in the substance, existence and essence of the person, in the first source of human dignity. Nevertheless, also this second dimension of personal dignity is inalienable as long as a person lives consciously or can reawaken. It does not depend on qualitative values, except in the weak sense explained below; the evil man possesses it just as the good man.

Most importantly, the second source of dignity of persons can never substitute the first and most foundational source of personal dignity.

---

‡‡ In rejecting this dignity founded on the very substance of the person, Singer makes an excellent point regarding abortion:[43]

> I will only point out that if we believe it is the potential of the infant that makes it wrong to kill it, we seem to be committed to the view that abortion, however soon after conception it may take place, is as seriously wrong as infanticide.

§§ One of the first sharp critics of the identification of death with brain death was Hans Jonas.[44,45]

4.4. A Third Sense of Dignity: A Positive Content of Consciousness (Quality of Life) and Moral Dignity

Thirdly, there is a human dignity which is the result only of the good actualizations of the person. One could list here pleasure as against displeasure, happiness as distinct from unhappiness, etc. But while these qualities stand in the center of today's quality of life discussions, they do not truly constitute a new sense of "dignity." Not even intellectual development and education or the actualization of other potentialities in a person bestow *per se* new dignity on a human being, at least not in the full sense. Rather, the new third sense of dignity is much rather constituted through higher values, most of all through moral values such as justice, love of truth, kindness, etc. The mere actualization of personhood and a consciously lived life are not sufficient to ground this moral level of human dignity which also gives rise to many new human rights.

As a matter of fact, a minimal level of dignity in the third sense is a condition for the fullness of possession of the rights which are rooted in the second level of human dignity. For the criminal can be deprived of such elementary rights as the right to freedom of movement, or of the right to educate his own children. But he must never be deprived of other human rights which are rooted entirely and solely in the second level of personal dignity: for example the right to a fair trial, to defend himself in court, to the freedom of conscience and religion. Thus some of the human rights rooted in the consciousness of living persons are absolutely inalienable as long as they live, and whatever crime they commit. Others rooted in the second level can be lost when their abuse is pushed too far or when the justice of punishment require it.

About this level of the moral dignity of the person Gabriel Marcel says rightly that it is a conquest and not a possession, which would not be true about the first ontological level of personhood and only true in a very restricted sense about the second level and source of the dignity of a person. Through education, through refinement of thought, but much more through the acquisition of moral values this dimension of personal dignity is increased.

The totally new dimension and source of personal dignity emerges before our minds when we contemplate the fact that it depends on the good use of freedom. This dignity is not inalienable nor does it automatically belong to us as persons. It is the fruit of morally good acts and thus radically distinct from the first type of dignity. It also has a distinct and unique quality which, as Kant rightly points out, culminates in holiness. This dignity differs from the purely ontological dignity of persons in that it knows opposites: it constitutes the radical contrast to the moral indignity and wretchedness, or maliciousness of a Hitler who loses any moral dignity through his actions. Evilness thus can make a person lose temporarily or for ever this dignity. Moreover, a completely new type of esteem is owed to a person in virtue of this third source of human dignity, an esteem in a far more literal and deep sense than the one which is the adequate response of respect owed to any being which bears human countenance, the *visage* in the sense of Lévinas.[46]

The third level of personal dignity, in this respect like the second one, can undergo innumerable degrees. New human rights, such as the right to a good moral reputation and to one's honor being preserved in society, have their root in this dignity and cannot be equally claimed or at all claimed where this dignity in the moral sense of the term is largely or wholly lost.

4.5. A Fourth Source of Human Dignity: Bestowed Dignity

A fourth source of dignity does not depend on the substantial being of a person nor on the use of his freedom but on gifts which neither each person possesses nor each person who has them possesses in the same degree. With regard to these gifts fundamental inequalities exist so that a fraternal recognition of the value of each person seems to stand in contrast to the claim to total equality, as thinkers such as Gabriel Marcel or Erik Kühnelt-Leddin have pointed out. The cry for *fraternité* of the French Revolution in some understanding of the term contradicts the cry for *egalité* of the same revolution. Max Scheler sought to demonstrate that a claim to equality that denies the existing differences of talents and gifts of various kinds results from a root of *ressentiment*[47] rather than from an insight into a universal equality of human nature. For, quite simply put, with respect to the fourth, as well as with respect to the second and third source of human dignity human beings definitely are not equal.[***]

Nevertheless, on the level of the first source of human dignity through the substantial character of a human person, we must uphold a universal equality of all human persons with regard to their dignity. For this dignity has no other conditions besides human nature and does not come in degrees. And we are obliged therefore to oppose any attempts to deny this level of *equal ontological human dignity*.

The gifts which endow either all men or some men with a special dignity can be natural gifts immanent in persons, such as beauty or intelligence, genius or charm, strength of character, etc. These latter gifts constitute a special dignity of the genius, of the artist, etc. They can also be gifts received through relations with other persons, such as St. Exupéry has in mind when he speaks in the *Le Petit Prince* of "être apprivoisé" (being tamed) by which term he refers to a uniqueness a rose or an animal receives in virtue of its being loved by persons. Symbolically, this stands for similar values of uniqueness received by human beings through becoming the object of the love of other persons. Of course, personal uniqueness already precedes and motivates human love, but a new dimension of uniqueness results from being loved. To this source of dignity belong several values of the individual that result from intersubjective relations and from being loved by others: for example the value of a person *as member of a community*, as well as the value the child receives through his or her social acceptance, *as friend of other persons* or through the child's acceptance through parents or society.

Gabriel Marcel's "existentialist foundation" of human dignity also refers to this fourth source of dignity. Marcel sees dignity founded in the gift of a common fatherhood, which also those non-believers in God who defend human dignity unconsciously may accept in a faith deeper than their actual beliefs and opinions. Marcel thinks that only such a reference to a common father renders intelligible the sense of brotherhood which also many atheists feel to other men and which is conceived by Marcel as including an acceptance of differences and therefore as entailing some contradiction to the cry for equality. Thus both brotherhood and the ultimate human dignity presuppose the relationship of the bearer of human dignity to a Thou.[48,49]

[***] This does not exclude that the fourth source of dignity through extrinsic gifts could in principle also be bestowed upon all men.

Also, for example, the office of the judge bestowed upon a man or woman by society gives rise to a new dignity and to a new human right: the right to the independence of the judge.

The human dignity which proceeds from gifts which go beyond the immanent rational nature of persons can also refer to a religious-theological and simultaneously ontological dimension of such gifts. We can not only affirm in religious faith that such a dignity exists but also see in a philosophy of religion that this deepest religious source of dignity of persons is possible and constitutes, if it exists, quite another type of dignity in comparison with the dignity which simply immanently proceeds from a person's nature.[†††]

As the second and third, so also this fourth source of human dignity knows many diverse forms and can be lost because it is—at least in most of its dimensions—not simply immanent to the being and life of a person. We do not deny that the existence and life of a person, too, can be rightly interpreted as gifts, but they do not have the peculiar character of extrinsic gifts intended here. Existence and life of the person are not elements which go beyond the possession of human being and nature.

If we look at this fourfold root of human dignity, we see that the notion often brought up in lectures and discussions throughout the preceding Symposium on Coma and Death (2000), as if human life deserved respect only in view of consciousness, is mistaken, reduces the source of human dignity to the second level, and confuses the different sources and kinds of human dignity. Notwithstanding the immense value and dignity of a positive content of human consciousness, the human person possesses an underlying and inalienable dignity which cannot be reduced to that which proceeds from his conscious life or its positive content. Since the human person cannot be reduced to a product of brain activity and since the value of human life and the dignity of the person has three other roots besides consciousness, not only the argument in favor of brain death from the loss of integration (an argument that received its final blow, particularly through D. Alan Shewmon's work of comparison with other similar forms of integration/disintegration in clearly living organisms), but also the philosophical argument equating brain death with actual death because of the irreversible loss of consciousness and the alleged "irreversible loss of quality of life and consciousness" in "brain dead" individuals prove untenable.

5. REFERENCES

1. Eccles J, Robinson DN. *The Wonder of Being Human. Our Brain and Our Mind.* New York-London: Springer, 1984.
2. Hildebrand DV. *Die Idee der sittlichen Handlung.* 2nd ed. Darmstadt: Wissenschaftliche Buchgesellschaft, 1969, pp. 1–126, pp. 8 ff.
3. Hildebrand DV. *Moralia. Nachgelassenes Werk. Gesammelte Werke Vol. V.* Regensburg: Josef Habbel, 1980:208 ff.
4. Hildebrand DV. *Ästhetik.* 1. Teil. Gesammelte Werke, Vol. V. Stuttgart: Kohlhammer, 1977:pp. 32–40; 49–57.
5. Hildebrand DV. *Ethics.* 2nd ed. Chicago: Franciscan Herald Press, 1978:191 ff.
6. Hildebrand DV. Das Cogito und die Erkenntnis der realen Welt. Teilveröffentlichung der Salzburger Vorlesungen Hildebrands (Salzburg, Herbst 1964): "Wesen und Wert menschlicher Erkenntnis" (7. und 8. Vorlesung). In: *Aletheia* VI, 1993–1994:2–27.

[†††] Philosophy cannot assert the *existence* of this supernaturally (divinely) bestowed dignity, but a philosophy of religion can understand in some measure its *essence* and *its different types.*

7. Wojtyìa K. *The Acting* Person. Boston: Reidel, 1979.
8. Seifert J. Karol Cardinal Karol Wojtyìa (Pope John Paul II) as Philosopher and the Cracow/Lublin School of Philosophy. In: *Aletheia* II, 1981:130–199.
9. Seifert J. *Back to Things in Themselves. A Phenomenological Foundation for Classical Realism.* London: Routledge, 1987, pp. 144 ff., 176 ff., 181–198, 249 ff., 286 ff.; see also the passages in the Index under "consciousness" and "constitution."
10. Seifert J. *Leib und Seele. Ein Beitrag zur philosophischen Anthropologie.* Salzburg: A. Pustet, 1973, pp. 45 ff.; 71–89.
11. Seifert J. *Erkenntnis objektiver Wahrheit. Die Transzendenz des Menschen in der Erkenntnis.* 2nd ed. Salzburg: A. Pustet, 1976, pp. 59 ff., 65 ff., 118 ff., 212 ff., 233 ff., 150 ff., 161 ff., 203 ff.
12. Leibniz GW. *Monadologie.* In: *Leibniz, Die Hauptwerke,* zusammengefaßt und übertragen von Gerhard Krüger. Stuttgart: Alfred Kröner Verlag, 1958.
13. Eccles J, Popper KR. *The Self and Its Brain.* Berlin/Heidelberg/London/New York: Springer-Verlag International, 1977/ corrected printing 1981.
14. Chisholm RM. *The First Person.* Minneapolis, 1981.
15. Chisholm RM. Is There a Mind-Body Problem? *Philosophic Exchange 1978;* 2: 25:32.
16. Seifert J. Essence and Existence: A New Foundation of Classical Metaphysics on the Basis of "Phenomenological Realism" and a Critical Investigation of "Existentialist Thomism." In: *Aletheia* I,1, 1977: pp. 17–157; I,2 (1977), pp. 371–459.
17. Seifert J. *Sein und Wesen.* Heidelberg: Universitätsverlag C. Winter, 1996.
18. Plato. *Phaedo,* 94b–95a.
19. Plato. *Phaedrus,* 245d–246a.
20. Plato. *Nomoi,* 895–896.
21. Plato. *Politeia,* IV, 439.
22. Plato. *Kleitophon,* 408.
23. Plato. *Kratylos,* 400.
24. Boethius. *Contra Eutychen et Nestorium,* cap. 3.
25. Boethius. *Patrologia Latina,* 64, 2343.
26. Kant I. *Vorlesungen über die Metaphysik.* K.H.L. Pölitz, ed. and Introduction. Erfurt, Keysersche Buchhandlung, 1821:201–202.
27. Kant I. *Critique of Pure Reason.* Norman K. Smith, trans. Toronto/New York: Macmillan and St Martin's Press, 1929, 1965.
27a. Kant I. *Critique of Practical Reason* VI, 33.
28. Seifert J. *What is Life? On the Originality, Irreducibility and Value of Life.* Rodopi, 1997.
29. Seifert J. Gibt es ein Leben nach dem Tod?. In: *Forum Katholische Theologie* 4/1989.
30. Seifert J. *Überwindung des Skandals der reinen Vernunft. Die Widerspruchsfreiheit der Wirklichkeit – trotz Kant.* Freiburg/München: Karl Alber, 2001.
31. Heisenberg W. *Physics and Philosophy. The Revolution in Modern Science.* First Harper and Row edition. New York: Harper and Row, 1962.
32. Hölscher L. *The Reality of the Mind. St. Augustine's Arguments for the Human Soul as Spiritual Substance.* London: Routledge and Kegan Paul, 1986.
33. Conrad-Martius H. *Die Seele der Pflanze.* In: H. Conrad-Martius, *Schriften zur Philosophie.* Eberhard Avé-Lallement ed., Vol. 1. München: Kösel, 1963:276–362.
34. Bergson H. *Matière et mémoire. Essai sur la relation du corps à l' esprit.* Bibliothèque de philosophie contemporaine. Paris: Alcan, 1896:221–222.
35. Singer P. *Muß dieses Kind am Leben bleiben*? Erlangen: Harald Fischer Verlag, 1993.
36. Alexander of Hales, *Glossa* 1, 23, 9.
37. Kant I. *Grundlegung zur Metaphysik der Sitten.* In: *Kants Werke,* Akademie-Textausgabe, Bd. IV. Berlin: Walter de Gruyter & Co., 1968.
38. Kant I. *Metaphysik der Sitten.* In: *Kants Werke,* Akademie-Textausgabe, Bd. VI. Berlin: Walter de Gruyter & Co., 1968.
39. Seifert J. Absolute Moral Obligations towards Finite Goods as Foundation of Intrinsically Right and Wrong Actions. A Critique of Consequentialist Teleological Ethics: Destruction of Ethics through Moral Theology?. In: *Anthropos* 1, 1985:57–94.
40. Richard of St. Viktor. *Trin.* 4, 22; 4, 25.
41. Schwarz S. *The Moral Question of Abortion.* Chicago: Loyola University Press, 1990:100–113.
42. Scheler M. *Die Stellung des Menschen im Kosmos* (1928). In: Max Scheler, *Gesammelte Werke,* Bd. IX. Manfred Frings, ed. Bern: A. Francke Verlag, 1976:pp. 7–72; p. 39.
43. Singer P. Unsanctifying Human Life. In: Peter Singer, *Ethical Issues Relating to Life and Death.* Melbourne, 1979, pp. 41–61; pp. 41 ff., 59; 43, 50.

44. Jonas H. Against the Stream: Comments on the Definition and Redefinition of Death. In: Hans Jonas, *Philosophical Essays: From Ancient Creed to Technological Man.* Englewood Cliffs, N.J.: Prentice-Hall, 1974:132–140.
45. Seifert J. Is "Brain Death" actually Death?. *The Monist 1993;*76:175:202.
46. Lévinas E. *Humanisme de l'Autre Homme.* 2nd ed. Montpellier: Fata Morgana, 1978.
47. Scheler M. *Ressentiment.* Tr. William W Holdheim. First edition: Lewis A. Coser, ed., with an Introduction. New York: Free Press of Glencoe, 1961; reprint: New York: Schocken Books, 1972. Second edition: Introduction by Manfred S. Frings. Marquette University Press, 1994.
48. Marcel G. *Die Menschenwürde und ihr existentieller Grund.* Frankfurt am Main: Josef Knecht, 1965; *The Existential Background of Human Dignity.* Cambridge, Mass.: Harvard University Press, 1962:139–162; 163 ff.
49. Kühnelt-Leddin E. *Liberty or Equality: The Challenge of Our Time.* London: Hollis and Carter, 1952.

THE DEATH OF DEATH

James J. Hughes*

1. INTRODUCTION

The current definitions of brain death are predicated on the prognostic observation that brain dead patients would quickly die even with intensive care. But this is now shown to be untrue.[1-4] Neuroremediation technologies and advances in intensive care will make it increasingly possible to keep alive the bodies of patients who would currently be classified as brain dead, and recover much of the memories and capabilities that we currently consider irrecoverable.

The on-going redefinition of death is the result of the technological deconstruction of dying. Instead of a relatively instantaneous, binary process, death is now more like a "syndrome," a cluster of related attributes, with a probabilistic diagnosis.[2] This disaggregation requires that we decide how many of these attributes are required before we begin treating someone as "dead," just as physicians must decide how many psychiatric traits are required before making a diagnosis of "schizophrenia." Electroencephalograms can only determine if there is a cessation of electrical activity on the surface of the brain, not in the deeper structures, and cannot determine if the electrically quiescent brain tissue is irrecoverable. Many of those who are diagnosed as brain dead in fact have clear evidence of functioning midbrains and brainstems, and are not necessarily irreversible.[3] A key argument in favor of whole brain death criteria over neo-cortical death, that the brain provides integrative functions that the body needs to survive, has also been shown to be fallacious since patients meeting the current clinical criteria for whole brain death have survived for years.[2]

Below I review some of the emergent technologies which will allow us to not only keep "brain dead" patients alive, but also repair their brain damage, requiring a re-specification of death from the whole-brain standard to the neo-cortical standard and beyond. The development of these life support and neuroremediation technologies will force us to finally accept a personhood-based "neo-cortical" position on brain death. But instead of legitimating the simple termination of life support for PVS patients, diagnostic protocols used to determine permanent vegetative state and brain death will have to be supplemented by therapeutic trials of neuroremediation, and predictions about the results

* James Hughes Ph.D., Public Policy Studies, Trinity College, 71 Vernon St., Hartford CT, 06106, USA. Email: jhughes@changesurfer.com.

Brain Death and Disorders of Consciousness, Edited by Machado and Shewmon
Kluwer Academic/Plenum Publishers, New York 2004

of such trials. The legal identity of the reconstituted personalities resulting from these therapies will be contested, and oblige the clarification of the legal status of discontinuous personalities in the law.

Ultimately, the nanotechnological neuro-prosthetics that we develop to remediate brain injuries will also lend themselves to the sharing and backing up of memories, thoughts and personalities. That point may be recognized as the "death of death."

2. STEM CELLS, NEUROGENESIS AND NEUROTROPHICS

Research is proceeding rapidly in the repair of brain damage with stem cell transplants and neurotrophics to stimulate neurogenesis.[4] Research has demonstrated that there are stem cells in the adult brain, and that they can locate and differentiate to repair damage in the brain, restoring function. Jeffrey Gray has repaired ischemic brain damage in rats with injections of rat stem cells.[5] Macklis and his team at Harvard have successfully induced neural stem cells present in adult brains of mice to differentiate and repair sections of damaged mouse brain.[6] Research is also rapidly progressing on gene therapy on neurons in the brain, to increase dopamine production in Parkinsons and boost nerve growth factors.[7,8]

In the near future, with combinations of stem cells, neurotrophic drugs and gene therapies, we will be able to remediate damage to brains that would meet current clinical criteria for brain death. Brain repairs might reconnect undamaged parts of the brain which still encode memories, skills and personality traits.

3. NEURAL-COMPUTER PROSTHESES

Methods to stimulate neurogenesis not only offer means of regenerating brain tissue for those who have suffered severe brain injuries or brain death, but also to facilitate the integration of brain prostheses. Many centers around the world are perfecting direct two-way communication between the brain and neuro-computing prostheses.[9,10,11]

The first step has been direct brain control of external computing and prostheses. For instance Mojarradi et al. at NASA have designed microarrays for the motor cortex to wirelessly control arm or leg prostheses.[12] Krishna Shenoy's lab at Stanford[13] is working on implanted electrodes which drive external computing. Roy Bakay and Phil Kennedy at Emory University have put computer chips into the motor control regions of the brains of patients in the locked-in state. The patients' neurons grow into contact with the chip permitting two-way communication, and when the patients think about moving various muscles they send a radio signal to an external computer. Slowly the patterns of neuronal firing against the chips' electrodes allow the patients to control computer cursors by thought alone.[14] Totally paralyzed people are now surfing the web, sending email and controlling their environment by thought alone.

The next step is to implant computing media in the brain able to act as prosthetic replacements for damaged brain structures. Berger and his lab at the University of Southern California have successfully modeled the neuronal inputs and outputs of the hippocampus, and are conducting trials of computerized prosthetic hippocampi.[15] Rapid advances in neural network software and the reverse engineering of the brain's neuronal interactions will allow these prosthetics to be adaptive and record new memories and pathways.[16] Eventually brain prostheses will replace the parts of a severely damaged brain which cannot be repaired or regrown using stem cells and neurogenesis.

4. CRYONIC SUSPENSION

Another technology that may eventually challenge our death concept is cryonic suspension, the freezing of heads or whole bodies for eventual reanimation.[17] All diagnostic protocols for the determination of brain death call for hypothermia to be ruled out when patients appear to be dead. But what if the patient is thoroughly frozen, and can't be safely thawed for many decades? If future nanomedical devices could repair the cellular damage caused by freezing, the ontological status of the frozen patient would be the same as the current hypothermically dead or the future, remediable brain dead patient: missing and probably dead, but possibly retrievable.

Unfortunately those who wish to be cryonically preserved must be legally dead first, else those who assist their freezing are considered accessories to suicide or murder. Cryonicists believe that future reanimation would be more successful if they could initiate the freezing before somatic failure, and certainly before cerebral death. Cryonics firms have already been accused (and acquitted) of murder for having failed to have a physician pronounce death before they began the suspension procedure. In 1993, the California Supreme Court ruled that a man with a terminal brain tumor could not have his head removed and frozen before he died.

Clearly the frozen do not meet a consciousness-based definition of life based on continuous, waking sentience, much less self-aware personhood. On the other hand, if such a standard is applied too rigidly, people who are in deep sleep, who are hypothermically suspended[18] but revivable, or who have been placed in temporary therapeutic states of arrest, would also be dead. The key question for the status of the cryonically preserved is the assessment of the likelihood that they can be revived. We have a very high likelihood that the sleeping will awake as their former selves, and at least we know of cases when hypothermic patients have fully recovered. Since we have never seen a reconstituted frozen mammal, the possibility of cryonic recovery is harder to accept. But the rapidly developing field of nanotechnology is creating an expectation that we will be able to repair and revive frozen bodies in this century.

5. NANONEUROLOGICAL REPAIR AND PROSTHESES

Cryonicists acknowledge that the freezing process results in the rupture of many cellular membranes and that micro-cellular repair will be the principal challenge of future reanimators. Cryonicists have therefore enthusiastically embraced the new field of nanotechnology which promises to eventually create molecular-scale, self-replicating robots capable of moving through frozen tissue without further disrupting cell walls, identifying damaged tissue and repairing it. Cryonicists expect this level of nanotechnology to be available within the next forty to hundred years.

The rapid advance of nanomedicine certainly gives reason for such optimism. The August 30, 2003 issue of the Lancet editorializes "Nanomedicine is a discipline whose time has come."[19] Similarly in 2002 the US National Science Foundation (NSF) convened a large project to examine the consequences of the convergence of nanotechnology, biotechnology, information technology and cognitive science (NBIC) for "improving human performance." The project drew together more than a hundred experts on science and technology from government, academia and business. The enormous 2002 report of the NBIC project includes detailed plans for nanotechnology to be intimately woven into the body and brain in the very near future, including a variety of projections of brain prostheses that the converging technologies will make possible in the coming decades. Computer scientist Ray Kurzweil projects that by 2030 the accelerating, converging and shrinking technologies will make possible a complete intracranial network of nanorobots, communicating with each neuron, and permitting immersive virtual reality, parallel processing, and backing up of memories and thoughts.[20] The introduction of such a nano-neuro network would have obvious therapeutic potential for the reconstruction of personality, skills and memory in a damaged brain.

6. THE MISSING AND THE DEAD

As our technologies for brain repair and prosthetic support of brain functions improve, so also do the odds that significant personhood can be recovered from a "dead" brain. In some sense, although we know the whereabouts of their body, the brain damaged person is a missing person. They are in a condition from which they may or may not return. There is the possibility that they have suffered information loss beyond the ability of technology to restore, or they may not. We have to decide whether to mount a search on the basis of our assessment of the likelihood that they are still alive.

Our ability to predict the likelihood of recovering some semblance of the original person will be important for families who are considering attempting reanimation. If recovery is unlikely, the person's advance directives or surrogate decision-makers should specify whether reanimation should be attempted when the likely result is a new person. Once the person is revived, they will need to be assessed for whether they are in fact sufficiently identical to the prior person. If the reanimated person is a substantially new person there will be a strong case for them to be considered a successor or relative of the deceased.

The prognostic prediction is then a part of the determination of the deadness of the patient, just as it is in the determination of the permanence of the "permanently vegetative state" (PVS). American practice toward PVS is based on the recommendations of the Multi-Society Task Force on the Permanent Vegetative State that patients should be considered permanently unconscious if they are unaware for three months after non-traumatic injury (such as chemical overdose) or 12 months after a traumatic brain injury. Once classified as permanently vegetative, caregivers and physicians are given much more latitude for conservative treatment, often in effect allowing death to "take its course."

Another cognate situation is the missing person. When someone goes missing on the high seas, or doesn't return from war, common law has long held that these missing persons be declared dead for practical reasons. If the circumstance of the missing person's death is uncertain, and there is some possibility they have been shipwrecked, taken hostage, or hiding, the court imposes a waiting period of some years before death may be declared. The Uniform Probate Code, adopted by 18 states and more or less in effect in the rest, declares death to have occurred after a five year waiting period. If there continues to be strong evidence that the missing person may be living, judges may put off declaring death even after this waiting period. Once the absentee is declared dead, most states protect the heirs and those who declared the person dead from liability for wrongly distributing their property, and otherwise harming their interests.

The deadness of the missing person is also partly determined by our decision to mount a search mission or not, which is also true with the therapeutic situation of the "do not resuscitate" order. The person in arrest with a DNR is much deader than the person in arrest subject to resuscitation. The non-heart-beating donor controversy has also made explicit what was implicit with DNRs; the intention to resuscitate a heart/breathing-arrested person partly determines when in the dying process a person is declared dead.

Similarly the intention to treat the brain dead as alive determines their status under New Jersey's brain death law. New Jersey permits Orthodox Jews to define and treat their brain dead relatives as alive. Also the U.S. approach to the fetus can show a similar social determination, as in the proposed US laws which would punish as a murderer someone whose assault leads to the miscarriage of a desired fetus, but would still permit a mother to have an abortion of an undesired fetus.

So the intention not to attempt identity recovery is itself part of the determination of death. As technology erodes the brain death standard, the future operational definition of "dead enough" will become something like: "The patient cannot be revived to self-awareness, with continuity of previous memories and personality, because they have irretrievably lost that information, or we are unable to recover them, or the patient or their surrogate decision-makers do not wish them to be revived."

7. THE STATUS OF RECONSTRUCTED PERSONALITIES

The beneficiaries of these therapies will probably continue to be disabled in many ways, and have lost much of their memory, skills and personality. The neurogenesis required to restore self-awareness may actually destroy identity-critical memories, as is suggested by Tsien et al.[21] The question these technologies will raise is how much of one's personality and memories one can lose for the new person grown in one's brain to be considered a new legal person. At the extreme McMahan asserts that the complete replacement of the cerebral tissues would clearly constitute a new person:

> *"Replacement of the (cerebral) tissues through the transplantation of new hemispheres might make consciousness possible, but this would not count as receiving the same mind, even if the new hemispheres were perfect duplicates of those destroyed. There would be a new and different mind."*[22]

But what of a new person who had forgotten everything of their former self except their taste in food or music? Would the family, society and the law consider this a successful recovery of the original person?

I believe there are three steps we should take to address the legal and identity ambiguities of personalities reconstituted from destroyed brains. First, we should develop a set of criteria to distinguish the legal personhood of the prior person and any future person who occupies their body but lacks sufficient psychological continuity to be the same person. The prior person should be declared dead, all their obligations and contracts should cease, and the subsequent person should become an heir of the prior person. The prior person's property should automatically be ceded to a trust to pay off debts and disburse to beneficiaries of the will, but funds should be reserved to provide for the subsequent person's welfare. The patient's family should have custody of the new person until they are returned to full autonomy.

Second we need to develop a thorough predictive model of the destruction of personality and memory in the patient's brain, and the likely results of remediation, before initiating therapy. This model will allow the family and other surrogate decision-makers to decide if there will likely be enough of their loved one left to allow continuity of personal identity. Above a certain threshold of prognostic certainty and level of identity recovery the family and physicians will be obliged to attempt the recovery. If the likelihood that the person will be recovered falls below a certain threshold the patient's advance directives and surrogate decision-makers will be allowed to determine whether there should be a trial of neuroremediation. If the person is sufficiently recovered, they would assume their former identity, but if they are judged a different person, they would be an heir.

Third, advance directives will need to be amended to elicit patients' wishes about reconstituting partial selves. Some may fear the prospect of a partially reconstituted self, while others may welcome any continuity, even if it is just physical continuity with a new person inheriting their body.

Some will certainly argue that legal personhood is identical to body continuity, and that any reconstituted personality would be identical to the prior person. Others may limit identity to those who recover identity-critical information and self-awareness without becoming significantly dependent on non-biological prostheses. For instance Youngner and Bartlett addressed the possibility of future mechanical remediation of brain injuries, and accepted that such remediation would be unproblematic for brainstem functions. But they go to great lengths to reject the possibility of mechanically-mediated cognition.

> *"It is...easy to imagine a patient's integrated vegetative functions being fully assumed by complex machines or well-trained health professionals. Any problems in such a takeover would be of a purely technical nature. In contrast, conceptual problems surround the replacement of a patient's consciousness and cognition. We believe it is impossible for a person's thoughts and feelings to be replaced by a mechanical device and still retain their essential nature. If the replacement is successful, the thoughts and feelings would no longer be those of a human; if they remained essentially unchanged, the replacement was not successful. This point shows the essential, conceptual connection between higher brain functions and the continued life of the person. No comparable problem arises with the replacement of vegetative operations...If a living person is to exist, the thoughts and conscious processes must be those of a human, not a machine."*[23]

Youngner's distinction between continuity of personhood maintained by organic therapy and continuity maintained by non-organic prostheses has no ethical justification. Nor is such prejudice likely to persist as people with powerful brain prostheses become increasingly common.

8. BEYOND PERSONAL IDENTITY

> *"Despite our every instinct to the contrary, there is one thing that consciousness is not: some entity deep inside the brain that corresponds to the 'self,' some kernel of awareness that runs the show, as the 'man behind the curtain' manipulated the illusion...in The Wizard of Oz. After more than a century of looking for it, brain researchers have long since concluded that there is no conceivable place for such a self to be located in the physical brain, and that it simply doesn't exist."* [24]

Just as technology drives us to clarify that we value continuous, discrete self-aware persons more than the biological platforms they come on, so it will also force us to acknowledge that continuous, discrete personhood is a fiction. Neuroremediation technology and brain-computer interfaces will erode the apparent boundaries and continuity of the self, and the autonomy of the individual and her decisions.

Threats to the self will develop in many areas. Our control over the brain will slowly make clear that cognition, memory and personal identity are actually many processes that can be disaggregated. We will have increasing control over our own personalities and memories. Full nanorobotic replication of the mental process opens the possibility of identity cloning, distributing one's identity over multiple platforms, sharing of mental components with others, and the merging of several individuals into one identity.

8.1. Technological Threats to Liberal Individualistic Assumptions about Continuous Discrete Personhood

- *Identity Malleability:* Parental, social and personal control of memory, identity and personality
- *Posthuman Persons:* Radically enhanced minds
- *Identity Sharing:* Sale or sharing of memories, thoughts and skills,
- *Identity Cloning:* Persons multiply copied into new biological or computing media
- *Distributed Identity:* Distinct persons distributed over, or sharing, a set of bodies and machines
- *Group Identity:* Multiple bodies and machines integrated into a collective identity, without clear personal identity, e.g. hive minds, the Borg

When we get to the point where neurological functions can be controlled, designed, cloned, shared, sold, and turned on and off, the fact that the continuous, autonomous self is an illusion will become more obvious. This will also pose a fundamental challenge to liberal democracy as noted by the futurist thinker Alexander Chislenko: "When one can easily modify, borrow, or drop, merge with others, and separate, any of their external and internal features ... there won't be distinct lines between individuals anymore." [25]

Once we are forced to drop this fundamental predicate of Enlightenment ethics, the discrete, continuous, autonomous individual, we are beyond the ethical frameworks of liberal democratic law and bioethics. Perhaps we will also then be beyond death, since it will be possible to backup and preserve all aspects of personal identity and experience, either as discrete individuals, or in distributed parts.

9. CONCLUSIONS

The current definition of death, worked out twenty years ago to address the technology of the respirator, is falling apart. Some suggest we dispense with "death" as a unitary marker of human status, while others are pushing for the recognition of a neocortical standard. This century will begin to see a shift toward consciousness and personhood-centered ethics as a means of dealing not only with brain death, but also with extra-uterine feti, intelligent chimeras, human-machine cyborgs, and the other new forms of life that we will create with technology. The status of these forms of life will be one important element of the struggle between the "transhumanists," advocates of non-anthropocentric personhood and post-human technological possibilities, and their opponents the "biofundamentalists" or "bioLuddites." Each proposal to extend human capabilities beyond our "natural" and "God-given" limitations, or to blur the boundaries of humanness with brain prostheses, will be fought politically and in the courts. Prominent bioethicist George Annas suggests for instance that brain implants and human-machine cyborgs, among many other technologies, be made "crimes against humanity" by international treaty.[26]

Nonetheless I believe we will soon need to scrap the brain death standard in favor of a much more tentative, probabilistic, information-theoretic understanding of death, as the loss of identity-critical information. The preservation of identity-critical information, regardless of whether on an organic or inorganic platform, will be considered continuity of legal personhood. Then, we will, eventually, face the challenge of a society in which death becomes completely meaningless and individual identity itself begins to unravel.

10. REFERENCES

1. Shewmon D. Recovery from 'brain death': a neurologist's apologia. *Linacre Q* 1997;**64:**31-96.
2. Emanuel L. Reexamining death: the asymptomatic model and a bounded zone definition. *Hastings Cent Rep* 1995;**25(4):**27-35.
3. Truog R. Is it Time to Abandon Brain Death? *Hastings Cent Rep* 1997;**27(1):**29-31.
4. Kozorovitskiy Y, Gould E. Adult Neurogenesis: A Mechanism for Brain Repair? *J Clin Exp Neuropsychol* 2003;**25(5):**721-732.
5. Hodges H, Veizovic T, Bray N, et al. Conditionally immortal neuroepithelial stem cell grafts reverse age-associated memory impairments in rats. *Neuroscience* 2000;**101(4):**945-955.
6. Arlotta P, Magavi SS, Macklis JD. Induction of adult neurogenesis: molecular manipulation of neural precursors in situ. *Ann N Y Acad Sci* 2003;**991:**229-236.
7. Fitzsimons HL, Bland RJ, During ML. Promoters and regulatory elements that improve adeno-associated virus transgene expression in the brain. *Methods* 2002;**28(2):**227-236.
8. Poulsen DJ, Harrop JS, During MJ. Gene therapy for spinal cord injury and disease. *J Spinal Cord Med* 2002;**25(1):**2-9.
9. Helms Tillery SI, Taylor DM, Schwartz AB. Training in cortical control of neuroprosthetic devices improves signal extraction from small neuronal ensembles. *Rev Neurosci* 2003;**14(1-2):**107-119.
10. Wolpaw JR, Birbaumer N, McFarland DJ, et al. Brain-computer interfaces for communication and control. *Clin Neurophysiol* 2002;**113(6):**767-791.
11. Taylor DM, Tillery SI, Schwartz AB. Direct cortical control of 3D neuroprosthetic devices. *Science* 2002;**296(5574):**1817-1818.
12. Mojarradi M, Binkley D, Blalock B, et al. A miniaturized neuroprosthesis suitable for implantation into the brain. *IEEE Trans Neural Syst Rehabil Eng* 2003;**11(1):**38-42.
13. Shenoy KV, Meeker D, Cao S, et al. Neural prosthetic control signals from plan activity. *Neuroreport* 2003;**14:**591-596.
14. Herberman E. Mind over Mater: Controlling Computers with Thoughts. *ALS News [serial online]* 1999; Mar 5. Available at: http://www.rideforlife.com/n_thought030899.htm. Accessed: September 1, 2003.
15. Alataris K, Berger TW, Marmarelis VZ. A novel network for nonlinear modeling of neural systems with arbitrary point-process inputs. *Neural Netw* 2000;**13(2):**255-66.
16. Mussa-Ivaldi FA, Miller LE. Brain–machine interfaces: computational demands and clinical needs meet basic neuroscience. *Trends Neurosci* 2003;**26(6):**329-334.
17. Alcor. 2000. *Cryonics Reaching for Tomorrow [on line]*. Available at: http://www.alcor.org/CRFTnew/crft08.htm. Accessed: September 1, 2003.
18. Safar P. Suspended animation for delayed resuscitation from prolonged cardiac arrest that is unresuscitable by standard cardiopulmonary-cerebral resuscitation. *Crit Care Med* 2000; **11(Suppl):**N214-N218.
19. Nanomedicine: grounds for optimism, and a call for papers. *Lancet* 2003;**362**(9385).

20. Kurzweil R. *The Coming Merging of Mind and Machine. In Sci Am. [serial online] 1999;* Fall. Available at: http://www.sciam.com/article.cfm?articleID=0007EE6E-F71F-1C72-9B81809EC588EF21. Accessed: September 1, 2003.
21. Feng R, Rampon C, Tang YP, et al. Deficient neurogenesis in forebrain-specific presenilin-1 knockout mice is associated with reduced clearance of hippocampal memory traces. *Neuron* 2001;**32(5):**911-926.
22. McMahan J. The metaphysics of brain death. *Bioethics* 1995;**9:**91-126.
23. Youngner S, Bartlett E. Human death and high technology: the failure of whole-brain formulations. *Ann Intern Med* 1983;**99:**252-258.
24. Nash M, Park A, Wilworth J. Glimpses of the mind. *Time* 1995;**July 17:**44-52.
25. Chislenko E. Interview with Alexander Chislenko. *Thing Reviews [serial online] 1997.* Available at: http://old.thing.net/ttreview/marrev97.03.html. Accessed: September 1, 2003.
26. Annas GJ. The man on the moon, immortality, and other millenial myths: the prospects and perils of human genetic engineering. *Emory Law J* 2000;**49(3):**753-782.

THE SEMIOTICS OF DEATH AND ITS MEDICAL IMPLICATIONS

D. Alan Shewmon and Elisabeth Seitz Shewmon*

Die Grenzen meiner Sprache bedeuten die Grenzen meiner Welt.
[The limits of my language signify the limits of my world.]
(Ludwig Wittgenstein, *Tractatus Logico-Philosophicus* 5.6)

In any discussion pertaining to this world, human beings are bound to the medium of language. We are often unaware of the degree to which language, as the physical shape of all thought, exerts its influence on the distinctions we make and consequently on the formation of our notions and ideas. The way we think shapes the way we speak, but also conversely, the language we speak shapes the thoughts we think. The languages we live in are the result of all preceding evolution of thought, cultural interaction and the communication thereof within a linguistic community, already presenting us with a pre-formed way of structuring and interpreting the world (*Weltanschauung*) at the time of primary language acquisition. We are not able to step out of the medium of language, but we are able to step out of one particular language into another, thereby critically evaluating the distinctions, terms and notions we usually take for granted. Surprising new perspectives open up on seemingly well-known objects of debate, and from the meta-linguistic point of view, more often than not what appeared to be a problem on the object level turns out to be an inherent feature of the language we use to discuss it. A case in point is the phenomenon called death.

1. INTRODUCTION

This presentation will (1) elaborate on the ambiguity of the concept corresponding to the word "death" and (2) suggest that the "dead donor rule" has caused us to focus all along on the wrong ethical question surrounding transplantation. The first part will be

* D. Alan Shewmon, MD, Professor of Neurology and Pediatrics, David Geffen School of Medicine at UCLA, Los Angeles, CA. Email: ashewmon@mednet.ucla.edu. Elisabeth Seitz Shewmon, Ph.D., Linguist, Los Angeles, CA. Formerly Assistant Professor of Slavic Linguistics at Eberhard-Karls-University, Tübingen, Germany. Email: eshewmon@socal.rr.com.

structured according to three levels of linguistic consideration as they relate to the death debate: (1) respecting word-usage conventions for the sake of communication (and the ill-formulated process-vs.-event debate), (2) the role of vocabulary in reflecting importance and focusing attention (and the multiplicity of death-related events), and (3) the dynamic interaction between language and thought (how much might our unitary concept of death be conditioned by the language we think and communicate in). The final section will consider the implications of all this for transplantation.

But first let us recall the traditional hierarchical schema (concept-criterion-tests) for discussing the definition and diagnosis of death and the associated assumption that "beginning at the beginning" means beginning with a concept of death. This structured approach was logical and served the field well for many years; nevertheless, the best point of intellectual departure may actually not be a concept, but rather observation – not only of various death-scenarios but also of how people speak and think about death.

1.1. Concept-Criterion-Tests: Sounds Nice but Gets Us Nowhere

In the four-decade-old debate over the definition and diagnosis of death, virtually everyone has accepted the conceptual framework eloquently outlined by Bernat and colleagues in their seminal paper of 1981,[1] namely (1) to distinguish clearly between three levels of discussion – concept, criterion, and tests – and (2) to respect the flow of logic in just that direction: from the abstract to the concrete. It has seemed obvious to most participants in the debate that the only way to get a proper handle on "death" is to begin with a correct (or at least agreed upon) definition or concept of death (a purely philosophical matter), then determine what anatomical criterion instantiates that concept (a hybrid philosophical/medical matter), and finally determine what clinical signs or tests reliably indicate that the criterion is fulfilled in a concrete case (a purely medical matter). After all, how can one devise tests for an unstated criterion, or how can one formulate a criterion for an undefined or vague concept?

For many years the first author also accepted this modus operandi as the only rational way to attack the problem of diagnosing death,[2] but he is no longer so sure. First of all, there are reasons to doubt whether a valid "concept of death" can even be clearly formulated. It is inextricably bound up with the concept of "life": death is the cessation of life. In his book "What is Life?" philosopher Josef Seifert convincingly argues that "life" is intrinsically undefinable.[3] It is an "ur-phenomenon," conceptually fundamental in its class; no more basic concepts exist to which it can be reduced. It can only be intuited from our experience of it, both around us (in nature and human society) and especially within us (our own personal psycho-physical life). Not even the thermodynamical concept of life as localized anti-entropy,[4-7] which appeals especially to scientists, adequately captures the essence of life in general. To cite one of Seifert's examples, a spore can remain inert for centuries but still possess life. If there can be no totally adequate definition of "life," neither can we presume to begin the logical cascade of concept-criterion-tests with a totally adequate "concept of death."

Many commentators have pointed out that, depending on the extent to which a concept relates to the physical domain, it may or may not specify a physically definite moment, process or state, and hence be correspondingly more or less relevant to the clinical diagnosis of death. For example, "cessation of the organism as a whole" is both a concept of death and the presumed description of a physical event (even if unobservable directly). (The precision with which it can be pinpointed in practice is a separate matter.) Some philosophical concepts of "soul" (e.g., the Aristotelian-Thomistic "substantial form of the body" officially endorsed by the Catholic Church) correspond to "organism as a whole" in the biological domain; thus, "departure of the soul" (so understood) corresponds in principle to a definite, physically definable moment. Other concepts of soul, however, (e.g., in Platonism and eastern religions) do not imply anything in the physical domain related to its coming or going. What becomes of the three-level schema if a philosophically favored "concept of death" happens to imply no clear empirical consequences for biology or medicine?

In some socio-medical contexts the very "concept" of death is what the three-level schema would call a "test." For example, in an anticipated home death, when the family at the bedside relates the onset of death to final apnea, their implicit concept of death is "expiration," quite different from, but no less appropriate than, "loss of the organism as a whole," "irreversible loss of the capacity for consciousness," "loss of potential for cardiac autoresuscitation," or any of the other popular candidate concepts in the current medical-ethical debate. The "concept" of death is similarly at the level of "tests" when a physician declares (again, appropriately) the official time of death of a monitored, non-donor patient with a do-not-resuscitate order to be when the EKG goes flat – although one could argue that this is not really a concept (but what more genuine concept is it a "test" for?) or that it is merely a legal fiction.

Veatch rightly distinguishes two basic types of concept of death: normative and ontological.[8] Those who favor a unitary, objective concept tend to think of death ontologically, whereas those who favor a context-dependent notion tend to think of it normatively. We submit that a normative concept implies a tacit ontological concept: if death is defined as the moment after which, or state when, it is ethical to treat the person as no longer alive, there is an already understood ontological concept of death. Moreover, in what follows we shall suggest (contrary to the first author's previous writings) that there is no singular, clear ontological concept of death, as the single word "death" tempts us to imagine, but rather a legitimate context-dependency even in the ontological order.

Therefore, instead of proceeding via logical deduction from some preconceived, abstract "concept of death," we propose a phenomenological approach: to step back from linguistically constrained notions (to the extent possible) and simply look afresh at the spectrum of realities surrounding what people call and have called "death" – as though one had never heard of the term "death" and feels no compulsion to label any particular phenomenon "death."

This mental exercise could prove invaluable and liberating, if in fact it is a delusion that our concepts of life and death are simply generalized from everyday experience and labeled by linguistic convention with the words "life" and "death," when in actuality those very concepts have been shaped by our language, and more importantly *constrained* by it (particularly by the poverty of a lexicon that offers only the single dichotomy "life" vs. "death" to apply to a rich and complex portion of reality). If our language deceives us into thinking that there must be only one reality of death merely because we have the one word "death," no wonder that we end up arguing endlessly and uselessly over which of the various events or processes is the *real* "death." No wonder also that the death-debate is so full of logical inconsistencies and failures to communicate, because people use the same word to express different concepts, not quite realizing (precisely *because of* the single word) that the concepts *are* different.

2. WATCH YOUR LANGUAGE! THE POINTLESS DEBATE OVER PROCESS VS. EVENT

Even the most cursory re-examination of death-related phenomena reveals the existence of various processes and events, all of which are equally "real." The debate whether death is a process or an event has dragged on for over three decades since the famous kick-off pair of articles by Morison[9] and Kass.[10] Nothing substantially new is ever said, and proponents of each side simply talk past each other. As traditionally formulated, the debate is intrinsically unresolvable, because it is based on terminology that is linguistically incorrect and fails to adequately convey the real issue.

A variation on the "process" theme is Linda Emanuel's "asymptotic model," which rejects a dichotomy between life and death and focuses instead on dy*ing* as "a bounded zone of residual states of life."[11] One problem with the asymptotic metaphor is that by definition an asymptote never reaches its limit. That means that even with rigor mortis and putrefaction, the corpse may be dead "for all intents and purposes" but still isn't finally and totally dead (and never will be). If the approaching curve actually reached its limit, it would not be properly speaking an "asymptote"; moreover, there would then be an "event" that could legitimately be called "death" or "the moment of death." A more serious problem with the model is the axis labeled "life" or the multiple axes proposed for the expanded metaphor ("personhood" and "biological life," each of which could in turn be divided into various sub-axes: "self-awareness," "ability to love," etc., at the "organic, cellular and molecular" levels). To treat such things as quantitative parameters representable as orthogonal axes in multidimensional space is more poetic than scientific. What could possibly be meant for any such axis by a quantitative value along the continuum of real numbers, and how would such a value be determined? If "life," "personhood," etc. are not meaningfully quantifiable, neither is the asymptote a helpful mathematical metaphor.

2.1. Precedents for State-Discontinuity Brought About by Continuous Change in Parameters

Process advocates have universally failed to comprehend (and event advocates have failed to point out) that invisible though real discontinuities of state can result from continuous changes in observable parameters. Continuity and discontinuity are mutually contradictory only in the same domain; system-state and parameter values are in different domains. To deny state discontinuities simply on the grounds of continuity of underlying processes is unscientific and anachronistic; it ignores some of the most interesting developments in modern science and biology.

2.1.1. Quantum Mechanics and Eigenstates

Quantum mechanics has something important to teach in this regard. Take the most elementary quantum mechanical system, a hydrogen atom. There is an infinite continuum of possible positions in space where the electron might be found, but only discrete possible energy levels and corresponding probability distributions of position (specified by the Schrödinger equation). No matter where the electron is in space, its position is mathematically compatible with either a free or a bound state, and if bound, with any of an infinite number of discrete energy levels. Consider a free electron shooting towards a free proton; suddenly a quantum of energy is emitted and we now have a hydrogen atom. Despite the continuity of the electron's trajectory, a dramatic yet unobservable event takes place: one moment there are two free particles and an infinitesimal moment later there is one atom.

If this is the case for something as simple as a hydrogen atom, how much more should we expect to find state discontinuities in more complex quantum systems and even in macroscopic systems whose differential equations, like the Schrödinger equation, yield discrete eigenvalues (the mathematical basis for the "quanta" of quantum mechanics). Varela made a good case that the dynamics of living systems must be mathematically modeled by enormous, complex sets of differential equations whose solution sets, as in quantum mechanics, contain discrete eigenvalues, corresponding to discrete eigenstates.[12] When a continuous change in one or more parameters occasions a change from one state to another, the state-change is necessarily discontinuous (like the transition between energy levels of a hydrogen atom). If that is the case within life, it should hardly be surprising or incomprehensible at the transition between life and non-life.

2.1.2. Nonlinear Dynamics and Bifurcations

The burgeoning field of nonlinear dynamics makes the process-event debate seem almost antediluvian. Apart from eigensolutions, nonlinear differential equations often occasion a type of state-discontinuity known as bifurcation. A classical example is the Hopf bifurcation, which describes the behavior of an oscillator subjected to a particular pattern of external influence. It can be represented by the following pair of equations:[13]

$$dx/dt = -y + x\{\mu - (x^2+y^2)\}$$
$$dy/dt = +x + y\{\mu - (x^2+y^2)\}$$

The qualitative behavior of the system is determined by the control parameter μ. For $\mu<0$, the system undergoes a damped oscillation asymptotically approaching an equilibrium point. For $\mu>0$, however, the system asymptotically approaches a "limit cycle" of self-sustained, spontaneous oscillation with radius $\sqrt{\mu}$ and period 2π. As μ is varied continuously from negative to positive, when it passes the bifurcation value $\mu=0$, a qualitative state-change occurs: an infinitesimal instant earlier the system tended to static equilibrium, and an infinitesimal instant later it tends to spontaneous oscillation. This radical discontinuity in system property is very real, though at the moment quite unnoticeable by anyone merely watching the oscillator and following the continuously changing values of x, y, and μ.

Bifurcations of this and other sorts are ubiquitous throughout nature, especially biological nature.[14-18] The transition between homeostasis (characterizing life) and lack thereof (characterizing decay) is a bifurcation not unlike the Hopf bifurcation, except on an infinitely more complex scale.

2.2. Linguistic Respect

The whole process-event question is simply mis-posed as necessarily "either-or." Between the beginning of dying and the end of decaying, continuous processes and discontinuous events occur (some directly observable, others not). Whether we assign the word "death" to one or more of these processes or events is primarily a linguistic question, not a biological one.

Death has always been understood as the cessation of life. But within this general notion the word "death," like many words, can have more than one nuance of meaning in common parlance, or in Saussurian terms, a range of *signifiés*. It can refer to (1) the state of the remains of a previously living thing, (2) the state of the remains of a previously physically living thing, which still lives in some spiritual sense, or (3) the moment of transition from living to one of these two states. "State" and "moment" are not opposed concepts of death but merely acceptable alternative usages of the word "death." When we employ it in the third sense, e.g., "death occurred at 2:15," this is merely another way of saying "[the state of] death began at 2:15." By contrast, the adjectival forms "alive" and "dead" apply only to states.

The referent of the verb "to die," however, depends on its form. The simple present or past tense is usually understood as referring to the momentary act of transition from life to death, e.g.: "He died at 2:15," or (last words of a martyr) "I die for the cause of justice!" (referring to the immediate future). The present perfect, "has died," indicates a present state of death that began at the moment of death. The progressive form, present participle, and gerund indicate, by their very linguistic nature, a process: "dying," "is dying," etc. can only refer to the gradual fading of life processes, tending toward [the moment of] death (= ending with the beginning of [the state of] death). (For an extensive analysis of the aspectual qualities of "die" verbs in 18 languages, see Botne.[19])

We therefore fail to see the novelty of James DuBois's proposal of death as a "state," as opposed to an event or process.[20] Of course it is a state, which begins with an instantaneous transition-event that linguistic convention allows us also to call by the same name "death." What is provocative about DuBois's proposal is rather that the state of death need not, by definition, be irreversible – but more on that below.

There is nothing particularly esoteric about all of this. There are processes and events related to death, and the English language provides us with the means of communicating accurately about it all and being understood. Nothing is to be gained, and only confusion sown, by insisting that a word which on linguistic grounds refers to a moment or a state must henceforth be used to refer to a process.

Some antithetical pairs logically entail a continuum of gradations, such as light and dark, rich and poor, beautiful and ugly. Any transition between the two is necessarily gradual; there is no intelligible, non-arbitrary point along the spectrum of illumination-intensity at which "light" becomes "dark." By contrast, there can be no intelligible continuum between logical contradictories, such as pregnant and not-pregnant. Properties traditionally associated with life and death form such pairs: being vs. non-being, unity vs. multiplicity, endogenous opposition to entropy vs. tendency toward increasing entropy, self-sustained limit-cycle vs. damped oscillation tending to static equilibrium (in the Hopf bifurcation analogy). Between any of these and its opposite the transition must, by logical necessity, be instantaneous and discontinuous, as must also be the case with the intimately related dichotomy life vs. non-life (i.e., death).

It would be linguistically possible, though uninsightful, to claim that "life" and "death" have been misconceived all along, that there are only graded realities like "light" and "dark," and that there is no discrete event or moment that separates the two, only a fuzzy transition. But it is simply an abuse of language – which biologists, physicians and ethicists have no right to change – to insist that "death" is a "process" when one could just as well use the term "dying" instead.

This foolish debate may be partly language-specific. Among death-related expressions in German, for example, there is the noun *Tod* (death) and the verb *sterben* (to die), with its nominalized form *das Sterben*, referring to the act or process of dying. The difference between the two nouns is clearly reflected when they are juxtaposed in a title, such as *"Tod und Sterben"* (corresponding to the English "On Death and Dying"). The debate whether death is an event or a process would already for that reason not make sense in German, because if one wants to speak of a process one uses *Sterben*, and if one wants to speak of an event one uses *Tod*. It would be incomprehensible even to ask whether *Tod* is a process or *Sterben* an event. Since English has only the one noun "death," and since the nominalized form of "to die" derives etymologically from the same root, it is easier to fall into wondering unwittingly whether "death" is a process or an event, but the question is still just as malformulated linguistically as it would be in German regarding *Tod*.

3. MARK MY WORD: A VOCABULARY OF DEATH-EVENTS

The lexicon of any culture or society reflects what is important for them; conversely, children learn what distinctions are important in the world around them partly by the language they learn. Pilots and weather forecasters have many terms for what the general public calls "clouds." Their specialized vocabulary is not merely descriptive; it is essential to the skilled and safe performance of their task. Sylvan societies have an expanded vocabulary for plants, serving not only a communicative function (reflecting the interests of adults) but also a pedagogical one (helping children, through learning the names, to learn the distinctions signified by the names: what is safe to eat, how to make various medicines and paints).

Eskimos and skiers have a multiplicity of words for frozen water. One reference on West Greenlandic, for example, identifies at least 49 different words/lexemes referring to ice and snow.[21(pp.366-367)] The linguistic controversies over the exact number, the distinction between "word" and "lexeme," etc. do not concern us here. The point is that what to us are incidentally different qualities or locations of "snow" (snow on the ground, snow falling in air, air thick with snow, feathery clumps of falling snow, newly fallen snow…) are perceived and conceived by the natives of western Greenland as significantly different things. It is logical that distinctions important for way of life and survival should be reflected in linguistic distinctions; for those of us for whom snow is but a wintry nuisance or source of fun, the words "snow," "slush" and "ice" probably suffice.

Now imagine that pioneers from the tropics migrated to the arctic and suddenly had to learn how to survive there. Their very language, which allows them to see only "snow" and "ice" all around, is itself detrimental to survival. They would do well to learn the Eskimo vocabulary in order to focus attention on important aspects of their new reality, critical distinctions regarding that white stuff they so carelessly refer to as "snow."

This is precisely the situation we find ourselves in regarding "death." We have migrated through human history into the modern ICU, bringing with us the linguistic baggage of a relatively simple concept of death for which the one word had always sufficed. Now we find ourselves in a situation for which medically real and ethically critical distinctions lack words in the common vocabulary. The best we can do is to speak in awkward paraphrases, such as "the point in time beyond which cardiac auto-resuscitation is impossible." To ask which of these technological mumbo-jumbos is *really* "death" may perhaps be as linguistically and epistemologically inappropriate as asking an Eskimo which of *sullarniq, aput, qaniit, nittaalaq, ...* is *really* "snow."

Having established above that system-level events can and must occur along the process of dying/decaying, and following a phenomenological approach, let us simply examine some sample events, each in its own proper context, putting aside the question as to which is the one "true" death. (Why must we assume *a priori* that there is only one?) Because our language lacks simple names for the technologically defined events, let us try to free our minds to the extent possible by simply calling the candidate events E1, E2, and so on, as defined below. (Such neutralized nomenclature is similar in approach to what Lynn and Cranford have labeled "Time 1," "Time 2," etc.[22] Some events are directly observable, such as E1 and E2; others can only be inferred as having happened already or not yet. Depending on the type of event, such inference could be based on something observable (e.g., with E5, inferring unconsciousness from unresponsiveness) or on cumulative knowledge from scientific studies (e.g., E3, E4, E6, E7).

Let us begin with common notions of death as an event. It could be instructive to recall what death-scenes were like before ventilators made possible a dissociation of the chief physiological components and before transplantation became a utilitarian motive for changing traditional concepts of death.[23,24(pp.20-37),25] Given how foreign an environment modern ICUs are to most people, perhaps the average non-medical person's understanding of death might not be that different today from what it was in pre-ICU times, once we get beyond the uncomprehending parroting of certain "brain-death" rhetoric picked up from the media. The first two E's to be considered are of this type.

3.1. E1 – Expiration ("Ex-Spiration")

A common pre-ICU-era death scenario was the terminally ill person lying in bed at home, surrounded by family members keeping vigil. Breathing becomes shallower and finally stops. As soon as the onlookers realize that the rhythm of breathing has halted and that a next breath is not going to come, they infer that with the end of that final exhalation the loved one died. Someone closes the deceased's eyes (if not already closed) and the family begins mourning. There is no checking for pulses or EKG electrical activity or brain waves. In such a context final apnea constitutes the most glaring (perhaps the only evident) discontinuity between the process of dying and the process of decaying. Let us call this event E1.

Of course, as Emanuel points out, actual death scenarios aren't always so straightforward.[11] There may be irregularly timed agonal gasps. What the family thinks is the last may turn out not to be so. Such a pattern of terminal respiration may temporarily complicate matters for the family in deciding when E1 occurs, but it does not undermine the concept of E1. No matter how many times the family might be fooled into thinking that a certain breath was the last, eventually there will be a truly last breath, which will in retrospect be known as such with certainty. The end of that final exhalation is E1.

It was common to declare, in colloquial theological terms, that at this moment the person "gave up the ghost." In Jewish tradition, cessation of breathing has generally been considered a more critical and essential marker of death than cessation of circulation.[26(pp.241-254),27,28] One of the Yiddish expressions for "to die" is *oyshoykhn di neshome*, in which *neshome* means "soul" and *oyshoykhn* means "breathe out" (i.e., one's final breath).[29(p.222)]

Christian tradition likewise contains an E1-precedent for the moment of death in the narration of Jesus's own death on the cross. All four gospels link his final crying out with the giving up of his spirit and illustrate the close relationship between breath and soul in English, Latin and Greek (Mt 27:50, Mk 15:37, Lk 23:46, Jn 19:30). In modern English, "to expire," meaning "to die," is etymologically derived from the Latin *exspirare*, "to breathe out." Further instances of the conceptual connection between "life" and "breath" are the English words "(to) animate" and "animation," which derive from the Latin *anima* or *animus*, "soul" or "spirit." In ancient Greek these correspond to two words, each of which is related to both life and breath: *Psychê* (ψυχή) is translated most often as life or soul, but also as ghost, departed spirit, conscious self, personality; whereas the verb form ψύχω means "to breathe," "to blow." *Pneuma* (πνεῦμα) signifies "breath," "spirit," "mind"; the verb πνέω "to blow," "to breathe out." The perfect πέπνυμαι means "to possess a soul," "to be animated," "to be alive," and "to have understanding"; whereas ἡ πνέουσα is a favorable wind.

Of course we know from modern medicine that in scenarios where death is understood as apnea (or at least coinciding with apnea), had the pulse been measured, it would have been noted to persist some tens of seconds after the final exhalation. Ineffectual EKG electrical activity might persist some minutes afterwards; all brain function would not be irreversibly lost for some minutes more – and who knows how long after that the spiritual soul *really* departs to enter the afterlife. In his previous unitary-death mind-set, the first author would have said that E1 was a convenient fiction for non-medical people in a non-medical setting, and that the "giving up of the ghost" at that moment was forgivable pious poetry, but that the "real" moment of death occurred some time later (whether defined philosophically in terms of loss of the substantial form of the body, or systems-dynamically in terms of loss of integrative unity of the organism as a whole, or thermodynamically as the transition from endogenous opposition to entropy to a giving-in to entropy – all of which were assumed to necessarily coincide, because they were assumed to be merely different ways of looking at the same phenomenon).

But does our medical and philosophical understanding mean that we should intrude on the family's death-bed scene to explain that they ought not to begin grieving yet, because their loved one isn't *really* dead yet, and won't be for some unknown number of minutes more? Does the absurdity of such an imaginary intrusion derive simply from its impoliteness and insensitivity, or rather from some deeply seated intuition that perhaps the family was right all along to understand that their loved one really did die and "give up the ghost" with that final exhalation? Why must an ICU context be the only "correct" one for understanding death?

Some authors have argued that irreversibility is not, and has never been, intrinsic to the notion of death.[20,30] Resurrection stories in religious history may be miraculous or mythical, but they do not do linguistic violence to the concept of death. Citing some fascinating studies on near death experiences, DuBois suggests that it may be reasonable after all to regard such patients as transiently, reversibly in the state of death, and that the European term "reanimation" might be literally more correct than our "resuscitation" after all.[20] The German term *Wiederbelebung* expresses the idea even more explicitly (literally "the giving back of life" to somebody). DuBois would not declare the state of death until all three major bodily systems have shut down (at first reversibly, soon irreversibly): breathing, circulation, and brain function.[20] But why must even this be so?

Let us take his proposal one step further. In the kind of scenario we are talking about, the person "falls asleep" (a common poetic expression for "dies") with the last breath: consciousness is lost, even if all brain functions are not yet lost. Why should a residual heartbeat for the next half-minute or so preclude this state from being death already, any more than it should preclude "brain death" from being death? For those who accept the notion of soul as life-principle, why could it not be that, rather than asystole causing or occasioning the soul to depart, the heart ends up stopping precisely *because* the soul has already departed? If hypothetically the person were successfully "reanimated" by doctors, this would not prove that the soul had not yet departed and that the person was not really dead; it could merely indicate that the soul had returned.

The more one gets into the role of the devil's advocate about all this, the more it becomes obvious how, in the home death-bed context, the various physiologically and systems-dynamically defined "events" are relatively abstract, hypothetical, unobservable, and without the least practical consequence, compared to that dramatic final exhalation. Who are we to say that the person didn't *really* die then and that the family was "wrong" (even if forgivably so) to initiate death-behaviors at that moment? Would it not rather be like forcing a square peg through a round hole to insist that one of the several candidate ICU-understandings of death be applied to this context?

3.2. E2 – Final Asystole

Nowadays the death-bed scene often takes place in a hospital, with the patient connected to various monitors. Suppose a do-not-resuscitate order is in place for a terminally ill patient. At one point the person stops breathing, and everyone watches the cardiac monitor until the rate slows, the pulse weakens, and finally the EKG goes flat. Let us call this latter event E2. At that moment the physician or nurse notes the official "time of death" to be entered on the death certificate, and the family begins its grieving process. (If the scenario were different such that E2 preceded E1, as in sudden cardiac arrest, it would of course be more appropriate to call E1 death.)

It might happen that some other pulseless cardiac rhythm preceded asystole, such as ventricular fibrillation or electromechanical dissociation. It is a moot point whether the "official" moment of death for purposes of the death certificate and pronouncement to the family is at the onset of such rhythm or of asystole. If desired, we could subdivide E2 into two cardiac-related events: the onset of pulselessness (E2a) and the onset of asystole (E2b); if bradycardia slows to asystole without any other terminal arrhythmias intervening, then E2a and E2b will coincide.

Are staff and family "wrong" in considering E2 the moment of death, given that, in principle, for the next few minutes resuscitative efforts could succeed in restoring cardiac function? If the state of death need not be intrinsically irreversible, as DuBois suggests, the answer is clearly "no." Or would they be "wrong" in the other direction, given that, had the same terminal illness ended at home unmonitored, they would have identified E1 as the moment of death? We do not think so. Since each "moment of death" seems intuitively appropriate within its own context, why must one trump all but one with the assumption of a unitary concept, as though correctness of E1 implied erroneousness of E2 and vice versa? Why not rather give priority to the intuitive correctness of both over the unitary-concept assumption?

3.3. E3 – Loss of Potential for Cardiac Auto-Resuscitation

The moment of loss of potential for cardiac auto-resuscitation (let's call it E3) seems to be of interest only in the context of non-heart-beating organ donors (NHBD). Some critics of the Pittsburgh protocol question whether two minutes of asystole is sufficient for determining E3.[31] This is an empirical question having nothing to do with the conceptual point of discussion here. It is interesting that such critics never take issue with the families declaring home-death at E1, or with medical staff for declaring non-transplant-related hospital death at E2. Whether or not one agrees with the ethics of NHBD in general or the details of some protocol in particular, three observations are pertinent. (1) Because E3 marks a point beyond which organ removal will not in any way alter the physical course of events already set in motion in the dying/decaying body, it is at least a reasonable candidate for a moral-legal "moment of death" in the context of NHBD. (2) This reasonableness does not invalidate the reasonableness of E1 and E2 in the more ordinary, non-NHBD contexts. (3) By not objecting to E1 and E2 in their own contexts, those who insist on the moral importance of E3 in the NHBD context (including ourselves) implicitly espouse a heterogeneous, context-dependent-event notion of death.

3.4. E4 – Loss of Potential for Interventional Resuscitation

The NHBD literature abounds with debate over the various senses of "irreversibility" of apnea and asystole. The "ethical" interpretation ("irreversible" because one should not and will not try to reverse it) corresponds to E1-E3,[32] whereas the "physiological" interpretation (loss of physical potential for resuscitation despite all technological means) corresponds to E4. Even E4 can be subdivided into "strong" and "weak" interpretations of irreversibility ("not reversible now with existing technology" vs. "can never be reversed despite all future technology").[30,33] Unlike E1 and E2, which are distinctly observable events, E3 and E4 are unobservable (based on subtle molecular-level changes), but no less real and distinct, moments along the continuum of physiological changes following hypoxic-ischemic damage to the heart.

3.5. E5 – Onset of Permanent Loss of Consciousness

This is the analog of E2 in the neurological domain, occurring around 5 to 10 seconds or so after (permanent) cessation of blood flow to the brain. If there is no moral obligation to try to restore consciousness (e.g., through restoring circulation of oxygenated blood), and especially if there is an obligation *not* to try, then E5 is also analogous to E3. Objections to NHBD on the basis that the donors are not yet "brain dead," i.e., that they are at E5 but not yet E6 or E7 (*vide infra*), are analogous to the cardiac-domain objections that donors are at E3 but not yet E4. Both domains of objection focus on the wrong question ("Are they dead?" vs. "Will they be killed or harmed?" – see final section below) and gratuitously absolutize one particular E for no apparent reason other than to build a case against NHBD.

3.6. E6 – Loss of Potential for Recovery of Consciousness

E6 corresponds to the notion of "irreversible coma" or (in the case of damage primarily to the cerebral hemispheres with a relatively intact brain stem) "permanent vegetative state." As with the cardiac E3 and E4, our ability or inability to determine in practice when E6 occurs, to such-and-such degree of precision, is an important empirical issue but is beside the point of the present conceptual discussion. If what sets the death-events in motion is primarily neurological rather than cardio-respiratory (i.e., all the sorts of etiologies that typically lead to "brain death" or persistent vegetative state), then the sequencing of E5-E7 relative to E2-E4 will be reversed, and the sequencing relative to E1 will depend on the precise temporal-anatomical pattern of brain insult.

3.7. E7 – Irreversible Loss of All Brain Function

This is how the Universal Determination of Death Act and most US state laws formulate the neurological arm of the bifurcated statutory definition of death, a condition commonly but unfortunately[6] known as "brain death." The fact that present diagnostic standards do not actually require strict fulfillment of this statutory definition (by allowing for some residual brain functions dismissed as "insignificant" or not true "functions" – cf. Shewmon[7,34]) is beside the point that E7 is a real moment along the continuum of death-related physiological changes. If desired, we could subdivide E7 according to the various schools of "brain-death" advocacy, such as E7a for the British "brainstem death" and E7b, E7c, ... for irreversible loss of all "critical" brain functions, depending on how one chooses to define "critical".[35,36]

3.8. E8, E9, ...

We could think of many other potentially definable, real physiological events along the course of dying/decaying and give each an E-label. Who knows whether we would come up with as many clinically or morally relevant ones as the 49 types of ice and snow? For purposes of this paper, the principle has been sufficiently illustrated with E1-E7. For each of these events we could coin a name more descriptive than "E3" and less of a mouthful than "loss of potential for spontaneous cardiac auto-resuscitation," just as the West Greenlanders say *"siirsinniq"* instead of "S45" or "ice swelling over partially frozen river, etc. from water seeping up to the surface." But that exercise can be saved for a more propitious occasion.

3.9. Special Contexts

In certain rare contexts, for example decapitation, some of these E's may not apply and other E's become uniquely relevant, such as the moment of head-severing. In other rare contexts, such as instant vaporization from a nuclear blast, all the E's occur simultaneously. Although of possible theoretical interest, such unusual death-circumstances have little relevance to the "dead donor rule," the "physiological decapitation" analogy of "brain death" notwithstanding.

4. WE THINK WHAT WE SAY: IMPLICATIONS FOR A UNITARY CONCEPT OF DEATH

The dynamic interaction between language and thought goes much deeper than merely focusing attention by naming. What if our assumption that there must be a clear, unitary, objective, correct concept of death, is derived not so much from intellectual insight as from an accident of the language we think in: the singularity of the word "death"? What if our very lexicon is a set-up for the interminable and seemingly unresolvable debates about the nature and determination of death, as well as for the incoherent thinking about death that abounds among not only the general public but health professionals as well?

In the following, we will briefly survey some selected languages and their lexical material referring to the phenomenon of "death" and "dying". For this purpose, the authors' own linguistic competence was combined with a little field study among native speakers and an inquiry addressed to members of the Linguist List, an internationally acknowledged online forum for linguistic discussion. Naturally, a systematic study investigating all semantic and stylistic aspects involved would be desirable, but beyond the scope of this paper.

Most languages possess a single-word equivalent to the English "death," suggesting that there is indeed a corresponding singular concept or "ur-phenomenon" universally understood across societies down through history. This makes sense, because up until the very recent advent of "life support" in developed countries, the set of candidate death-events was fairly limited (final breath decapitation...). Moreover, nothing critical hinged on the exact timing of death (so long as it had surely occurred prior to burial). But modern developed countries now find themselves with death-situations unknown and inconceivable throughout the millennia over which languages developed. Therefore, just because we grew up learning to speak and think with the one word "death," it does not follow that we must also think with the same singular concept in the context of modern ICUs. (Neither does the new context necessarily imply that we *shouldn't* think in terms of a singular death-concept; it simply raises the question, which we believe is answered in the course of this paper.)

Wilhelm von Humboldt (1767-1835), one of the "fathers" of modern linguistics, developed as an integral part of his work a theory of the dynamic mutual interaction between how we think and how we speak. We use language to express ideas, but our range of ideas is also shaped and limited by the language we grow up in as we learn to think. This limitation can, of course, be transcended by reflecting about the preconditions provided by our native languages. Some of the more spectacular examples of the degree to which languages can differ structurally are the Algonquin language, which has no word for time,[37] and Classical Hebrew, which operates without the category of tense (for the tense/aspect discussion, see Cook[38]), as do Chinese languages. Although it is not true that speakers of these languages do not understand the concept of time, it has to be inferred that such structural details of a language do shape the world view of its speakers.

Distinctions in language derive from the awareness of distinctions in reality; but the reverse is also true: the perception of distinctions in the real world is facilitated when a linguistic distinction already exists. Speakers of languages with a reduced color lexicon, for example, consider as mere variations in a single color what others regard as fundamentally different colors.[39] In Russian there are two different colors, both called "blue" by English speakers (*sinij* for dark blue and *goluboj* for light blue), and there is no word corresponding to our genus "blue."

To what extent does this phenomenon influence our thinking about death? In other words, do we have the single word "death" to express a primarily intuited, unitary concept of death, or is our concept of death unitary because up to now we have had only the one word for it (apart from pejorative or euphemistic synonyms)? By examining death-words in other languages and cultures, we might get a sense of the extent to which the (largely English-speaking) debate about death might be conditioned or constrained by the English language.

Some languages have no equivalent for the English word "death." For example, in the Kovai language of Papua New Guinea, the verb *um* means "to die," but the noun formed from it, *umong*, means not only "death" but also mere "sickness" (not necessarily fatal). There is no other obvious word for death or sickness. This may be quite common in Papua New Guinean languages [personal communication, Michael Johnstone, Cambridge University]. In Tok Pisin (English-based creole of Papua New Guinea) "he dies/is dead" is rendered *em i dai*, which can also mean that he is unconscious. To indicate what we call death they add an aspectual qualifier: *em i dai pinis* (which can also mean something like "he is already dead" and which is not available for the future tense) or *dai olgeta* ("die altogether") [personal communication, Eva Lindström, Department of Linguistics, Stockholm University]. These people's very language seems to reflect a world-view in which the demarcation between life and death lies more in the direction of life than we tend to think.

A similar thing occurs in Quechua:

> "My sister-in-law is dying!" This, in Quichua, may mean anything from a headache to a snakebite. If one is in excellent health, he is "living." Otherwise, he is "dying." "What is the matter with your sister-in-law?" "She is causing a child to be born. Will you come?"[40(pp.42-43)]

Such a linguistic difference reflects a profound difference in world-view, in which death is viewed not as the end of life but as a kind of extreme of illness, after which the spirits of the dead continue to live (physically) in a different place, eating, sleeping, working, etc., from whence they may return periodically to speak about their present life to family members in dreams.

A parallel phenomenon occurs in Nivkh (older name: Gilyak; spoken on Sakhalin and opposite mainland by fewer than 1,000 people), in which the verb corresponding most closely to "die" is *mu-*. The interesting feature of this verb is that it also corresponds to the English "become." There is thus no verb for "die" in Nivkh which is really equivalent to the English [personal communication, Daniel Abondolo, School of Slavonic and East European Studies University College, London, UK]. What might this imply for the world-view of this culture, in contrast to ours, where, conversely, living is a constant "becoming" and dying is a ceasing to be?

The opposite phenomenon obtains in languages with two or more non-synonymous words which we would translate indiscriminately as "death." Italian, for example, has three words for death. *Morte* (= death, neutral) and *decesso* (= death at the end of illness) refer to essentially the same concept under different aspects, but *trapasso* (= passage from this life (on earth) to a better one (in paradise)) expresses a religious concept of death that the other words do not, namely that after bodily *morte*, personal life continues in a spiritual form. This is similar to the English euphemism "to pass on," which implies passage to somewhere else, as opposed to "pass away," which carries more overtones of disappearance from existence.

Tamil has three nouns for death. *Irappu* means departing from the world, shedding life (the opposite of *pirappu*, which means "birth," "coming into the world with life"); it is a process. *Chavu* refers to the state of no life, and can refer to plants and animals as well as people. *Maranam* refers to the event of death and is used only in reference to humans [personal communication, A. S. Sundar, Etymologist, Ponicherry, India].

The latter example is also a particularly clear instance of another way that death concepts and the whole death debate can be influenced by language: namely, the distinction that many languages, but not English, make between human and non-human death. Polish, for example, has two nouns for death. The most frequently used is *śmierć* (of common Slavonic origin), a non-marked equivalent of the English "death." It is used in everyday speech as well as in legal and medical language, e.g. *śmierć kliniczna* (clinical death), *śmierć mózgu* (death of the brain). The other term, *zgon*, is not only more formal and medical, but means specifically "death of a person": one may not say *"zgon mózgu"* (zgon/death of the brain) but only *"śmierć mózgu"* (śmierć/death of the brain); afterwards a doctor may declare *zgon* of the patient. Polish also has two separate words for "to die": *umrzeć*, used only in reference to human beings, and *zdechnąć*, used in reference to animals or derogatorily in reference to humans. Similarly, the German verb *sterben* applies only to humans (and personalized animals like pets); for animals one uses *verenden,* and for plants or animals, *eingehen.*

Hindi has three words for death: two from the same root, *mrityu* (Sanskrit) and *maut* (modern form of *mrityu*), and one from a different one, *dehant* (meaning literally "end of the body"). Although the three can be used relatively interchangeably, the interesting thing is that, in contrast to the aforementioned languages which have separate words for human vs. non-human death, these Hindi death-words apply equally to humans and animals, but not to plants. In English we can say that a plant "died," but in Hindi they can only say that it "dried up" or "withered." Thus the Hindi language both reflects and reinforces the Hindu belief in reincarnation among sentient forms of life.

The insistence by some, including the first author in his earlier thinking, following Bernat, that the correct concept of death must be species-nonspecific, applicable to all living things, might therefore derive more from a linguist accident (that English has only one proper word for death) than from anything else. The debate over whether human death should be conceived as primarily personal vs. biological might not even take place in languages that distinguish the two concepts of death by different words. (Whether "personal" can be reduced to "mental" is another whole issue.) English is surprisingly impoverished in this area, compared to its general richness of vocabulary and capacity for nuance of expression.

This view represents a radical departure from the first author's previous understanding of death as a species-nonspecific, unitary phenomenon of a fundamentally biological nature, to which other disciplines ought to refer. Defending "brain death" at the Pontifical Academy of Sciences in 1989, for example, he stated (in a subsection entitled "Death is a unitary phenomenon"): "The only alternative [to postulating multiple "kinds of death"], which is intuitively more reasonable anyway, is to regard human death as a singular reality, which can be considered from a variety of perspectives [biological, clinical, philosophical, theological, legal]."[6(p.28)] Even after his subsequent rejection of the traditional "organism as a whole" or "integrative unity" rationale for "brain death,"[2,7,34, 41-43] he continued to subscribe to the "unitary phenomenon" assumption, up until the recent rethinking expounded here.

All that time he had assumed that the essence, in the biological realm, of the one phenomenon of death (cessation of life, regardless of form or species) could be expressed in at least two ways (two candidate "concepts of death," in the tripartite terminology of Bernat et al.[1]): loss of endogenous opposition to entropy, and loss of integrative unity of the organism as a whole. (Maintaining the parallel with the empirical events labeled "E_", let us call these conceptual moments C1 and C2, respectively.) It was taken as axiomatic that these two concepts were merely different ways of describing the same thing and therefore necessarily pointed to the same moment in time. Such "concepts" of death are not so abstract as others like "loss of the life principle," which do not imply a physically definable moment along the biological time line. The single moment corresponding to these two physicalistic "concepts" was taken to be the "true" moment of death, which one should not assume *a priori* to correspond necessarily to any of the E's listed in the previous section, some of which could at best be tolerated as token "moments of death" for the sake of cultural sensitivity or convenient legal fiction.

Now we must question whether either C1 or C2 is determinable or even meaningful on the empirical level; much less can their temporal coincidence be taken as axiomatic. Moreover, there is no *a priori* reason to assume that either of these concepts is intrinsically "truer" or more "objective" than the death-concepts in cultures with no linguistic equivalent for the English word "death."

4.1. C1 – Loss of Endogenous Opposition to Entropy

As noted in the beginning, "life" – like "unity," "being," and "consciousness" – is an "ur-phenomenon," a fundamental concept which we come to know only through direct experience, not through more fundamental concepts to which it is reducible, of which there are none.[3] Localized opposition to the general tendency in nature to increasing entropy is an extremely important aspect of physical life, but, as Seifert points out, it does not apply to all instances of life (e.g., spores, frozen goldfish), nor does it encompass the entire essence of life even in instances where it applies. If we accept a richer, undefinable but intuitable notion of life that cannot entirely be reduced to thermodynamics, we should expect a corresponding variety of types and manifestations of death, not entirely reducible to thermodynamics.

Let us take Seifert's thesis one step further. It is simply a gratuitous assumption that entropy is a measurable – or even meaningful – property of living systems. It is a quantitative abstraction with different meanings in different contexts. Thermodynamic entropy is the amount of thermal energy unavailable to do work in a closed system. The Second Law of Thermodynamics states that entropy in a closed system can never decrease (i.e., available energy can never spontaneously arise out of nowhere). Logical or informational entropy is a measure of disorganization or disorder. A common-sense analog of the Second Law states that order does not spontaneously arise from disorder.

Without entering into the fascinating debate about entropy and the origin and evolution of life on earth, let it be simply noted that for any individual living organism there is no method or formula for calculating its entropy (whether thermodynamic or informational). It is not at all evident that such an abstraction, invented for the physics of gases, has any quantitative meaning in biology. Moreover, entropy (even if it could be measured) and its rate and direction of change are surely not homogeneously distributed within a dying organism; there are no doubt pockets or subsystems already breaking down and increasing in "disorder" while other parts continue to resist that tendency with greater or lesser success. There is simply no coherent way, even in a thought experiment, to quantitate the "entropy" of such a system. Moreover, in what sense is a terminally ill, irreversibly dying (but not yet "dead") organism still "anti-entropic" anyway?

The entropy metaphor from physics sounds attractive and explanatory at first, but when its application to the transition from life to death is carefully examined, it is seen to be precisely a metaphor, not a more accurate description. There is no operationally meaningful way, even in principle, to determine a moment along the time course of dying/decaying when "endogenous opposition to entropy" instantaneously changes to "a tendency to increasing entropy." C1 is nothing more than a mental construct, which for logical and linguistic reasons must refer to an imaginary moment in time and not a duration, but which lacks any clear physical correlate.

4.2. C2 – Cessation of the Organism as a Whole

In a recent article the first author proposed an operational definition for "integrative unity," intended as a first pass which hopefully others would help to fine-tune.[7] It was based on the implicit assumption, taken universally for granted in the "brain-death" debate, that unity ("organism as a whole") vs. multiplicity (collection of organs and tissues) is a dichotomy translatable (at least in theory) from the philosophical to the physical domain. But if that assumption is incorrect, as we now suspect, then any attempt to operationally define "organism as a whole," with the goal of enabling unequivocal, nonarbitrary, dichotomous categorization of all cases, is an exercise in futility.

To begin with, there are difficult questions regarding exactly what constitutes "the organism." For example, what is the status (whether part of "the organism" or not) of an abscess, a cyst, a hamartoma, a benign tumor, an oocyte in an ovary or Fallopian tube, a conjoined twin with a common heart, the tip of a hair or fingernail, stones in the gall bladder, "normal" bacteria in the gut necessary for digestion, a transplanted organ, a molecule of ethanol in the bloodstream, a molecule of botulinum toxin in the bloodstream, antibodies causing automimmune disease, a partially amputated finger dangling from the hand by a thread of skin, etc. We are increasingly skeptical that there can be some all-encompassing, philosophically sound criterion for determining what, in medical practice, constitutes "the organism," especially during its dying phase as various parts become dysfunctional and necrose at different times and rates.

That article cited a litany of properties "at the level of the whole," some or many of which are present in at least some "brain-dead" bodies, from which it was concluded (and we would still conclude) that those bodies must therefore be "organisms as a whole." The problem is how to apply this concept to bodies dying in the more usual cardio-pulmonary way, in which there is a relatively long window of time during the dying/decaying process when no holistic function is any longer evident but neither can it be demonstrated or inferred that all are absent (especially those that are not directly observable but exist as potencies). Take, for example, "capacity for wound healing," an empirical test for which would take days, but the dying/decaying process might predictably reach the death-state with certainty within a few hours. How, then, can one determine at what moment the holistic property "capacity for wound healing," along with many others like it, is lost?

The situation seems something like (though not on the same basis as) Gödelian unprovability. For a given set of axioms (viz., the death-concept axiom, in the logical domain, of intrinsic discontinuity between unity and multiplicity, between anti- vs. pro-"entropy"), there is an infinite number of statements that are neither provable nor disprovable (an infinite number of moments along the course of dying/decaying in a concrete case that can neither be proved nor disproved to correspond to the moment of discontinuity in the logical domain). Or it is like the fuzzy fractal border of certain mathematical structures such as the Mandelbrot set, for which there may be infinitely many points intrinsically impossible to determine whether inside or outside the set.[44(pp.138-140)]

A possible reaction to this unsatisfying lack of clear projection from a moment in the logical-conceptual domain to a moment in the physico-temporal domain might be to reject the philosophical notion of unity altogether – bringing us face to face with the deepest controversies of ontology. We find it far preferable to say that healthy, living organisms are obviously integrated unities, that decomposing corpses are obviously not unities, and that there is a fuzzy area in between that is intrinsically undecidable. This is similar to how in color charts or prismatic spectra, two "adjacent" colors, e.g., green and blue, are clearly distinct and identifiable when their central areas are compared, but points along the transition zone are undecidable. The understanding of certain comparisons as dichotomous, when viewed on a large scale, such as between a healthy organism and a putrefying corpse, is not invalidated by the fuzziness of the transition viewed on a small scale, any more than the representational meaning of an impressionist painting is vitiated by the fact that close-up one sees only brush strokes.

4.3. Death as a Contextually Defined Event

Unlike E1-E7, which are clear, physically definable moments, C1 and C2 ultimately are abstract philosophical notions that do not translate or project operationally to the physical level. Although they may be good (possibly the best) candidates for an absolute, unitary concept of "death," they are ironically no better than "departure of the soul" for leading practically to criteria and tests. Since neither the cessation of "anti-entropy" nor the cessation of the "organism as a whole" occurs at a determinable moment, *a fortiori* there is no reason to insist axiomatically that these two indeterminable moments should coincide. And (for those interested in the soul) what basis is there to suppose that the cessation of the soul's "informing" the body corresponds to C1 or C2, if it seems intuitive that it ought to correspond to both, but the two may not necessarily correspond to each other? And if it might not correspond to one or the other, why must it necessarily correspond to either, and not to E1, E2, or ..., depending on the context?

Perhaps a more mature approach to the question of the moment of death would be to recall Humboldt's theory of the dynamic interaction between language and thought, and language and culture: perhaps our notion of death as a unitary phenomenon is after all not a self-evident "given" but an oversimplification reinforced by the fact that our language contains just a single word for death (namely, "death"). Much futile argumentation could be spared by recognizing that "death" is as rich and multi-layered a concept as "life" and equally undefinable. We should abandon the search for criteria for the universally "true" moment of death, as there is no single, context-independent, "true" moment of death. Rather, there are various moments of state-discontinuity, not all of which necessarily occur in a given case, and not all of which are equally striking to the senses and intellect of an observer. All of these state-discontinuities are equally "real" and "valid" phenomena in themselves, and there is no *a priori* reason that one of them must be singled out for the designation "death" while the others slip into conceptual obscurity for want of a word.

Once we recognize the restrictions that our language tends to impose on our ways of thinking about death, we can attempt to transcend them through expanding the vocabulary to correspond to the more enlightened understanding. We could invent words for E1, E2, etc., that were distinct enough not to create a false impression that they were all species of the same conceptual genus "death," but simply different moments of state-discontinuity resulting from changes in observable parameters along the continuous process known as dying and decaying.

Depending on the clinical circumstances of the dying person and the behavioral or ethical issues for those standing round, one of these state-discontinuities will stand out as particularly observationally striking and/or ethically determining. Such an E could legitimately be labeled "death" for purposes of that clinical-ethical context. This approach is somewhat similar to proposals that we should "unbundle" what Veatch calls "death behaviors," focusing not on when "death" occurs but on when such-and-such "death behavior" becomes appropriate or licit.[45-48,49(p.26),50] We suggest that the candidates for death-moments be defined not purely by ethical propriety, but also by physical circumstances. Some moments might involve the same set of death behaviors but are legitimately distinguished from each other as equally valid "moments of death" or "death events" within their own contexts (e.g., E1 and E2).

5. IMPLICATIONS FOR TRANSPLANTATION

What implications does all this have for the "dead donor rule" and transplantation? Among the various types of "death behavior," removal of vital organs is no doubt the one for which precision of timing is most critical. For non-transplant issues such as grieving, inheritance, burial, etc., it doesn't really matter whether "death" is considered to occur at E1 or E2. If the timing is off even by hours, there are no practical consequences, except possibly in bizarre, highly improbable scenarios purposefully constructed to create a hypothetical consequence. Neither does the choice of E matter for burial, because by the time funeral services are arranged and burial actually takes place, it is typically a day or more after all of the Es. For ordinary purposes, it is reasonable to label as "death" (and to define as "legal death") whichever E entails the most obvious state-discontinuity: in a home death that would probably be E1, whereas in a hospital death it could be E2.

Regarding transplantation of unpaired vital organs (or both of paired vital organs), the "dead donor rule" reflects the belief of many that the critical ethical question is: "Is the donor dead?" But if there is no one, true "moment of death" in an absolute sense, but rather a multiplicity of moments, any one of which might serve as a reasonable demarcation for a particular context, how does one decide which is most appropriate for the context of transplantation? We agree with Veatch and others that the proper choice of E is essentially not an ontological but a moral issue. "Is the patient dead?" is not only the wrong question to ask on the practical, physical level; it is not even a meaningful one when asked on a microscopic time-scale at the transition between life and death (like zooming in on the prismatic spectrum midway between green and blue, and demanding that someone not only identify it unequivocally as either "green" or "blue" but also have a convincing, logical rationale for doing so).

The question that really matters is: If we extirpate such-and-such organ(s) in such-and-such a way, do we kill or harm the patient? Although the verb "to kill" implicitly involves a dichotomous notion of life and death, it also involves causality and intentionality. The latter aspects make it possible in some situations to bypass the fuzzy, intrinsically undecidable border-zone between life and death, so that one can be morally certain of "not killing" even without first having to determine which side of the life-death boundary the donor is on at the time.

Truog, who agrees that a dead brain *per se* does not constitute a dead organism, maintains that removal of vital organs (especially the heart) from a heart-beating donor actively, iatrogenically kills the donor, but that it is nevertheless ethical because being killed supposedly does not "harm" the donor.[51,52] He reaches this startling conclusion through a subjective, first-person-experiential (essentially Cartesian) notion of "harm," according to which permanently unconscious patients are by definition beyond "harm."

We do not subscribe to such a concept of either person or harm, and we believe that one can justify many transplantation actions without having to resort to such a redefinition, fully respecting the traditional injunction *primum non nocere* even in the sense of physical harm. It is beyond the present scope to enter into the philosophical debate over the fundamental principles of ethics. Let us therefore simply state without argument that we believe, *contra* Truog, that there is a profound and critical difference between killing and letting-die, as well as between intending and foreseeing death, even if in some pairs of examples the physical acts or omissions might look outwardly identical. We also believe that the principle of double effect is valid and necessary for bioethics, as applied eloquently in DuBois's justification of certain controversial practices involving NHBD.[53]

DuBois zeroed in on the real issue in the very title of his article, "Is organ procurement causing the death of patients?",[53] in contrast to its counterpoint piece, "The importance of being dead: Non-heart-beating organ donation."[54] DuBois is right to suggest in his abstract "that many of the debates over death can be bypassed by changing the term of the debate: ..." But what follows ("what matters is not whether death is a process or an event, but death as a state") falls short of his own most fundamental point. It should rather be said that what matters is not whether death is a process or an event or a state, but whether "organ procurement caus[es] the death of patients," as his title asks.

Let us restrict discussion to the ideal NHBD context: assuming the legitimacy of stopping life support (independent of transplant considerations), truly informed consent, lack of conflict of interest, medical certainty that apnea will supervene once the ventilator is discontinued, etc. When the ventilator is withdrawn in the operating room, the first E to occur will be E1, final apnea; then will ensue E2, E3, E4, and E6, in that order (presumably E5, the onset of loss of consciousness has already taken place, whether from primary brain damage or from sedation for the procedure). One should not base the timing of organ retrieval on which E represents "true" death, because this is an improperly posed question. Rather one should ask, "Beyond which of these events does the removal of organs X, Y, Z... neither kill nor harm the patient (even in the physical sense of accelerating the dying process)?" The answer depends on which organs one is talking about.

In the case of all organs except heart and lungs, removing them even before E1 will neither cause nor hasten death (however defined), because, in the *ideal* NHBD context under discussion (the ideality supposition must be repeatedly emphasized), by the time loss of those organs might exert even the tiniest systemic effect, all the E's will have supervened long before anyway. It takes days or weeks to die from renal or hepatic failure or intestinal non-absorption or pancreatic insufficiency; for the first several hours, absence of such organs has no significant adverse effect on the body. Between discontinuation of the ventilator and even the most conservative choice of E, the effects of hypoxia-ischemia will totally overshadow whatever infinitesimal adverse effects might theoretically result from an incipient lack of kidney or liver functioning, etc. Thus, for transplantation of non-cardiopulmonary organs, it is utterly irrelevant ethically whether "brain death" is "really death," or whether the Pittsburgh protocol's 2 minutes of asystole is "really death," or whether any other physical event is "really death." Such questions are both malformulated and ethically beside the point.

For heart and/or lungs, the moral requirement "to do no harm" by the extirpation is trickier. A solution suggested by some would be to place the patient on cardiopulmonary bypass, remove heart and lungs without affecting systemic circulation or oxygenation, then declare the bypass machine an ethically inappropriate ("extraordinary" or "disproportionate") means of life support (just as the original ventilator was), and disconnect it. Critics argue that this seems nothing more than ethical sleight of hand to pseudo-justify something that is morally equivalent to simply cutting out the beating heart. Even if such a roundabout, complicated and more expensive way of doing things really made a moral difference between unethical killing and ethical "letting die" (which we seriously doubt), such a procedure would still create the appearance of something shady. We doubt whether such an approach would do anything to relieve the moral stress and emotional burn-out experienced by operating room personnel involved in transplantation.[55-57]

An alternative approach that seems to respect traditional ethical principles and sensibilities could be the following. If, as we have been assuming for the ideal case, it is known for certain that (1) the patient will not breathe spontaneously off the ventilator and (2) it is moral to disconnect the ventilator, then it makes no physical or moral difference whether the ventilator is disconnected before or after opening the chest cavity. Therefore, go ahead and disconnect it with the chest cavity open, perhaps after already having removed some of the non-cardiopulmonary organs discussed above. Be prepared to instill directly into the heart and/or pulmonary circulation some cool, tissue-preserving fluid as soon as final asystole, E2, can be determined with moral certainty to be truly final. The organs could also have been pretreated with some tissue-preserving medication infused into the blood stream, as long as it wouldn't diminish cardiopulmonary functioning and informed consent had been granted.

This approach to heart/lung retrieval does not cause or hasten death, because once circulation has effectively ceased due to the effect of progressive hypoxia on the heart, the dying or decaying process continues just the same regardless whether the nonbeating heart and nonfunctioning lungs remain physically in the circulationless body or not. The phrase "circulation has effectively ceased" is intentionally chosen, because what is important here, both physiologically and ethically, is the circulation of blood (even as its oxygen content progressively diminishes), not the QRS complex of the electrocardiogram, nor even a possible surprise "truly final" beat or two after (say) half a minute of asystole, beats which would not produce an "effective circulation" that would change in any significant way the process of dying/decaying already set in motion. Thus, it does not really matter exactly when E3 (loss of potential for cardiac auto-resuscitation) occurs or whether 2 or 5 minutes or some other duration of asystole suffices to provide certainty that E3 has already passed. This is because any potential cardiac auto-resuscitation would necessarily be very transient, by virtue of the ongoing apnea and severe hypoxemia; moreover, a very brief and weak circulation of anoxic blood would do nothing to counteract the inexorable process of dying/decaying already set in motion. In such a context, it makes no physiological difference in the dying process of the pulseless, hypoxemic body whether the potential for cardiac auto-resuscitation is briefly actualized or not.

Therefore, as soon as (1) the blood becomes hypoxemic enough that any brief resumption of its circulation would make no physiological difference *and* (2) the cessation of effective heartbeat lasts long enough to make an experienced physician suspect that the last (in the sense of previous) heartbeat will prove to be the last (in the sense of final) physiologically effective one, then (a) there is no need to wait further for E3 in order to avoid causing or hastening death by cardiac removal, and (b) there is no moral problem in making the suspected E2 definitively E2 (via pre-E3 organ removal).

Transplant surgeons might complain that having to wait for E2 would only complicate their already difficult task and might risk ischemic damage to the donated organ(s). Perhaps the old adage "necessity is the mother of invention" applies here: transplant surgeons never developed such a technique for heart-lung retrieval, primarily because the "brain-death" fiction convinced them that there was no need to. We believe that a historically honest and physiologically enlightened appraisal of "brain death" makes it an ethical requisite. With a little creativity, the complication of waiting for E2 can surely be effectively dealt with – and for the sake of everyone's consciences, it is a complication worth accepting and dealing with.

Statutory law does not require donors to be dead, but it does contain strong injunctions against voluntary homicide. In the transplant setting, these laws can be respected without needing to gerrymander a special statutory definition of "death" in order to seemingly fulfill a "dead donor rule" that isn't even a law. Statutory law would benefit by refocusing from the wrong medical-ethical question ("When is the patient dead?") to the right one, which it already has ample provision for ("Is homicide being committed?").

5.1. Disclaimer

To avoid potential misinterpretation, let it be stated explicitly that the foregoing discussion does not constitute positive advocacy of any particular transplantation protocol, whether heart-beating or non-heart-beating, present or future. The above analysis addresses the very precise and limited question whether it is possible in principle to remove vital organs without causing or hastening death or violating the time-honored injunction *primum non nocere*. Our conclusion is: yes, it is possible in principle. But before deciding whether it would be prudent to put this principle into practice in today's society, many other factors must be considered which are outside the scope of this discussion – such as whether donor consent can be guaranteed to be truly informed and free, whether in the case at hand apnea off life-support can be predicted with medical and moral certainty, whether such eviscerating procedures respect human dignity even if they might not cause or hasten death, whether the risk of public misperception that this is utilitarian killing can be minimized, etc. If the answer to one or more of these "whethers" is "no," as critics of NHBD protocols claim,[58] then it behooves us to hold off implementing the otherwise intrinsically ethical procedure until all the circumstantial details are worked out. The foregoing section is merely an appeal for clarity of terminology and logicality of thought, absolute prerequisites for fruitful discussion and valid ethics.

6. CONCLUSION

Linguistic considerations surrounding the term and concept "death" suggest that society has traditionally assumed a univocal notion of "death," in large part because, up until very recently in human history, there was no need for a more nuanced notion. Thus, the English language developed with only a single word for death (namely, "death" and its relatives "dead," "to die," etc. – euphemisms, contextual and stylistic variants excluded). What served humankind well linguistically for most of history now tends to restrict thinking when applied to situations uniquely occasioned by modern medicine.

It is *not* a contradiction in terms to speak of state-discontinuities brought about by continuous changes in observable parameters; bifurcations in state-space are ubiquitous throughout nature, especially animate nature. The reality of death-processes therefore does not preclude the reality of death-events. There is no logical continuum between life and non-life, being and non-being, unity and multiplicity, and it is simply incorrect even to pose the tiresome question whether "death" is an event or a process. Linguistically it can be understood only as an event; there are other words for the process. It is time to expand our death-vocabulary to facilitate the recognition of multiple events, all equally real, along the process from declining health to decomposition.

Depending on the context, some of these death-related events may constitute a more obvious discontinuity than others and may more justifiably be considered "death" within that context. It may also be more appropriate emotionally and/or morally to begin certain kinds of "death behavior" at one of these moments and not others, depending on the clinical context and the behavior in question. There is no reason to assume *a priori* that there must be an overarching, unitary conception of death from which all diagnostic criteria and tests must derive.

Regarding organ transplantation, the important and truly meaningful question is not "When is the patient dead?" but rather "When can organs X, Y, Z... be removed without causing or hastening death or harming the patient in any way?" Perhaps some of the general public's confusion and incoherence surrounding the "dead donor rule" results from a mismatch between people's intuitive understanding of death in the era of modern medicine and the limited lexicon that our colloquial language imposes on us for articulating that intuitive understanding.

7. ACKNOWLEDGMENTS

The first author thanks Alan Garfinkel, Ph.D. for introducing him to nonlinear dynamics and the example of the Hopf bifurcation; and Stuart Youngner, M.D. and Laura Siminoff, Ph.D. for catalyzing this latest stage in his understanding of death. Both authors thank the many colleagues from the "Linguist List" (http://www.linguistlist.org/) who replied to our query about "death words" in other languages.

8. REFERENCES

1. Bernat JL, Culver CM, Gert B. On the definition and criterion of death. *Ann Intern Med* 1981;**94(3):**389-394.
2. Shewmon DA. "Brain-stem death", "brain death" and death: a critical re-evaluation of the purported evidence. *Issues Law Med* 1998;**14(2):**125-145.
3. Seifert J. *What is Life? The Originality, Irreducibility, and Value of Life*. Amsterdam: Rodopi, 1997.
4. Schrödinger E. *What is Life? The Physical Aspect of the Living Cell*. New York: The Macmillan Company, 1946.
5. Korein J. The problem of brain death: development and history. *Ann N Y Acad Sci* 1978;**315:**19-38.
6. Shewmon DA. "Brain death": a valid theme with invalid variations, blurred by semantic ambiguity. In: White RJ, Angstwurm H, Carrasco de Paula I, eds. *Working Group on the Determination of Brain Death*

and its Relationship to Human Death. 10-14 December, 1989. (Scripta Varia 83). Vatican City: Pontifical Academy of Sciences, 1992:23-51.

7. Shewmon DA. The brain and somatic integration: insights into the standard biological rationale for equating "brain death" with death. *J Med Philos* 2001;**26(5):**457-478.
8. Veatch RM. The conscience clause: how much individual choice in defining death can our society tolerate? In: Youngner SJ, Arnold RM, Schapiro R, eds. *The Definition of Death: Contemporary Controversies.* Baltimore, MD: Johns Hopkins University Press, 1999:137-160.
9. Morison RS. Death: Process or event? *Science* 1971;**173:**694-698.
10. Kass LR. Death as an event: a commentary on Robert Morison. Attempts to blur the distinction between a man alive and a man dead are both unsound and dangerous. *Science* 1971;**173:**698-702.
11. Emanuel LL. Reexamining death: the asymptotic model and a bounded zone definition. *Hastings Cent Rep* 1995;**25(4):**27-35.
12. Varela FJ. *Principles of Biological Autonomy.* New York: North Holland, 1979.
13. Hilborn RC. *Chaos and Nonlinear Dynamics. An Introduction for Scientists and Engineers.* New York: Oxford University Press, 1994.
14. Aihara K, Matsumoto G. Chaotic oscillations and bifurcations in squid giant axons. In: Holden AV, ed. *Chaos.* Princeton: Princeton University Press, 1986:257-269.
15. Garfinkel A. A mathematics for physiology. *Am J Physiol* 1983;**245(Regulatory Integrative Comp Physiol 14):**R455-R466.
16. Garfinkel A, Chen P-S, Walter DO, Karagueuzian HS, Kogan B, Evans SJ, Karpoukhin M, Hwang C, Uchida T, Gotoh M, Nwasokwa O, Sager P, Weiss JN. Quasiperiodicity and chaos in cardiac fibrillation. *J Clin Invest* 1997;**99(2):**305-314.
17. Glass L, Guevara MR, Shrier A, Perez R. Bifurcation and chaos in a periodically stimulated cardiac oscillator. *Physica* 1983;**7D:**89-101.
18. Strogatz SH. *Nonlinear Dynamics and Chaos, with Applications to Physics, Biology, Chemistry, and Engineering.* Reading, MA: Addison-Wesley, 1994.
19. Botne R. *To die* across languages: Toward a typology of achievement verbs. *Linguistic Typology* 2003;**7(2):** 233–278.
20. DuBois. Non-heart-beating organ donation: A defense of the required determination of death. *J Law Med Ethics* 1999;**27:**126-136.
21. Fortescue M. *West Geenlandic.* London: Croom Helm, 1984.
22. Lynn J, Cranford R. The persisting perplexities in the determination of death. In: Youngner SJ, Arnold RM, Schapiro R, eds. *The Definition of Death: Contemporary Controversies.* Baltimore, MD: Johns Hopkins University Press, 1999:101-114.
23. Potts M, Byrne PA, Nilges RG. *Beyond Brain Death. The Case Against Brain Based Criteria for Human Death.* Dordrecht: Kluwer Academic Publishers, 2000.
24. Singer P. *Rethinking Life & Death. The Collapse of Our Traditional Ethics.* New York, NY: St. Martin's Press, 1995.
25. Youngner SJ, Arnold RM, Schapiro R. *The Definition of Death: Contemporary Controversies.* Baltimore, MD: Johns Hopkins University Press, 1999.
26. Rosner F. *Modern Medicine and Jewish Ethics.* New York: Yeshiva University Press, 1986.
27. Rosner F. The definition of death in Jewish law. In: Youngner SJ, Arnold RM, Schapiro R, eds. *The Definition of Death: Contemporary Controversies.* Baltimore, MD: Johns Hopkins University Press, 1999:210-221.
28. Steinberg A. Ethical issues in nephrology - Jewish perspectives. *Nephrol Dial Transplant* 1996;**11:**961-963.
29. Yofe YA, Mark Y. Groyser verterbukh fun der yidisher shprakh. New York: Komitet farn Groysn Verterbukh fun der Yidisher Shprakh, 1961.
30. Cole DJ. The reversibility of death. *J Med Ethics* 1992;**18(1):**26-30; discussion 31-33.
31. Lynn J. Are the patients who become organ donors under the Pittsburgh protocol for "non-heart-beating donors" really dead? In: Arnold RM, Youngner SJ, Schapiro R, Spicer CM, eds. *Procuring Organs for Transplant: The Debate over Non-Heart-Beating Cadaver Protocols.* Baltimore: The Johns Hopkins University Press, 1995:91-101.
32. Tomlinson T. The irreversibility of death: reply to Cole. In: Arnold RM, Youngner SJ, Schapiro R, Spicer CM, eds. *Procuring Organs for Transplant: The Debate over Non-Heart-Beating Cadaver Protocols.* Baltimore: The Johns Hopkins University Press, 1995:81-89.
33. Cole D. Statutory definitions of death and the management of terminally ill patients who may become donors. In: Arnold RM, Youngner SJ, Schapiro R, Spicer CM, eds. *Procuring Organs for Transplant:*

The Debate over Non-Heart-Beating Cadaver Protocols. Baltimore: The Johns Hopkins University Press, 1995:69-79.

34. Shewmon DA. Spinal shock and 'brain death': somatic pathophysiological equivalence and implications for the integrative-unity rationale. *Spinal Cord* 1999;**37(5):**313-324.
35. Bernat JL. How much of the brain must die in brain death? *J Clin Ethics* 1992;**3(1):**21-26; discussion 27-28.
36. Bernat JL. A defense of the whole-brain concept of death. *Hastings Cent Rep* 1998;**28(2):**14-23.
37. Pritchard ET. *No Word for Time: The Way of the Algonquin People*. Tulsa, OK: Council Oak Books, 1997.
38. Cook JA. The Hebrew verb: a grammaticalization approach. *Zeitschrift für Althebraistik (Stuttgart)* 2001; **14(2):**117-143.
39. Berlin B, Kay P. *Universality and Evolution of Basic Color Terms*. Berkeley, CA: University of California, Berkeley, Laboratory for Language-Behavior Research, 1967.
40. Elliot E. *Through Gates of Splendor*. New York: Harper & Brothers, 1957.
41. Shewmon DA. Recovery from "brain death": A neurologist's *Apologia*. *Linacre Q* 1997; **64(1):**30-96.
42. Shewmon DA. Chronic "brain death": meta-analysis and conceptual consequences. *Neurology* 1998;**51(6):** 1538-1545.
43. Shewmon DA. Chronic "brain death": meta-analysis and conceptual consequences [response to letters]. *Neurology* 1999;**53(6):**1369-1372.
44. Penrose R. *The Emperor's New Mind*. New York: Penguin Books, 1989.
45. Brody BA. How much of the brain must be dead? In: Youngner SJ, Arnold RM, Schapiro R, eds. *The Definition of Death: Contemporary Controversies*. Baltimore, MD: Johns Hopkins University Press, 1999:71-82.
46. Fost N. The unimportance of death. In: Youngner SJ, Arnold RM, Schapiro R, eds. *The Definition of Death: Contemporary Controversies*. Baltimore, MD: Johns Hopkins University Press, 1999:161-178.
47. Halevy A. Beyond brain death? *J Med Philos* 2001;**26(5):**493-501.
48. Halevy A, Brody B. Brain death: reconciling definitions, criteria, and tests. *Ann Intern Med* 1993;**119(6):** 519-525.
49. Veatch RM. *Death, Dying, and the Biological Revolution: Our Last Quest for Responsibility*. New Haven: Yale University Press, 1976.
50. Youngner SJ, Arnold RM. Philosophical debates about the definition of death: Who cares? *J Med Philos* 2001;**26(5):**527-537.
51. Truog RD. Is it time to abandon brain death? *Hastings Cent Rep* 1997;**27(1):**29-37.
52. Truog RD, Robinson WM. Role of brain death and the dead-donor rule in the ethics of organ transplantation. *Crit Care Med* 2003;**31(9):**2391-2396.
53. DuBois JM. Is organ procurement causing the death of patients? *Issues Law Med* 2002;**18(1):**21-41.
54. Menikoff J. The importance of being dead: Non-heart-beating organ donation. *Issues Law Med* 2002;**18(1):** 3-20.
55. Castelnuovo-Tedesco P. Cardiac surgeons look at transplantation - interviews with Drs. Cleveland, Cooley, DeBakey, Hallman and Rochelle. *Semin Psychiatry* 1971;**3(1):**5-16.
56. Lock M. *Twice Dead: Organ Transplants and the Reinvention of Death*. Berkeley, CA: University of California Press, 2001.
57. Youngner SJ, Allen M, Bartlett ET, Cascorbi HF, Hau T, Jackson DL, Mahowald MB, Martin BJ. Psychosocial and ethical implications of organ retrieval. *N Engl J Med* 1985;**313(5):**321-324.
58. Arnold RM, Youngner SJ, Schapiro R, Spicer CM. *Procuring organs for transplant: the debate over non-heart-beating cadaver protocols*. Baltimore: The Johns Hopkins University Press, 1995.

ABOUT THE CONTINUITY OF OUR CONSCIOUSNESS

Pim van Lommel*

1. INTRODUCTION

Some people who have survived a life-threatening crisis report an extraordinary experience. Near-death experiences (NDE) occur with increasing frequency because of improved survival rates resulting from modern techniques of resuscitation. The content of NDE and the effects on patients seem similar worldwide, across all cultures and times. The subjective nature and absence of a frame of reference for this experience lead to individual, cultural, and religious factors determining the vocabulary used to describe and interpret the experience. NDE can be defined as the reported memory of the whole of impressions during a special state of consciousness, including a number of special elements such as out-of-body experience, pleasant feelings, seeing a tunnel, a light, deceased relatives, or a life review. Many circumstances are described during which NDE are reported, such as cardiac arrest (clinical death), shock after loss of blood, traumatic brain injury or intra-cerebral haemorrhage, near-drowning or asphyxia, but also in serious diseases not immediately life-threatening. Similar experiences to near-death ones can occur during the terminal phase of illness, and are called deathbed visions. Furthermore, identical experiences, so-called "fear-death" experiences, are mainly reported after situations in which death seemed unavoidable like serious traffic or mountaineering accidents. The NDE is transformational, causing profound changes of life-insight and loss of the fear of death. An NDE seems to be a relatively regularly occurring, and to many physicians an inexplicable phenomenon and hence an ignored result of survival in a critical medical situation.

And should we also consider the possibility of conscious experience when someone in coma has been declared brain dead by physicians, and organ transplantation is about to be started? Recently several books were published in the Netherlands about what patients had experienced in their consciousness during coma following a severe traffic accident, following acute disseminated encephalomyelitis (ADEM), or following complications with cerebral hypertension after surgery for a brain tumour, this last patient being

* Pim van Lommel, Cardiologist, Division of Cardiology, Hospital Rijnstate, PO Box 9555, 6800 TA Arnhem, The Netherlands. Email: pimvanlommel@wanadoo.nl.

Brain Death and Disorders of Consciousness, Edited by Machado and Shewmon
Kluwer Academic/Plenum Publishers, New York 2004

declared brain dead by his neurologist and neurosurgeon, but the family refused to give permission for organ donation. All these patients reported, after regaining consciousness, that they had experienced clear consciousness with memories, emotions, and perception out of and above their body during the period of their coma, also "seeing" nurses, physicians and family in and around the ICU. Does brain death really means death, or is it just the beginning of the process of dying that can last for hours to days, and what happens to consciousness during this period? Should we also consider the possibility that someone who is clinically dead during cardiac arrest can experience consciousness, and even whether there could still be consciousness after someone really has died, when his body is cold? How is consciousness related to the integrity of brain function? Is it possible to gain insight in this relationship? In my view the only possible empirical approach to evaluate theories about consciousness is research on NDE, because in studying the several universal elements that are reported during NDE, we get the opportunity to verify all the existing theories about consciousness that have been discussed until now. Consciousness presents temporal as well as everlasting experiences. Is there a start or an end to consciousness?

In this paper I first will discuss some more general aspects of death, and after that I will describe more details from our prospective study on near-death experience in survivors of cardiac arrest in the Netherlands, which was published in the Lancet.[1] I also want to comment on similar findings from two prospective studies in survivors of cardiac arrest from the USA[2] and from the United Kingdom.[3] Finally, I will discuss implications for consciousness studies, and how it could be possible to explain the continuity of our consciousness.

2. ABOUT DEATH

First I want to discuss death. The confrontation with death raises many basic questions, also for physicians. Why are we afraid of death? Are our concepts about death correct? Most of us believe that death is the end of our existence; we believe that it is the end of everything we are. We believe that the death of our body is the end of our identity, the end of our thoughts and memories, that it is the end of our consciousness. Do we have to change our concepts about death, not only based on what has been thought and written about death in human history around the world in many cultures, in many religions, and in all times, but also based on insights from recent scientific research on NDE?

What happens when I am dead? What is death? During our life 500000 cells die each second, each day about 50 billion cells in our body are replaced, resulting in a new body each year. So cell death is totally different from body death when you eventually die. During our life our body changes continuously, each day, each minute, each second. Each year about 98% of our molecules and atoms in our body have been replaced. Each living being is in an unstable balance of two opposing processes of continual disintegration and integration. But no one realizes this constant change. And from where comes the continuity of our continually changing body? Cells are just the building blocks of our body, like the bricks of a house, but who is the architect, who coordinates the building of this house. When someone has died, only mortal remains are left: only matter. But where is the director of the body? What about our consciousness when we die? Is someone his body, or do we "have" a body?

3. SCIENTIFIC RESEARCH ON NEAR-DEATH EXPERIENCE

In 1969 during my rotating internship a patient was successfully resuscitated in the cardiac ward by electrical defibrillation. The patient regained consciousness, and was very, very disappointed. He told me about a tunnel, beautiful colours, a light and beautiful music. I have never forgotten this event, but I did not do anything with it. Years later, in 1976 Raymond Moody first described the so-called "near-death experiences", and only in 1986 I read about these experiences in the book by George Ritchie entitled "Return from Tomorrow," which relates what he experienced during a period of clinical death of 6-minutes duration in 1943 during his medical study.[4] After reading his book I started to interview my patients who had survived a cardiac arrest. To my great surprise, within two years about fifty patients told me about their NDE.

My scientific curiosity started to grow, because according to our current medical concepts, it is not possible to experience consciousness during a cardiac arrest, when circulation and breathing have ceased.

Several theories on the origin of an NDE have been proposed. Some think the experience is caused by physiological changes in the brain such as brain cells dying as a result of cerebral anoxia, and possibly also caused by release of endorphins, or NMDA receptor blockade.[5] Other theories encompass a psychological reaction to approaching death[6] or a combination of such reaction and anoxia.[7] But until now there was no prospective, meticulous and scientifically designed study to explain the cause and content of an NDE. All studies had been retrospective and very selective with respect to patients. In retrospective studies 5-30 years can elapse between occurrence of the experience and its investigation, which often prevents accurate assessment of medical and pharmacological factors. We wanted to know if there could be a physiological, pharmacological, psychological or demographic explanation why people experience consciousness during a period of clinical death. The definition of clinical death was used for the period of unconsciousness caused by anoxia of the brain due to the arrest of circulation and breathing that happens during ventricular fibrillation in patients with acute myocardial infarction.

We studied patients who survived cardiac arrest, because this is a well-described life threatening medical situation, where patients will ultimately die from irreversible damage to the brain if cardio-pulmonary resuscitation (CPR) is not initiated within 5 to 10 minutes. It is the closest model of the process of dying.

So, in 1988 we started a prospective study of 344 consecutive survivors of cardiac arrest in ten Dutch hospitals with the aim of investigating the frequency, the cause and the content of an NDE.[1] We did a short standardised interview with sufficiently recovered patients within a few days of resuscitation, and asked whether they could remember the period of unconsciousness, and what they recalled. In cases where memories were reported, we coded the experiences according to a weighted core experience index. In this system the depth of the NDE was measured according to the reported elements of the content of the NDE. The more elements were reported, the deeper the experience was and the higher the resulting score was.

Results: 62 patients (18%) reported some recollection of the time of clinical death. Of these patients 41 (12%) had a core experience with a score of 6 or higher, and 21 (6%) had a superficial NDE. In the core group 23 patients (7%) reported a deep or very deep experience with a score of 10 or higher. And 282 patients (82%) had no recollection of the period of cardiac arrest.

In the American prospective study of 116 survivors of cardiac arrest 11 patients (10%) reported an NDE with a score of 6 or higher; the investigators did not specify the number of patients with a superficial NDE with a low score.[2] In the British prospective study of 63 survivors of cardiac arrest only 4 patients (6.3%) reported an NDE with a score of 6 or higher, and 3 patients (4.8%) had a superficial NDE, a total of 7 patients (11%) with memories from the period of cardiac arrest.[3]

In our study about 50% of the patients with an NDE reported awareness of being dead, or had positive emotions, 30% reported moving through a tunnel, had an observation of a celestial landscape, or had a meeting with deceased relatives. About 25% of the patients with an NDE had an out-of-body experience, had communication with "the light," or observed colours, 13% experienced a life review, and 8% experienced a border.

What might distinguish the small percentage of patients who report an NDE from those who do not? We found that neither the duration of cardiac arrest nor the duration of unconsciousness, nor the need for intubation in complicated CPR, nor induced cardiac arrest in electrophysiologic stimulation (EPS) had any influence on the frequency of NDE. Neither could we find any relationship between the frequency of NDE and administered drugs, fear of death before the arrest, foreknowledge of NDE, religion or education. An NDE was more frequently reported at ages lower than 60 years, and also by patients who had had more than one CPR during their hospital stay, and by patients who had experienced an NDE previously. Patients with memory defects induced by lengthy CPR reported an NDE less frequently. Good short-term memory seems to be essential for remembering an NDE. Unexpectedly, we found that significantly more patients who had an NDE, especially a deep experience, died within 30 days of CPR ($p<0.0001$).

We performed a longitudinal study with taped interviews of all late survivors with NDE 2 and 8 years following the cardiac arrest, along with a matched control group of survivors of cardiac arrest who did not report an NDE.[1] This study was designed to assess whether the transformation in attitude toward life and death following an NDE is the result of having an NDE or the result of the cardiac arrest itself. In this follow-up research into transformational processes after NDE, we found a significant difference between patients with and without an NDE. The process of transformation took several years to consolidate. Patients with an NDE did not show any fear of death, they strongly believed in an afterlife, and their insight in what is important in life had changed: love and compassion for oneself, for others, and for nature. They now understood the cosmic law that everything one does to others will ultimately be returned to oneself: hatred and violence as well as love and compassion. Remarkably, there was often evidence of increased intuitive feelings. Furthermore, the long lasting transformational effects of an experience that lasts only a few minutes was a surprising and unexpected finding.

Several theories have been proposed to explain NDE. However, in our prospective study we did not show that psychological, physiological or pharmacological factors caused these experiences after cardiac arrest. With a purely physiological explanation such as cerebral anoxia, most patients who had been clinically dead should report an NDE. All 344 patients had been unconscious because of anoxia of the brain resulting from their cardiac arrest. Why should only 18% of the survivors of cardiac arrest report an NDE?

And yet, neurophysiological processes must play some part in NDE, because NDE-like experiences can be induced through electrical "stimulation" of some parts of the cortex in patients with epilepsy,[8] with high carbon dioxide levels (hypercarbia)[9] and in decreased cerebral perfusion resulting in local cerebral hypoxia, as in rapid acceleration during training of fighter pilots,[10] or as in hyperventilation followed by Valsalva maneuver.[11] Also NDE-like experiences have been reported after the use of drugs like ketamine,[12] LSD,[13] or mushrooms.[14] These induced experiences can sometimes result in a period of unconsciousness, but can at the same time also consist of out-of-body experiences, perception of sound, light or flashes of recollections from the past. These recollections, however, consist of fragmented and random memories unlike the panoramic life-review that can occur in NDE. Further, transformational processes are rarely reported after induced experiences. Thus, induced experiences are not identical to NDE.

Another theory holds that NDE might be a changing state of consciousness (transcendence, or the theory of continuity), in which memories, identity, and cognition, with emotion, function independently from the unconscious body, and retain the possibility of non-sensory perception. Obviously, consciousness during NDE was experienced independently from the normal body-linked waking consciousness.

With lack of evidence for any other theories for NDE, the concept thus far assumed but never scientifically proven, that consciousness and memories are localized in the brain should be discussed. Traditionally, it has been argued that thoughts or consciousness are produced by large groups of neurons or neuronal networks. How could a clear consciousness outside one's body be experienced at the moment that the brain no longer functions during a period of clinical death, with flat EEG?[15] Furthermore, blind people have also described veridical perceptions during out-of-body experiences at the time of their NDE.[16] Scientific study of NDE pushes us to the limits of our medical and neurophysiological ideas about the range of human consciousness and relationship of consciousness and memories to the brain.

Also Greyson[2] writes in his discussion: "No one physiological or psychological model by itself explains all the common features of NDE. The paradoxical occurrence of heightened, lucid awareness and logical thought processes during a period of impaired cerebral perfusion raises particular perplexing questions for our current understanding of consciousness and its relation to brain function. A clear sensorium and complex perceptual processes during a period of apparent clinical death challenge the concept that consciousness is localized exclusively in the brain." And Parnia and Fenwick[3] write in their discussion: "The data suggest that the NDE arises during unconsciousness. This is a surprising conclusion, because when the brain is so dysfunctional that the patient is deeply comatose, the cerebral structures, which underpin subjective experience and memory, must be severely impaired. Complex experiences such as are reported in the NDE should not arise or be retained in memory. Such patients would be expected to have no subjective experience [as was the case in the vast majority of patients who survive cardiac arrest in the three published prospective studies[1-3] or at best a confusional state if some brain function is retained. Even if the unconscious brain is flooded by neurotransmitters this should not produce clear, lucid remembered experiences, as those cerebral modules, which generate conscious experience, are impaired by cerebral anoxia. The fact that in a cardiac arrest loss of cortical function precedes the rapid loss of brainstem activity lends further support to this view. An alternative explanation would be that the observed experiences arise during the loss of, or on regaining consciousness. The transition from consciousness to unconsciousness is rapid, with the EEG showing changes within a few seconds, and appearing immediate to the subject. Experiences which occur during the recovery of consciousness are confusional, which these were not". In fact, memory is a very sensitive indicator of brain injury and the length of amnesia before and after unconsciousness is an indicator of the severity of the injury. Therefore, events that occur just prior to or just after loss of consciousness would not be expected to be recalled. And as stated before, in our study[1] patients with loss of memory induced by lengthy CPR reported significantly fewer NDE. Good short-term memory seems to be essential for remembering NDE.

4. SOME TYPICAL ELEMENTS OF NDE

Before I discuss in greater detail some neurophysiologic aspects of brain functioning during cardiac arrest, I would like to reconsider certain elements of the NDE, like the out-of-body experience, the holographic life review and preview, the encounter with deceased relatives, the return into the body and the disappearance of the fear of death.

4.1. The Out-of-Body Experience

In this experience people have veridical perceptions from a position outside and above their lifeless body. NDEers have the feeling that they have apparently taken off their body like an old coat and to their surprise they appear to have retained their own identity with the possibility of perception, emotions, and a very clear consciousness. This out-of-body experience is scientifically important because doctors, nurses, and relatives can verify the reported perceptions. This is the report of a nurse of a Coronary Care Unit:

> *"During night shift an ambulance brings in a 44-year old cyanotic, comatose man into the coronary care unit. He was found in coma about 30 minutes before in a meadow. When we go to intubate the patient, he turns out to have dentures in his mouth. I remove these upper dentures and put them onto the 'crash cart.' After about an hour and a half the patient has sufficient heart rhythm and blood pressure, but he is still ventilated and intubated, and he is still comatose. He is transferred to the intensive care unit to continue the necessary artificial respiration. Only after more than a week do I meet again with the patient, who is by now back on the cardiac ward. The moment he sees me he says: 'O, that nurse knows where my dentures are.' I am very surprised. Then he elucidates: 'You were there when I was brought into hospital and you took my dentures out of my mouth and put them onto that cart, it had all these bottles on it and there was this sliding drawer underneath, and there you put my teeth.' I was especially amazed because I remembered this happening while the man was in deep coma and in the process of CPR. It appeared that the man had seen himself lying in bed, that he had perceived from above how nurses and doctors had been busy with the CPR. He was also able to describe correctly and in detail the small room in which he had been resuscitated as well as the appearance of those present like myself. He is deeply impressed by his experience and says he is no longer afraid of death."*

4.2. The Holographic Life Review

During this life review the subject feels the presence and renewed experience of not only every act but also every thought from one's past life, and one realizes that all of it is an energy field which influences oneself as well as others. All that has been done and thought seems to be significant and stored. Insight is obtained about whether love was given or on the contrary withheld. Because one is connected with the memories, emotions and consciousness of another person, you experience the consequences of your own thoughts, words and actions to that other person at the very moment in the past that they occurred. Hence there is during a life review a connection with the fields of consciousness of other persons as well as with your own fields of consciousness (*interconnectedness*). Patients survey their whole life in one glance; time and space do not seem to exist during such an experience. Instantaneously they are where they concentrate upon (*non-locality*), and they can talk for hours about the content of the life review even though the resuscitation only took minutes. Quotation:

> *"All of my life up till the present seemed to be placed before me in a kind of panoramic, three-dimensional review, and each event seemed to be accompanied by a consciousness of good or evil or with an insight into cause or effect. Not only did I perceive everything from my own viewpoint, but I also knew the thoughts of everyone involved in the event, as if I had their thoughts within me. This meant that I perceived not only what I had done or thought, but even in what way it had influenced others, as if I saw things with all-seeing eyes. And so even your thoughts are apparently not wiped out. And all the time during the review the importance of love was emphasised. Looking back, I cannot say how long this life review and life insight lasted, it may have been long, for every subject came up, but at the same time it seemed just a fraction of a second, because I perceived it all at the same moment. Time and distance seemed not to exist. I was in all places at the same time, and sometimes my attention was drawn to something, and then I would be present there."*

Also a preview can be experienced, in which both future images from personal life events (sometimes remembered only later in the shape of "déja vu") as well as more general images from the future occur, even though it must be stressed that these surveyed images should be considered purely as possibilities. And again it seems as if time and space do not exist during this review. Quotation:

> *"I had a nice eye contact, they looked at me full of love, and then I surveyed a great part of my life to come; the care for my children, the terminal illness of my wife, the circumstances I would be mixed up with, in my job and besides. I surveyed it completely; and then I got the feeling that I had to decide now: 'I may stay here, or I have to go back,' but I had to decide now."*

4.3. The Encounter with Deceased Relatives

If deceased acquaintances or relatives are encountered in an otherworldly dimension, they are usually recognized by their appearance, while communication is possible through thought transfer. Thus, during an NDE it is also possible to come into contact with fields of consciousness of deceased persons (*interconnectedness*). Sometimes persons are met whose death was impossible to have known; sometimes persons unknown to them are encountered during an NDE. Quotation:

> *"During my cardiac arrest I had a extensive experience (...) and later I saw, apart from my deceased grandmother, a man who had looked at me lovingly, but whom I did not know. More than 10 years later, at my mother's deathbed, she confessed to me that I had been born out of an extramarital relationship, my father being a Jewish man who had been deported and killed during the second World War, and my mother showed me his picture. The unknown man that I had seen more than 10 years before during my NDE turned out to be my biological father."*

4.4. The Return into the Body

Some patients can describe how they returned into their body, mostly through the top of the head, after they had come to understand through wordless communication with a Being of Light or a deceased relative that "it wasn't their time yet" or that "they still had a task to fulfil." The conscious return into the body is experienced as something very oppressive. They regain consciousness in their body and realize that they are "locked up" in their body, meaning again all the pain and restriction of their disease. They also realize that a part of their consciousness with deep knowledge and understanding as well as the feeling of unconditional love and acceptance have been taken away from them again. Quotation:

> *"And when I regained consciousness in my body, it was so terrible, so terrible... that experience was so beautiful, I never would have liked to come back, I wanted to stay there... and still I came back. And from that moment on it was a very difficult experience to live my life again in my body, with all the limitations I felt in that period."*

4.5. The Disappearance of Fear of Death

Nearly all people who have experienced an NDE lose their fear of death. This is due to the realization that there is a continuation of consciousness, even when you have been declared dead by bystanders or even by doctors. You are separated from the lifeless body, retaining the ability of perception. Quotation:

> *"It is outside my domain to discuss something that can only be proven by death. For me, however, the experience was decisive in convincing me that consciousness lives on beyond the grave. Death was not death, but another form of life."*

Another quotation:

> *"This experience is a blessing for me, for now I know for sure that body and mind are separated, and that there is life after death."*

Following an NDE people know of the continuity of their consciousness, retaining all thoughts and past events. And this insight causes exactly their process of transformation and the loss of fear of death. Man appears to be more than just a body.

5. NEUROPHYSIOLOGY IN CARDIAC ARREST

All these elements of an NDE were experienced during the period of cardiac arrest, during the period of apparent unconsciousness, during the period of clinical death! But how is it possible to explain these experiences during the period of temporary loss of all functions of the brain due to acute pancerebral ischemia?

We know that patients with cardiac arrest are unconscious within seconds. But how do we know that the electroencephalogram (EEG) is flat in those patients, and how can we study this? Complete cessation of cerebral circulation is found in cardiac arrest due to ventricular fibrillation (VF) during threshold testing at implantation of internal defibrillators. This complete cerebral ischemic model can be used to study the result of anoxia of the brain.

In VF complete cardiac arrest occurs, with complete cessation of cerebral flow, resulting in acute pancerebral anoxia. The middle cerebral artery blood flow, V_{mca}, which is a reliable trend monitor of the cerebral blood flow, decreases to 0 cm/sec immediately after the induction of VF.[17] Through many studies in both human and animal models, cerebral function has been shown to be severely compromised during cardiac arrest, and electrical activity in both cerebral cortex and the deeper structures of the brain has been shown to be absent after a very short period of time. Monitoring of the electrical activity of the cortex (EEG) has shown that ischemia produces a decrease of power in fast activity and in delta activity and an increase of slow delta I activity, sometimes also an increase in amplitude of theta activity, progressively and ultimately declining to isoelectricity. More often initial slowing and attenuation of the EEG waves is the first sign of cerebral ischemia. The first ischemic changes in the EEG are detected an average of 6.5 seconds after circulatory arrest. With prolongation of the cerebral ischemia, progression to isoelectricity occurs within 10 to 20 (mean 15) seconds from the onset of cardiac arrest.[18-21]

After defibrillation the V_{mca}, measured by transcranial Doppler technique, returns rapidly within 1-5 seconds after a cardiac arrest of short duration. However, in the case of a prolonged cardiac arrest of more than 37 seconds, the V_{mca} shows an initial overshoot upon reperfusion, a transient global hyperaemia, followed by a significant decrease in cerebral blood flow up to 50% or less of normal.[22] This results also in an initial overshoot of cerebral oxygen uptake (hyperoxia) with a fast decrease in cerebral oxygen uptake to borderline values for a considerable time due to delayed hypoperfusion.[18,22] In the case of a prolonged cardiac arrest the EEG recovery also takes more time, and normal EEG activity may not return for many minutes to hours after cardiac function has been restored, depending on the duration of the cardiac arrest, despite maintenance of adequate blood pressure during the recovery phase. Additionally, EEG recovery underestimates the metabolic recovery of the brain, and cerebral oxygen uptake may be depressed for a considerable time after restoration of circulation.[18] In acute myocardial infarction the duration of cardiac arrest (VF) in the Coronary Care Unit (CCU) is usually 60-120 seconds, on the cardiac ward 2-5 minutes, and in out-of-hospital arrest it usually exceeds 5-10 minutes. Only during threshold testing of internal defibrillators or during electrophysiologic stimulation studies will the duration of cardiac arrest rarely exceed 30-60 seconds.

Anoxia causes loss of function of our cell systems. However, in anoxia of only some minute's duration this loss may be transient; in prolonged anoxia cell death occurs, with permanent functional loss. During an embolic event a small clot obstructs the blood flow in a small vessel of the cortex, resulting in anoxia of that part of the brain, with loss of electrical activity. This results in a functional loss of the cortex like hemiplegia or aphasia. When the clot is dissolved or broken down within several minutes the lost cortical function is restored. This is called a transient ischemic attack (TIA). However, when the clot obstructs the cerebral vessel for minutes to hours, it will result in neuronal cell death, with a permanent loss of function of this part of the brain, with persistent hemiplegia or aphasia, and the diagnosis of cerebrovascular accident (CVA) is made. So transient anoxia results in transient loss of function.

In cardiac arrest global anoxia of the brain occurs within seconds. Timely and adequate CPR reverses this functional loss of the brain, because definitive damage of the brain cells, resulting in cell death, has been prevented. Long lasting anoxia, caused by cessation of blood flow to the brain for more than 5-10 minutes, results in irreversible damage and extensive cell death in the brain. This is called brain death, and most patients will ultimately die.

From these studies we know that in our prospective study[1] as well as in the other studies[2,3] of patients who have been clinically dead (VF on the ECG), total lack of electric activity of the cortex of the brain (flat EEG) must have been the only possibility, but also the abolition of brain-stem activity, such as the loss of the corneal reflex, fixed and dilated pupils, and the loss of the gag reflex, is a clinical finding in those patients. However, patients with an NDE can report a clear consciousness, in which cognitive functioning, emotion, sense of identity, and memory from early childhood was possible, as well as perception from a position out and above their "dead" body. Because of the occasional and verifiable out-of-body experiences, like the one involving the dentures in our study,[1] we know that the NDE must happen during the period of unconsciousness, and not in the first or last seconds of this period. There is also a well documented report of a patient with constant registration of the EEG during surgery for an gigantic aneurysm at the base of the brain, operated with a body temperature between 10 and 15 degrees Celsius. She was connected to a heart-lung machine, with VF, with all blood drained from her head, with a flat line EEG, with clicking devices in both ears, with eyes taped shut, and this patient experienced an NDE with an out-of-body experience, and all details she perceived and heard could later be verified.[15]

So we have to conclude that NDE in our study,[1] as well as in the American[2] and the British study,[3] was experienced during a transient functional loss of all functions of the cortex and of the brainstem. How could a clear consciousness outside one's body be experienced at the moment that the brain no longer functions during a period of clinical death, with a flat EEG? Such a brain would be roughly analogous to a computer with its power source unplugged and its circuits detached. It couldn't hallucinate; it couldn't do anything at all. As stated before, up to the present it has generally been assumed that consciousness and memories are localized inside the brain, that the brain produces them. According to this unproven concept, consciousness and memories ought to vanish with physical death, and necessary also during clinical death or brain death. However, during an NDE patients experience the continuity of their consciousness with the possibility of perception outside and above one's lifeless body. Consciousness can be experienced in another dimension without our conventional body-linked concept of time and space, where all past, present and future events exist and can be observed simultaneously and instantaneously (*non-locality*). In the other dimension, one can be connected with the personal memories and fields of consciousness of oneself as well as others, including deceased relatives (*universal interconnectedness*). And the conscious return into one's body can be experienced, together with the feeling of bodily limitation, and also sometimes the awareness of the loss of universal wisdom and love they had experienced during their NDE.

6. NEUROPHYSIOLOGY IN A NORMAL FUNCTIONING BRAIN

For decades, extensive research has been done to localize consciousness and memories inside the brain, so far without success. In connection with the unproven assumption that consciousness and memories are produced and stored inside the brain, we should ask ourselves how a non-material activity such as concentrated attention or thinking can correspond to an observable (material) reaction in the form of measurable electrical, magnetic, and chemical activity at a certain place in the brain,[23-25] even an increase in cerebral blood flow is observed during such a non-material activity as thinking.[26] Neurophysiological studies have shown these aforesaid activities through EEG, magnetoencephalography (MEG), magnetic resonance imaging (MRI) and positron emission tomography (PET) scanning. Specific areas of the brain have been shown to become metabolically active in response to a thought or feeling. However, those studies, although providing evidence for the role of neuronal networks as an intermediary for the manifestation of thoughts, do not necessary imply that those cells also produce the thoughts. Direct evidence of how neurons or neuronal networks could possibly produce the subjective essence of the mind and thoughts is currently lacking. It is also not well understood how to explain that in a sensory experiment, the subject stated that he was aware (conscious) of the sensation a few thousands of a second following the stimulation, whereas neuronal adequacy in the subject's brain wasn't achieved until a full 500 msec following the sensation. This experiment has led to the so-called delay-and-antedating hypothesis,[27] and it is a challenge to our current neurophysiological theories, as well as phenomena like anticipatory activation, or presentiment,[28] with changes on MRI up to 3 seconds preceding emotional stimuli. [29]

The brain contains about 100 billion neurons, 20 billion of which are situated in the cerebral cortex. Several thousand neurons die each day, and there is a continuous renewal of the proteins and lipids constituting cellular membranes on a time-span basis ranging from several days to a few weeks.[30] During life the cerebral cortex continuously adaptively modifies its neuronal network, including changing the number and location of synapses. All neurons show an electrical potential across their cell membranes, and each neuron has tens to hundreds of synapses that influence other neurons. Transportation of information along neurons occurs predominantly by means of action potentials, differences in membrane potential caused by synaptic depolarization and hyperpolarization. The sum total of changes along neurons causes transient electric fields and therefore also transient magnetic fields along the synchronously activated dendrites. During cerebral activity, these electrical and magnetic patterns of the 100 billion neurons change each nanosecond. Neither the number of neurons, nor the precise shape of the dendrites, nor the position of synapses, nor the firing of individual neurons seem to be crucial for information processing properties, but the derivative, the fleeting, highly ordered 4-dimensional (space and time) patterns of the electromagnetic fields generated along the dendritic trees of specialized neuronal networks. These patterns should be thought of as the final product of chaotic, dynamically governed self-organization.[31]

The influence of external localized magnetic and electric fields on these constant changing electromagnetic fields during normal functioning of the brain should now be mentioned. Neurophysiological research is being performed using transcranial magnetic stimulation (TMS),[32] in the course of which localized magnetic fields are produced. TMS can excite or inhibit different parts of the brain, depending of the amount of energy given, allowing functional mapping of cortical regions and creation of *transient functional lesions*. It allows assessing the function in focal brain regions on a millisecond scale, and it can study the contribution of cortical networks to specific cognitive functions. TMS can interfere with visual and motion perception, by *interrupting* cortical processing for 80-100 milliseconds. Intracortical inhibition and facilitation obtained during paired-pulse studies with TMS reflect the activity of interneurons in the cortex. Also TMS can alter the functioning of the brain beyond the time of stimulation, but it does not appear to leave any lasting effect.[32]

Interrupting the electrical fields of local neuronal networks in parts of the cortex also disturbs the normal functioning of the brain. By localized electrical stimulation of the temporal and parietal lobe during surgery for epilepsy the neurosurgeon and Nobel prize winner Wilder Penfield could sometimes induce flashes of recollection of the past (never a complete life review), experiences of light, sound or music, and rarely a kind of out-of-body experience (OBE).[33,34] These experiences did not produce any life-attitude transformation.

The effect of the external magnetic or electrical stimulation depends on the intensity and duration of energy given. There may be no clinical effect; sometimes an effect occurs when only a small amount of energy is given. But during stimulation with higher energy, *inhibition* of local cortical functions occurs by extinction of their electrical and magnetic fields (personal communication Dr. Olaf Blanke, neurologist, Laboratory for Presurgical Epilepsy Evaluation and Functional Brain Mapping Laboratory, Department of Neurology, University Hospital of Geneva, Switzerland). Blanke recently described a patient with induced OBE by *inhibition* of cortical activity caused by more intense external electrical stimulation of neuronal networks in the gyrus angularis in a patient with epilepsy.[35]

We have to conclude that localized artificial stimulation with real photons (electrical or magnetic energy) disturbs and inhibits the constantly changing electromagnetic fields of our neuronal networks, thereby influencing and inhibiting the normal functions of our brain. Could consciousness and memories be the product or the result of these constantly changing fields of photons? Could these photons be the elementary carriers of consciousness?[31]

Some researchers try to create artificial intelligence by computer technology, hoping to simulate programs evoking consciousness. But Roger Penrose, a quantum physicist, argues that "Algorithmic computations cannot simulate mathematical reasoning. The brain, as a closed system capable of internal and consistent computations, is insufficient to elicit human consciousness."[36] Penrose offers a quantum mechanical hypothesis to explain the relation between consciousness and the brain. And Simon Berkovitch, a professor in Computer Science of the George Washington University, has calculated that the brain has an absolutely inadequate capacity to produce and store all the informational processes of all our memories with associative thoughts. We would need 10^{24} operations per second, which is absolutely impossible for our neurons.[37] Herms Romijn, a Dutch neurobiologist, comes to the same conclusion.[30] One should conclude that the brain has not enough computing capacity to store all the memories with associative thoughts from one's life, has not enough retrieval abilities, and seems not to be able to elicit consciousness.

7. QUANTUM MECHANICS AND THE BRAIN

With our current medical and scientific concepts it seems impossible to explain all aspects of the subjective experiences as reported by patients with an NDE during their period of cardiac arrest, during a transient loss of all functions of the brain. But science, I believe, is the search for explaining new mysteries rather than the cataloguing of old facts and concepts. So it is a scientific challenge to discuss new hypotheses that could explain the reported *interconnectedness* with the consciousness of other persons and of deceased relatives, to explain the possibility to experience instantaneously and simultaneously (*non-locality*) a review and a preview of someone's life in a *dimension without our conventional body-linked concept of time and space*, where all past, present and future events exist, and the possibility to have clear consciousness with memories from early childhood, with self-identity, with cognition, and with emotion, and the possibility of perception out and above one's lifeless body.

We should conclude, like many others, that quantum mechanical processes could have something critical to do with how consciousness and memories relate with the brain and the body during normal daily activities as well as during brain death or clinical death.

I would like now to discuss some aspects of quantum physics, because this seems necessary to understand my concept of the continuity of consciousness. Quantum physics has completely overturned the existing view of our material, manifest world, the so-called real-space. It tells us that particles can propagate like waves, and so can be described by a quantum mechanical wave function. It can be proven that light in some experiments behaves like particles (photons), and in other experiments it behaves like waves, and both experiments are true. So waves and particles are *complementary* aspects of light (Bohr).[38] The experiment of Aspect, based on Bell's theorem, has established *non-locality* in quantum mechanics (*non-local interconnectedness*).[39] Non-locality happens because all events are interrelated and influence each other.

Phase-space is an invisible, non-local, higher-dimensional space consisting of *fields of probability*, where every past and future event is available as a possibility. Within this phase-space no matter is present, everything belongs to uncertainty, and neither measurements nor observations are possible by physicists.[40] The act of observation instantly changes a probability into an actuality by collapse of the wave function. Roger Penrose calls this resolution of multiple possibilities into one definitive state "objective reduction".[35] So it seems that no observation is possible without fundamentally changing the observed subject; only *subjectivity* remains.

The phase-speed in this invisible and non-measurable phase-space varies from the speed of light to infinity, while the speed of particles in our manifest physical real-space varies from zero to the speed of light. At the speed of light, the speed of a particle and the speed of the wave are identical. But the slower the particle, the faster the wave-speed, and when the particle stops, the wave-speed is infinite. The phase-space generates events that can be located in our space-time continuum, the manifest world, or real-space. Everything visible emanates form the invisible.

According to Stuart Hameroff and Roger Penrose, microtubules in neurons may process information generated by self-organizing patterns, giving rise to coherent states, and these states could be the explanation of the possibility of experiencing consciousness.[42] Herms Romijn argues that the continuously changing electromagnetic fields of the neuronal networks, which can be considered as a biological quantum coherence phenomenon, possibly could be the elementary "carriers" of consciousness.[31]

Quantum physics cannot explain the essence of consciousness or the secret of life, but in my concept it is helpful for understanding the transition between the fields of consciousness in the phase-space (to be compared with the probability fields as we know from quantum mechanics) and the body-linked waking consciousness in the real-space, because these are the two *complementary* aspects of consciousness.[41] Our whole and undivided consciousness with declarative memories finds its origin in, and is stored in this phase-space, and the cortex only serves as a relay station for parts of our consciousness and parts of our memories to be received into our waking consciousness. In this concept consciousness is not physically rooted. This could be compared with the internet, which does not originate from the computer itself, but is only received by it.

Life creates the transition from phase-space into our manifest real-space; according to our hypothesis life creates the possibility to receive the fields of consciousness (waves) into the waking consciousness which belongs to our physical body (particles). During life, our consciousness has an aspect of waves as well as of particles, and there is a permanent interaction between these two aspects of consciousness. This concept is a complementary theory, like both the wave and particle aspects of light, and not a dualistic theory. Subjective (conscious) experiences and the corresponding objective physical properties are two fundamentally different manifestations of one and the same underlying deeper reality; they cannot be reduced to each other.[30] The particle aspect, the physical aspect of consciousness in the material world, originates from the wave aspect of our consciousness from the phase-space by collapse of the wave function into particles ("objective reduction"), and can be measured by means of EEG, MEG, MRI, and PET scan. And different neuronal networks function as interface for different aspects of our consciousness, as can be demonstrated by changing images during these registrations of EEG, MRI or PET scan. The wave aspect of our indestructible consciousness in phase-space, with non-local interconnectedness, is inherently not measurable by physical means. When we die, our consciousness will no longer have an aspect of particles, but only an eternal aspect of waves.

With this new concept about consciousness and the mind-brain relation all reported elements of an NDE during cardiac arrest could be explained. This concept is also compatible with the non-local interconnectedness with fields of consciousness of other persons in phase-space. Following an NDE most people, often to their own amazement and confusion, experience an enhanced intuitive sensibility, like clairvoyance and clairaudience, or prognostic dreams, in which they "dream" about future events. In people with an NDE the functional receiving capacity seems to be permanently enhanced.

When you compare this with a TV set, you receive not only Channel 1, the transmission of your personal consciousness, but simultaneously Channels 2, 3 and 4 with aspects of consciousness of others. This remote, *non-local communication* seems to have been demonstrated scientifically by positioning subject pairs in two separate Faraday chambers, which effectively rules out any electromagnetic transfer mechanism. A visual pattern-reversal stimulus is used to elicit visual evoked responses in the EEG registration of the stimulated subject, and this was *instantaneously* received by the non-stimulated subject resulting in an analogous neural event with a similar brain wave morphology, or transferred potentials, as revealed on the EEG.[43,44]

8. THE ROLE OF DNA

How should we understand the interaction between our consciousness and our functioning brain in our continuously changing body? As stated before, during our life the composition of our body changes continuously, as during each second 500000 cells are being replaced in our body. What could be the basis of the continuity of our changing body? Cells and molecules are just the building blocks. In assessing all the theories mentioned above, it seems reasonable to consider the person-specific DNA in our cells as the place of resonance, or the interface across which a constant informational exchange takes place between our personal material body and the phase-space, where all fields of our personal consciousness are available as fields of possibility.

DNA is a molecule, composed of nucleotides, with a double helix structure. In humans it is organized into 23 pairs of chromosomes, defines 30,000 genes, and contains about 3 billion base pairs.[45] About 95% of human DNA has a still unknown function, for which reason it is called "junk DNA," non-protein-coding DNA, or introns,[46] and the 5% protein-coding called exons. The more complex a species is, the more introns it has. Simon Berkovich assumes that this "junk DNA" could have an identifying purpose, comparable to a kind of "barcode" functionality. According to his hypothesis DNA itself does not contain the hereditary material, but is capable of receiving hereditary information and memories from the past, as well as the morphogenetic information, which contains the way the body will be built with all its different cell systems with specialized functions.[47] Person-specific DNA is in this model the receiver as well as the transmitter of our permanently evolving personal consciousness.

According to Erwin Schrödinger, a quantum physicist, DNA is an a-statistic molecule, and a-statistic processes are quantum mechanical processes which originate from phase-space.[48] In his theory DNA should function as a quantum antenna with non-local communication, and also Stuart Hameroff considers DNA as a chain of quantum bits (qubits) with helical twist, and according to him DNA could function in a way analogous to superconductive quantum interference devices. In his quantum computer model the 3 billion base pairs should function as qubits with quantum superposition of simultaneously zero and one.[49]

Following a heart transplant, the donor heart contains DNA material foreign to the recipient. In a few recent books it has been reported that sometimes the recipient experiences thoughts and feelings that are totally strange and new, and later it becomes obvious that they fit with the character and consciousness of the deceased donor.[50,51] The DNA in the donor heart seems to give rise to fields of consciousness that are received by the organ recipient. Unfortunately, until now scientific research on this has not been possible due to the reluctance of the transplant centers.

9. ANALOGY WITH WORLDWIDE COMMUNICATION

In trying to understand this concept of quantum mechanical mutual interaction between the invisible phase-space and our visible, material body, it seems appropriate to compare it with modern worldwide communication. There is a continuous exchange of objective information by means of electromagnetic fields for radio, TV, mobile telephone, or laptop computer. We are unaware of the vast amounts of electromagnetic fields that constantly, day and night, exist around us and through us, as well as through structures like walls and buildings. We only become aware of these electromagnetic informational fields at the moment we use our mobile telephone or by switching on our radio, TV or laptop. What we receive is not inside the instrument, nor in the components, but thanks to the receiver, the information from the electromagnetic fields becomes observable to our senses and hence perception occurs in our consciousness. The voice we hear over our telephone is not inside the telephone. The concert we hear over our radio is transmitted to our radio. The images and music we hear and see on TV are transmitted to our TV set. The internet is not located inside our laptop. We can receive what is transmitted with the speed of light from a distance of some hundreds or thousands of miles. And if we switch off the TV set, the reception disappears, but the transmission continues. The information transmitted remains present within the electromagnetic fields. The connection has been interrupted, but it has not vanished and can still be received elsewhere by using another TV set ("*non-locality*").

Could our brain be compared to the TV set, which receives electromagnetic waves and transforms them into image and sound, as well as to the TV camera, which transforms image and sound into electromagnetic waves? This electromagnetic radiation holds the essence of all information, but is only perceivable by our senses through suitable instruments like camera and TV set.

The informational fields of our consciousness and of our memories, both evolving during our lifetime by our experiences and by the informational input from our sense organs, are present around us, and become available to our waking consciousness only through our functioning brain (and other cells of our body) in the shape of electromagnetic fields. As soon as the function of the brain has been lost, as in clinical death or brain death, memories and consciousness do still exist, but the receptivity is lost, the connection is interrupted.

10. CONCLUSION

According to our concept, grounded on the reported aspects of consciousness experienced during cardiac arrest, we can conclude that our consciousness could be based on fields of information, consisting of waves, and that it originates in the phase-space. During cardiac arrest, the functioning of the brain and of other cells in our body stops because of anoxia. The electromagnetic fields of our neurons and other cells disappear, and the possibility of resonance, the interface between consciousness and physical body, is interrupted.

Such understanding fundamentally changes one's opinion about death, because of the almost unavoidable conclusion that at the time of physical death consciousness will continue to be experienced in another dimension, in an invisible and immaterial world, the phase-space, in which all past, present and future is enclosed. Research on NDE cannot give us the irrefutable scientific proof of this conclusion, because people with an NDE did not quite die, but they all were very, very close to death, without a functioning brain.

The conclusion that consciousness can be experienced independently of brain function might well induce a huge change in the scientific paradigm in western medicine, and could have practical implications in actual medical and ethical problems such as the care for comatose or dying patients, euthanasia, abortion, and the removal of organs for transplantation from somebody in the dying process with a beating heart in a warm body but a diagnosis of brain death.

There are still more questions than answers, but, based on the aforementioned theoretical aspects of the obviously experienced continuity of our consciousness, we finally should consider the possibility that death, like birth, may well be a mere passing from one state of consciousness to another.

11. REFERENCES

1. Van Lommel W, Van Wees R, Meyers V, Elfferich I. Near-death experience in survivors of cardiac arrest: a prospective study in the Netherlands. *Lancet* 2001;**358:**2039-2045.
2. Greyson B. Incidence and correlates of near-death experiences in a cardiac care unit. *Gen Hosp Psychiatry* 2003;**25:**269-276.
3. Parnia S, Waller DG, Yeates R, Fenwick P. A qualitative and quantitative study of the incidence, features and aetiology of near death experiences in cardiac arrest survivors. *Resuscitation* 2001;**48:**149-156.
4. Ritchie G.G. *Return from Tomorrow.* Grand Rapids, Michigan: Chosen Books of The Zondervan Corp., 1978.
5. Blackmore S. *Dying to Live: Science and the Near-Death Experience.* London: Grafton -- An imprint of Harper Collins Publishers, 1993.
6. Appelby L. Near-death experience: analogous to other stress induced physiological phenomena. *BMJ* 1989;**298:**976-977.
7. Owens JE, Cook EW, Stevenson I. Features of "near-death experience" in relation to whether or not patients were near death. *Lancet* 1990;**336:**1175-1177.
8. Penfield W. *The Excitable Cortex in Conscious Man.* Liverpool: Liverpool University Press, 1958.
9. Meduna LT. *Carbon Dioxide Therapy: A Neuropsychological Treatment of Nervous Disorders.* Springfield: Charles C. Thomas, 1950.
10. Whinnery JE, Whinnery AM. Acceleration-induced loss of consciousness. *Arch Neurol* 1990;**47:**764-776.
11. Lempert T, Bauer M, Schmidt D. Syncope and Near-Death Experience. *Lancet* 1994;**344**:829-830.
12. Jansen K. Neuroscience, Ketamine and the Near-Death Experience: The Role of Glutamate and the NMDA-Receptor, In: *The Near-Death Experience: A Reader.* Bailey LW, Yates J, eds. New York and London: Routledge, 1996:265-282.
13. Grof S, Halifax J. *The Human Encounter with Death.* New York: Dutton, 1977.
14. Schröter-Kunhardt M. Nah--Todeserfahrungen aus Psychiatrisch-Neurologischer Sicht. In: Knoblaub H, Soeffner HG, eds. *Todesnähe: Interdisziplinäre Zugänge zu Einem Außergewöhnlichen Phänomen.* Konstanz: Universitätsverlag Konstanz, 1999:65-99.
15. Sabom MB. *Light and Death: One Doctor's Fascinating Account of Near-Death Experiences: "The Case of Pam Reynolds." In chapter 3: Death: The Final Frontier.* Michigan: Zondervan Publishing House, 1998:37-52.
16. Ring K, Cooper S. *Mindsight: Near-Death and Out-Of-Body Experiences in the Blind.* Palo Alto: William James Center for Consciousness Studies, 1999.
17. Gopalan KT, Lee J, Ikeda S, Burch CM. Cerebral blood flow velocity during repeatedly induced ventricular fibrillation. *J Clin Anesth* 1999;**11(4):**290-295.

18. De Vries JW, Bakker PFA, Visser GH, Diephuis JC, A.C. Van Huffelen AC. Changes in cerebral oxygen uptake and cerebral electrical activity during defibrillation threshold testing. *Anesth Analg* 1998;**87:**16-20.
19. Clute H, Levy WJ. Electroencephalographic changes during brief cardiac arrest in humans. *Anesthesiology* 1990;**73:**821-825.
20. Losasso TJ, Muzzi DA, Meyer FB, Sharbrough FW. Electroencephalographic monitoring of cerebral function during asystole and successful cardiopulmonary resuscitation. *Anesth Analg* 1992;**75:**12-19.
21. Parnia S, Fenwick P. Near-death experiences in cardiac arrest: visions of a dying brain or visions. Review article. *Resuscitation* 2002;**52:**5-11.
22. Smith DS, Levy W, Maris M, Chance B. Reperfusion hyperoxia in the brain after circulatory arrest in humans. *Anesthesiology* 1990;**73:**12-19.
23. Desmedt JE, Robertson D. Differential enhancement of early and late components of the cerebral somatosensory evoked potentials during forced-paced cognitive tasks in man. *J Physiol* 1977;**271:**761-782.
24. Roland PE, Friberg L. Localization in cortical areas activated by thinking. *J Neurophysiol* 1985;**53:**1219-1243.
25. Eccles JC. The effect of silent thinking on the cerebral cortex. *Truth Journal, International Interdisciplinary Journal of Christian Thought* 1988:**Vol 2.**
26. Roland PE. Somatotopical tuning of postcentral gyrus during focal attention in man. A regional cerebral blood flow study. *J Neurophysiol* 1981;**46:**744-754.
27. Libet B. Subjective antedating of a sensory experience and mind-brain theories: Reply to Honderich (1984). *J Theor Biol* 1985;**144:**563-570.
28. Bierman DJ, Radin DI. Anomalous anticipatory response on randomised future conditions. *Percept Mot Skills* 1997;**84:**689-690.
29. Bierman DJ, Scholte HS. A fMRI brain imaging study of presentiment. *Journal of ISLIS* 2002;**20(2):**280-288.
30. Romijn H. About the origin of consciousness. A new, multidisciplinary perspective on the relationship between brain and mind. *Proc Kon Ned Akad v Wetensch* 1977;**100(1-2):**181-267.
31. Romijn H. Are Virtual Photons the Elementary Carriers of Consciousness? *Journal of Consciousness Studies* 2002;**9:**61-81.
32. Hallett M. Transcranial magnetic stimulation and the human brain. *Nature* 2000;**406:**147-150.
33. Penfield W. *The excitable cortex in conscious man.* Liverpool: Liverpool University Press, 1958.
34. Penfield W. *The Mystery of the Mind.* Princeton: Princeton University Press, 1975.
35. Blanke O, Ortigue S, Landis T, Seeck M. Stimulating illusory own-body perceptions. The part of the brain that can induce out-of-body experiences has been located. *Nature* 2002;**419:**269-270.
36. Penrose R. *Shadows of the mind.* Oxford: Oxford University Press, 1996.
37. Berkovich SY. On the information processing capabilities of the brain: shifting the paradigm. *Nanobiology* 1993;**2:**99-107.
38. Bohr N, Kalckar J, editors. *Collected Works. Volume 6: Foundations of Quantum Physics I (1926-1932).* Amsterdam, New York: North Holland, 1997:91-94.
39. Aspect A, Dalibard J, Roger G. Experimental tests of Bell's inequality using varying analyses. *Phys Rev Lett* 1982;**25:**1084.
40. Heisenberg W. *Schritte über Grenze.* Munchen: R. Piper & Co Verlag, 1971.
41. Walach H, Hartmann R. Complementarity is a useful concept for consciousness studies. A Reminder. *Neuroendrocrinol Lett* 2000;**21:**221-232.
42. Hameroff S, Penrose R. Orchestrated reduction of quantum coherence in brain microtubules. In: *Proceedings of the international neural Network Society, Washington DC*, Erlbaum, Hillsdale, NJ, 1995.
43. Thaheld F. Biological non-locality and the mind-brain interaction problem: comments on a new empirical approach. *Biosystems* 2003;**2209:**1-7.
44. Wackermann J, Seiter C, Keibel H, Walach H. Correlations between electrical activities of two spatially separated human subjects. *Neurosci Lett* 2003;**336:**60-64.
45. Ridley M. *Genome. The autobiography of a species in 23 chapters.* New York: Harper Collins Publishers, 2000.
46. Mantegna RN, *et al.* Linguistic features of non-coding DNA sequences. *Phys Rev Lett* 1994;**73:**31-69.
47. Berkovich SY. *On the "Barcode" Functionality of the DNA, or the Phenomenon of Life in the Physical Universe.* Pittsburgh: Dorrance Publishing CO, 2003).
48. Schrödinger E. *What is Life?* Cambridge: Cambridge University Press, 1944.
49. Hameroff S. Quantum computing in DNA. http://www.consciousness.arizona.edu/hameroff/New/Quantum_computing_in_DNA/index.htm.
50. Sylvia C, Novak W. *Change of Heart.* New York: Little, Brown, 1997.
51. Pearsall P. *The Heart's Code.* New York: Broadway Books, Bantam Doubleday Dell, Inc, 1998.

BRAIN DEATH AND ORGAN TRANSPLANTATION

Concepts and principles in Judaism

Z. Harry Rappaport and Isabelle T. Rappaport*

1. INTRODUCTION

When faced with mortality people tend to turn to their religious heritage, even if they are secularized. Ethical issues are deeply influenced by the religious background of our societies. In this essay we wish to examine the concept of death from a Jewish religious perspective. With the emergence of brain death criteria driven by the advances in medical technology, Jewish religious interpretation adjusted to take into account the possibilities of modern medical technology. While traditionalist elements cling to older interpretations, a process of ethical evolution, similar to that in other societies, has led to the widespread acceptance of brain death criteria for the purpose of organ donation in Israel.

2. BACKGROUND

The basis of Jewish religious jurisprudence is the Old Testament. The codification of a parallel oral tradition (*Talmud*) occurred between the third and fifth centuries in Babylon and is known as the *Mishna*. There followed a 200 year period of further elaboration of the codices in a body of work known as the *Gemara*. In the middle ages various commentators such as Rashi in France and Maimonides in Egypt laid the foundations for the modern interpretation of Jewish law (*Halakha*) encodified in the 17th century by Joseph Karo. Subsequent religious authorities would base their rulings on these foundations until the present era.

Criteria for the timing of death had not been a major issue for Judaism in the past. As a rule all vital systems (respiratory, circulatory, and neurological) would fail at the same time. Any individual organ system could not continue independently. With the significant advance in medical technology in the latter part of the twentieth century it became possible to artificially support the cardio-respiratory functions. This led to the need to

* Dept. of Neurosurgery, Rabin Medical Center, Tel-Aviv Univ. School of Medicine, 49100 Petah-Tiqva, Israel.

Brain Death and Disorders of Consciousness, Edited by Machado and Shewmon
Kluwer Academic/Plenum Publishers, New York 2004

ascertain with greater discrimination which systems were in fact indicative of life and which were indicative of death. The issue received urgency due to the scarcity of high tech resources and the need for allowing organ donation. For the Rabbis the questions arose of when was it permissible to discontinue a respirator without being responsible for murder, and if it was permissible to harvest a beating heart for organ transplantation. On the one hand, there was difficulty in declaring dead an artificially ventilated patient whose heart was still beating, and the corresponding concern that removing his heart would be tantamount to killing. On the other hand, if the potential donor was in fact already dead, there was a moral obligation to use every means to save the life of the potential organ recipient.

The sanctity of life is a principle of paramount importance in Judaism. One is obliged to use every means to preserve life, even if only for an instant. He who hastens the death of a human being is seen as a murderer. Therefore, the exact timing of the moment of death has a significant *Halakhic* significance. For only when the criteria of death, as enunciated by the *Halakha* are present, is one free of the obligation to use every available means in order to preserve life.

The first source that appears in the *Babylonian Talmud* concerning this issue deals with the situation of a body buried under a pile of rubble on a Sabbath[1]. Is it permissible to violate the law against performing manual labor on the Sabbath by excavating the body? The *Halakha* rules that the rescue effort must continue *"however remote the likelihood of rescuing life may be,"* and *"even if found crushed in such a manner that he cannot survive except for a short while."*[1] Judaism places an infinite value on human life. Life of a few seconds is as dear as that of one hundred years. However, if it were ascertained beyond a shadow of doubt that the buried body is dead, one should not continue to violate the Sabbath by unearthing it. Rather, one should stop the effort and continue only after the Sabbath has ended. The question then arises as to how much of the body must be uncovered before one can ascertain that the person is in fact dead. The first opinion states that one must continue to dig until the nose has been exposed. The person is pronounced dead in the absence of signs of respiration. The second opinion deals with the situation in which the body is uncovered from the feet up. When one reaches the thorax and cannot detect signs of a heartbeat the person is surely dead. This, however, remains a minority opinion. The authors felt that a weak heartbeat might be mistaken for an absent one. The absence of respiration was felt to be a much more robust sign. The centrality of cessation of respiration in determining death is based on the biblical sentence: "…all in whose nostrils is the breath of the spirit of life."[2] In fact the Hebrew word for soul, *neshama,* is closely related to the word for respiration, *neshima.*

The second source in the *Talmud* deals with issue of a Caeserian section on the Sabbath[3]. One is permitted to perform the operation on a Sabbath in order to save the life of an infant, whose mother had died in childbirth. A 17th century rabbinical authority, Rabbi Moshe Isserles, prohibited such a procedure, claiming that at his time there was insufficient medical expertise to determine the exact timing of the mother's death[4]. He feared that the operative intervention might hasten the demise of the mother, something that would be tantamount to murder. However, what would be the religious ruling in the case of a pregnant woman who was accidentally decapitated? The *Mishna* in Oholot[5] deals with the laws concerning ritual impurity due to contact with a cadaver. The *Mishna* rules that decapitation is an uncontestable sign of death even if the body still displays some form of motion. This motion is regarded as the twitching of an amputated lizard's tail, meaning a manifestation of cellular life after the death of the organism has occurred.

Apart from this rather hypothetical exception, the general consensus concerning the timing of death continued to rely exclusively on respiratory cessation. The heartbeat would stop several minutes after breathing stopped. Therefore, the absence of a heartbeat was viewed as an indication that respiration had irreversibly disappeared. The 11th century rabbinical authority Rashi, in his commentary on the Talmudic passage concerning the man buried under a pile of rubble, explained that, in order to proclaim him dead, the person must be without movement like an inanimate stone. This formed the basis of a triad of signs of death that has been current ever since in Jewish religious law:

1. The patient lies like an inanimate stone;
2. There is no pulse;
3. Respiration has ceased.

The presence of heart beat negates the first condition of being like an inanimate stone, and has therefore been used by some modern religious authorities as an argument against brain death.

3. CURRENT ATTITUDES

The necessity for religious authorities to deal with the issue of brain death achieved urgency following the first successful cardiac transplant by Christiaan Barnard in 1967. Prof. Morris Levy performed the first cardiac transplant in Israel, then the 100th in the world, a year later at our institution. He received special dispensation by the Chief Rabbinate of Israel to perform the surgery. However, the initial years of heart transplantation had a high mortality, and the operation was still considered experimental. Religious authorities in fact opined that the procedure amounted to double murder, that of the donor and that of the recipient. In the course of the years, as the procedure became more successful and established, religious authorities reappraised their attitude toward the concept of brain death.

The leading *Halakhic* authority at that time was Rabbi Moshe Feinstein, head of a Talmudic academy in New York. He was initially opposed to heart transplantation, and in fact was instrumental together with his Catholic colleagues in influencing the New York State legislature to adopt a religious exemption in their brain death legislation. Over time he adopted a more favorable approach to the concept of brain death, relying on the *Mishna* in Oholot that viewed a decapitated animal as conferring the ritual impurity of the dead. He submitted that a nuclear brain scan that showed cessation of blood flow to the brain in fact implied brain liquefaction, a situation tantamount to decapitation. Based on this interpretation of the religious literature, the Chief Rabbinate of Israel permitted organ transplantation based on brain death criteria in 1987. The establishment of the criteria was handed over to an expert medical committee with the participation of a member of the Rabbinate. At that time the following five points were included:

1. Knowledge of the cause of the illness;
2. Complete cessation of spontaneous ventilation;
3. Clinical demonstration of the destruction of the brain-stem;
4. Objective support of the clinical determination by brain-stem auditory evoked potentials;

5. Demonstration that absent respiration and brain-stem activity persist for at least 12 hours under full therapy.

The Chief Rabbinate accepted that the destruction of the lower brainstem was equivalent to irreversible cessation of respiration, the traditional criterion of death in the religious texts. They therefore did not include a negative brain scan as an additional requirement. This ruling opened the era of heart and liver transplantation in Israel, which had long stopped being experimental.

4. THE APPROACH OF JUDAISM TO DEATH IN CHILDREN AND INFANTS

Judaism does not in general differentiate between adults and children as to criteria for ascertaining death. If medical authorities would impose stricter parameters for arriving at a diagnosis of brain death in children, this would become the norm for the religious authorities. In its approach to feticide Judaism, however, differentiates itself from other religions, especially from that of the Catholic Church. The Scriptural basis for the Jewish attitude is to be found in *Exodus*[7]:

> *"And if men strive together and hurt a woman with child, so that her fruit depart, and yet no harm to follow, he shall be surely fined, according as the woman's husband lay upon him; and he shall pay as the judges determine. But if any harm follow, then shalt thou give life for life..."*[7]

This passage is interpreted in Judaism that, as long as there is no harm to the pregnant mother, there is no capital offence in the killing of the fetus. In early Christianity the passage "no harm to follow" was mistranslated by the Greek version of the Old Testament, the *Septuagint*, into *"[her child be born] imperfectly formed."* Tertullian and subsequent Church fathers made the distinction between an unformed and a formed fetus, branding the killing of the latter as a capital crime. This distinction was adopted by canon law and in Justinian Law[8]. Subsequently, in 1588, the Catholic Church came to view the killing of a fetus (abortion) as murder from the moment of conception.[9]

In Judaism the full title to life begins only at birth. In fact the *Talmud* sanctions the dismemberment of the fetus if it endangers the mother's life at birth:

> *"If a woman is in hard travail [and her life cannot otherwise be saved], one cuts up the child within her womb and extracts it member by member, because her life comes before that of [the child]. But if the greater part [or the head] was delivered, one may not touch it, for one may not set aside one person's life for the sake of another."*[10]

Later rabbinical responsa expanded on this subject. If a fetus is impacted during a breech delivery, it may be dismembered, even if the majority of the body has been delivered. This ruling applies to the situation when both mother and infant would die without this maneuver. During the first 30 days of life the infant is considered to be an entity that has yet to prove its viability. Thus parents may not *halakhically* morn the death of an infant during this time frame. There is therefore an inequality between the life of the mother, which is established, and that of the newborn, who has yet to prove his viability. This difference in status is usually of no significance, and does not sanction the killing of the infant to save the mother's life, except in the specific situation where both would be doomed without the intervention.

Whether this type of argumentation is applicable to the use of anencephalic newborns as organ donors is not yet clear. Jewish religious authorities in the past have, however, proven themselves able to deal with the ethical demands that modern medical innovations impose upon society's ethical traditions, without abandoning the framework of the traditional dialectic. It is the adaptability of traditional cultural norms to the exigencies of the modern situation that raises the hope for a society based on a shared morality in the future.

5. CONCLUSION

The need to deal with brain death criteria arose in Judaism as a result of the successes in organ transplantation. While the traditional criteria for death were based on the cessation of respiration, irreversible brain-stem nonfunction came to be accepted as its equivalent. The value placed upon the saving of a life (that of the organ recipient) allowed for innovation even in the face of opposition by religious traditionalists. The possibility of moral evolution within a rigorous religious framework is an example of placing humanistic values over dogma.

6. REFERENCES

1. Babylonian Talmud, Yoma 85a.
2. Old Testament, Genesis 7:22.
3. Babylonian Talmud, Erekhin, 7a.
4. M. Isserles, Orech Chaim, Question 8.
5. Babylonian Talmud, Mishna Oholot 1:6.
6. M. Feinstein, Igrot Moshe, Yore Deah III, no. 132, 1976.
7. Old Testament, Exodus 21:12
8. E. Westermarck, Christianity and Morals. Kegan Paul, Trench, London: Trubner & Co., 1939: 243.
9. A. Bonnar, The Catholic Doctor. New York: PJ Kenedy, 1938:78.
10. Babylonian Talmud, Mishna Oholot 7:6.

CUBA HAS PASSED A LAW FOR THE DETERMINATION AND CERTIFICATION OF DEATH

Calixto Machado,* Mayda Abeledo, Carlos Álvarez , Rosa M. Aroche, Irene Barrios, Alicia M. Lasanta, Ramón Beguería, Armando Cabrera, Berta L. Castro, Maria E. Cobas, Enma Cuspineda, Antonio Enamorado, Nicolás Fernández, Pedro Figueredo, Orlando D. García, Tania García, Nelson Gómez, Carlos González, Noel González, Jorge González, Armando González, Raúl Herrera, Jorge Lage, Alberto Martínez, Armando Pardo, Jesús Parets, Leonardo Pérez, Jesús Pérez, Margarita Pons, Desiderio Pozo, Ibis Rojas, José M. Román, Héctor Roselló, René Ruiz, Aquilino Santiago, Sofía Sordo, Roberto Suárez, Rene Zamora†

1. INTRODUCTION

During the last several decades physicians and the community have needed urgent changes in the legal codes for accepting brain death (BD) as death, to obtain organs from heart-beating donors. The "dead donor rule" requires that donors must be first declared dead.[1] For this reason, most codes legalizing BD are usually sections of transplant laws.[2] Thus, a conceptual and practical controversy emerged: if brain-dead cases were not useful as organ donors, they were usually kept on life support until cardiac arrest occurred.[2-4]

During the last several decades physicians and the community have needed urgent changes in the legal codes for accepting brain death (BD) as death, to obtain organs from heart-beating donors.

The National Commission for the Determination and Certification of Death has accepted a neurological view of death and remarked that any legal code of death should be completely separated from any norm governing organ transplants.[2,3]

* Calixto Machado, MD, PhD, President of the National Commission for the Determination and Certification of Death. Institute of Neurology and Neurosurgery, Havana, Cuba. Email: braind@infomed.sld.cu.

† Members of the Commission.

Brain Death and Disorders of Consciousness, Edited by Machado and Shewmon
Kluwer Academic/Plenum Publishers, New York 2004

2. CUBAN LAW FOR THE DETERMINATION OF DEATH

2.1. The Civil Code

The Cuban Civil Code has undergone several historical changes regarding the determination of death. [5]

2.1.1. Spanish Civil Code, Established in Cuba since 1889

This code remarked in its Article 31:

> *"Death occurs when the person is extinguished"*

2.1.2. New Civil Code (Cuban Law 1175, March 9, 1965)

This code stated:

> *"Death is the loss of every sign of life after a living birth"*

2.1.3. Development of Article 26.1 of the Present Civil Code (July 1987)

In its final version, in force since 1987, Article 26.1 stated:

> *"Physicians are the only professionals authorized to diagnose and certify death according to a norm established by the Ministry of Public Health."*

2.2. Distinguished Features of the Article 26.1 of the Current Cuban Civil Code

- It does not define death.
- It specifies that physicians are the only professionals authorized to diagnose and certify death.
- It specifies that the Ministry of Public Health must provide a norm for the determination of death (clinical and instrumental criteria).
- The Cuban Parliament gave full responsibility to the Ministry of Public Health to legalize the determination and certification of death.

2.3. Law within a Ministry

According to the Cuban Parliament, a "Resolution" is a law that legalizes a working standard within any Ministry and that could be signed and changed by the Minister in function.[2] Therefore, the Ministry of Public Health needed to respond to the present Civil Code, Article 26.1,[2] by writing a Resolution on this subject.

For this reason, the Ministry of Public Health organized the National Commission for the Determination and Certification of Death at the beginning of the 1990s. The Commission was composed by representatives from multiple disciplines, including various medical and legal specialties. After several versions, the Commission wrote the final resolution, signed by the Minister of Public Health on August 27, 2001.[3]

2.4. The Main Body of this Norm Contained the Following Points

- Physicians diagnose death by documenting the "signs of death."
- When the diagnosis of death is based on the irreversible loss of whole brain functions, medical specialists need to be accredited by the National Commission for the Determination and Certification of Death.
- Death is certified by the physician who diagnoses it. Physicians will register the moment of death upon completing the diagnostic procedure.
- The Commission will annually review the diagnostic criteria of death and will propose changes or amendments, according to medical and technological advances in this area.

2.5. Diagnosis of Death

The National Commission accepted only one kind of death, based on *the irreversible loss of brain functions*. As the Civil Code did not require a definition, the Commission followed this norm and presented not a concept, but the ways to diagnose death.[2] However, discussions left very clear that although there is only one kind of death, there are several ways of diagnosing it, according to the place and environment where death occurs.[2-8]

Nonetheless, the Commission adopted the view that the irreversible loss of cardio-circulatory and respiratory functions can only cause death when ischemia and anoxia are prolonged enough to produce an irreversible destruction of the brain.[2]

Physicians diagnose death by finding the "signs of death." Signs I and II correspond to the classical respiratory and cardio-circulatory functions. Signs III to VIII are related to forensic circumstances. Sign IX corresponds to the BD diagnosis.[2,3]

3. SIGNS OF DEATH

I. Irreversible loss of respiratory function
II. Irreversible loss of cardio-circulatory functions
III. Algor mortis (postmortem coldness)
IV. Livor mortis (postmortem lividity)
V. Rigor mortis (postmortem rigidity)
VI. Cadaveric spasm
VII. Loss of muscle contractions
VIII. Putrefaction
IX. Irreversible loss of brain functions

3.1. The Commission Enumerated Three Possible Situations for Diagnosing Death

- Outside the intensive care environment (without life support) physicians apply the cardio-circulatory and respiratory criteria.
- In forensic medicine circumstances physicians utilize cadaveric signs. (They don't even need a stethoscope.)
- In the intensive care environment (with life support) when cardio-circulatory and/or respiratory arrest occurs physicians utilize the cardio-circulatory and respiratory criteria. When physicians suspect an irreversible loss of brain functions in a heart-beating and ventilatory supported case, BD diagnostic criteria are applied.

The Commission proposed two groups of ancillary tests for an early diagnosis of irreversible loss of brain functions: those for testing the irreversible absence of intracranial cerebral blood flow, and those to document absence of bioelectrical activity.[2,9] Among the latter group, the Commission proposed a test battery composed by multimodality evoked potentials and electroretinography.[4,9-15]

4. REFERENCES

1. A definition of irreversible coma. Report of the Ad Hoc Committee of the Harvard Medical School to Examine the Definition of Brain Death. *JAMA* 1968;**205(6):**337-340.
2. Machado C. Resolution for the determination and certification of death in Cuba. *Rev Neurol* 2003;**36(8):**763-770.
3. Gaceta Oficial de la República de Cuba. Resolución No. 90 de Salud Pública. *Edición Ordinaria del 21 de Septiembre del 2001*. Ciudad de La Habana: Ministerio de Justicia. 2001.
4. Machado C, García-Tigera J, García OD, García-Pumariega J, Román JM. Muerte Encefálica. Criterios diagnósticos. *Rev Cubana Med* 1991;**30(3):**181-206.
5. Gaceta Oficial de la República de Cuba. Justicia. *Ley No. 59 del 16 de Julio de 1987 (Código Civil)*. Ciudad de La Habana:Ed. Ministerio de Justicia. Edición Extraordinaria No. 9, 16-7-1987; p. 39-81.
6. Machado C. Consciousness as a definition of death: its appeal and complexity. *Clin Electroencephalogr* 1999;**30(4):**156-164.
7. Machado C. Death on neurological grounds. *J Neurosurg Sci* 1994;**38(4):**209-222.
8. Machado C. Is the concept of brain death secure? In: Zeman A, Emanuel L. *Ethical Dilemmas in Neurology*. London: Ed. W. B. Saunders Company. Vol. 36, 2000:193-212.
9. Machado C, García OD. Guidelines for the determination of brain death. In: Machado C. *Brain Death* (Proceedings of the Second International Symposium on Brain Death). Amsterdam: Elsevier Science BV. 1995;75-80.
10. Machado C. Multimodality evoked potentials and electroretinography in a test battery for an early diagnosis of brain death. *J Neurosurg Sci* 1993;**37(3):**125-131.
11. Machado C, Pumariega J, García-Tigera J, Miranda J, Coutin P, Antelo J et al. A multimodal evoked potential and electroretinography test battery for the early diagnosis of brain death. *Int J Neurosci* 1989;**49:**241-242.
12. Machado C, Valdés P, Garcia T, Virues T, Biscay R, Miranda J et al. Brain-stem auditory evoked potentials and brain death. *Electroencephalogr clin Neurophysiol* 1991;**80(5):**392-398.
13. Machado-Curbelo C, Roman-Murga JM. Usefulness of multimodal evoked potentials and the electroretinogram in the early diagnosis of brain death. *Rev Neurol* 1998;**27(159):**809-817.
14. Machado C, Valdés P, García O, Coutin P, Miranda J, Román J. Short latency somatosensory evoked potentials in brain-dead patients using restricted low cut filter setting. *J Neurosurg Sci* 1993;**37(3):**133-140.
15. Machado C, Santiesteban R, García O, Coutin P, Buergo MA, Román J et al. Visual evoked potentials and electroretinography in brain-dead patients. *Doc Ophthalmol* 1993;**84(1):**89-96.

NEUROPROTECTION BECOMES REALITY

Changing times for cerebral resuscitation

Maxwell S. Damian*

1. INTRODUCTION

Sudden cardiac arrest (CA) is a leading cause of death worldwide, claiming over 1000 lives daily in the United States. Despite cardiopulmonary resuscitation (CPR) being one of the most dramatic and widely known medical procedures, its overall result remains unsatisfactory. One third of CPRs succeed in restoring spontaneous perfusion, but less than half of those patients ever regain consciousness, and less than 10 per cent of survivors can return to their former lifestyle.[1-4] The results of CPR are particularly dismal in subgroups such as the elderly or patients with primary asystole. Severe brain damage is the prime cause of death and disability in these patients, due to the exquisite sensitivity of the brain to anoxia.

Although vast efforts have been undertaken to improve the immediate response to CA by providing sufficient trained personnel, training the lay population and giving access to technology such as automatic defibillators, there was long little emphasis on treatment of brain damage after CPR. This was due in part to a misconception that the process damaging the brain is ended by providing adequate circulation, and to a fatalistic attitude seeing the concept of neuroprotection as purely theoretical science. Time and again, agents that showed neuroprotective potential on the bench have proved useless or worse at the bedside.[5-10] During the past few years, however, a substantial change in perception has occurred, with the recognition that reperfusion injury and delayed cerebral damage continue for many hours or even days after CPR. Major elements of this biochemical cascade have now been identified, bringing specific treatment strategies within reach.[2,11-13] Even more importantly, well-conducted clinical studies involving significant numbers of patients have finally confirmed that applying a neuroprotective strategy after CPR improves survival.[14,15] An active approach to the management of delayed posthypoxic brain injury has now become reality.

* Maxwell S. Damian, MD. Department of Neurology, University Hospitals of Leicester, UK. Ema maxwell.damian@uhl-tr.nhs.uk.

Brain Death and Disorders of Consciousness, Edited by Machado and Shewmon
Kluwer Academic/Plenum Publishers, New York 2004

2. PATHOPHYSIOLOGY OF POSTHYPOXIC BRAIN INJURY

The brain depends on uninterrupted oxidative metabolism not only to uphold neuronal function, but also for cellular detoxification and maintenance of membrane integrity. Under normal circumstances cerebral perfusion is 750-1000 ml/min, about 20% of the total cardiac output, to permit an oxygen consumption of 3.5ml/100g/min. There are virtually no stores of oxygen, and thus uninterrupted circulation is needed, even though some inefficient anaerobic metabolism is possible. Loss of consciousness ensues if oxygen supply is less than 2ml/100g/min, and the amount of neuronal damage sustained depends both on duration and severity of hypoxic ischemia. As even expert resuscitation can only provide a fraction of normal blood flow, the duration of CA and CPR determines the extent of immediate cell death, as well as the intensity of the metabolic cascade that leads to delayed neuronal damage.

The mechanisms of delayed neuronal injury are multifactorial, and their relative importance is undecided and probably varies from case to case. Certainly, profoundly abnormal cellular signal transduction through various pathways is central to reperfusion injury. One important factor is disruption of the blood-brain barrier, explained by an inflammatory response with leukocyte-derived mediators, reactive oxygen species (ROS) such as nitric oxide (NO) and peroxynitrite, and proteases in reperfused tissue. Inflammatory mediators activate matrix metalloproteinases (MMPs), which are toxic and increase vascular permeability. An interstitial fluid shift increases diffusion distances, which further impairs microvascular perfusion, and causes inefficient exchange of O_2 and toxic waste substances accumulated during CA. This impairs tissue function, causes delayed neuronal damage after reperfusion, and subsequent generalized mixed vasogenic and cytotoxic brain edema further reduce cerebral perfusion pressure.[4]

Ischemic upregulation of NO synthetase generates reactive oxygen species, which cause intracellular acidification, but also triggers complex mechanisms of neuronal injury by activating neuronal endonucleases and cystein preoteases.[13,16] ROS have been studied intensely, both as markers of neuronal damage and as targets for treatment.[17]

Anaerobic glycolysis produces excessive lactate, a major source of cellular acidification if not eliminated efficiently. Furthermore, ATP depletion removes the energetic basis for cellular ion homeostasis and Na-K-pump function, causing an intracellular sodium and water influx, and neuronal membrane depolarization. This is the signal for massive release of glutamate and other excitatory amino acids (EAAs). Overstimulation of glutamate receptors opens ion channels permeable to Ca^{++}, creating additional intracellular Ca^{++} influx. Cystein protease (caspase) activition, which can initiate programmed cell death after ischemia, may be modulated by metabotropic glutamate receptors.[11] Thus, glutamate release is an important stage in neuronal injury, and an attractive therapeutic target. However, the complex physiological functions of glutamate receptors have limited the usefulness of putative glutamate antagonists, especially in stroke trials.[10,18]

Ca^{++} overload carries various detrimental effects on neuronal integrity. Intracellular proteases are activated, causing further damage to the cytoskeleton. ROS synthesis is fuelled, and DNA fragmentation is facilitated [19]. Ca^{++} overload is toxic for mitochondria and through them serves to initiate pathways of programmed cell death. Although Ca^{++} influx is at the crossroads of several metabolic pathways to neuronal injury, the use of calcium antagonists as neuroprotective agents has been particularly disappointing, compared with their cardiovascular applications.[6-8]

Mitochondrial dysfunction has only recently been assigned a central role in this scenario. This has come together with a recognition of its central role in apoptosis. Caspase activation also alters mitochondrial permeability, and the importance of mitochondrial dysfunction in reperfusion injury has gained increasing recognition.[20] Mitochondria have even been called the "initiators of cell death",[13] because mitochondrial disruption is central to ATP depletion, collapse of the electrochemical gradient, intracellular acidification, and failure of oxygen free radical detoxification.[21] Apoptosis is fuelled by cytochrome c release through mitochondrial permeability transition pores and release of other apoptigenic molecules. Importantly, this is a delayed process that takes place over many hours following reperfusion, potentially offering a time window for intervention.[22,23] Recent studies suggest that the mitochondrial intermediate Coenzyme Q10 (= ubiquinone) is a potent neuroprotective agent that preserves the electrochemical gradient and reduces oxidative stress, also having effects on glutamate release and calcium influx. It improves oxidative metabolism in Huntington's disease and has recently been shown to slow the progression of Parkinson's disease.[24-26] Although treatment targeting mitochondrial metabolism has shown efficacy in neurodegenerative disorders, no data is available for severe hypoxia.

3. NEUROPROTECTIVE TREATMENT AFTER CARDIAC ARREST

Most trials of neuroprotection for acute disease have been in stroke and brain trauma; far less clinical experience exists with global ischemia. A number of drug trials are indicated above. Although they were all unsuccessful, the robust preclinical evidence for neuroprotection suggests that some factors that have a decisive influence on outcome may not have been recognized in these studies. One might be the importance of temperature control, and another is the still undecided duration of the therapeutic window for intervention. Hypothermia has several beneficial effects on the hypoxic brain: for instance, energy consumption decreases and high-energy phosphates are preserved; release of EAAs and formation of ROS is attenuated; brain swelling is reduced.[27-30] It is now considered that the therapeutic window for neuroprotection might be extended, avoiding a major handicap of neuroprotective drug administration in the clinical setting.

Therapeutic hypothermia is the first intervention shown to confer neuroprotection after acute hypoxia, now confirmed in two well-conducted randomized trials.[14,15] Bernard et al. reported survival to hospital discharge with good outcome in 21 out of 49 (49%) patients treated with hypothermia at 33°C for 12 hours vs. 9 out of 34 (26%) patients treated with normothermia after CPR.[14] Mortality itself was 51% in the hypothermia group and 68% in the normothermia group. The HACA study group maintained hypothermia at 32°-34°C for 24 hours after CPR, and reported a good Cerebral Performance Category rating in 75 out of 136 hypothermia patients (55%) vs. 54 out of 137 (39%) normothermia patients, as well as reduced mortality at 6 months (41% vs. 55%) [15]. The authors calculated that 6 patients would need to be treated to prevent an adverse outcome, and 7 to prevent one death. Neither study reported significant side-effects from hypothermia, and creatine kinase values were not higher in the hypothermia patients. In both studies, the initial cardiac rhythm was ventricular fibrillation; patients found asystolic at CPR were excluded. The different rates of survival in otherwise very similar patient groups, treated in a similar fashion, highlight the difficulty in comparing study results, but certainly the results using moderate hypothermia for the first day after

CPR are by far the most promising yet published. They underline the importance of temperature control and raise the question whether pharmacologic neuroprotection might not prove more successful if combined with hypothermia.

4. PRESENT STUDY

This study was undertaken to test the hypothesis whether a combination with pharmacologic neuroprotection could improve neuroprotection with hypothermia after severe hypoxia due to CA. CoQ10 was chosen as the study drug, because of its well-studied favourable side-effect profile, central role in mitochondrial metabolism, and proven effect in neurodegenerative disorders.

4.1. Patients and Methods

The study was performed at the University of Dresden, Germany. Patients entered within 6 hours after cardiac arrest fulfilled the following criteria: age between 18 and 80 years; witnessed cardiac arrest of cardiac origin; coma despite restored spontaneous perfusion; absence of neurological disease or pregnancy; previous heart failure not more than NYHA [New York Heart Association] class II. The trial was randomised and controlled with blind assessment of the outcome. Written consent was given by next of kin. All patients were cooled using a cooling mattress to a core temperature between 35° and 36°C, and rewarmed after 24 hours or earlier if instability was attributed to hypothermia; temperature remained below 37° in all cases. CoQ10 patients were given a loading dose of 250 mg lipophilic emulsion of CoQ10 (Sanomit^R, MSE Pharma, Bad Homburg, Germany) immediately after inclusion, followed by 150 mg t.i.d. Placebo patients were given identically coloured water from identical brown vials. Neurological assessments included clinical examination, 4 channel EEG and somatosensory evoked potentials (SSEP) on days 0, 1, 3 and 5 and 90 days after CPR; outcomes were classified according to the Glasgow Outcome Scale (GOS). Serum S100 protein was sampled on days 0, 1 and 5. The primary outcome variable was the 3-month survival; other outcome variables were: survival to discharge from ICU; GOS ratings after 3 months; S100 protein serum levels on days 0, 1 and 5; cortical SSEP amplitudes. We estimated that approximately 50 patients were necessary to provide sufficient power in a parallel design with one outcome variable. The SPSS 10.0 computer package was used, with a Kaplan-Meier analysis of survival, group comparisons by ANOVA and unpaired t-tests. Significance was defined as $p < 0.05$.

4.2. Results

Fifty consecutive eligible patients were included; there was one drop-out whose study medication was not administered, making 25 patients in the CoQ10 and 24 in the placebo group. There were no significant differences in age, medical history, sex, cause of arrest, delay to start CPR, duration of CPR or serum glucose at admission. 9 patients in the CoQ10 group, and 7 in the placebo group were found with asystole. There were no breaches of protocol or significant differences in treatment such as sedative or inotrope doses. There were no side effects attributable to study medication. Lactate and pH at

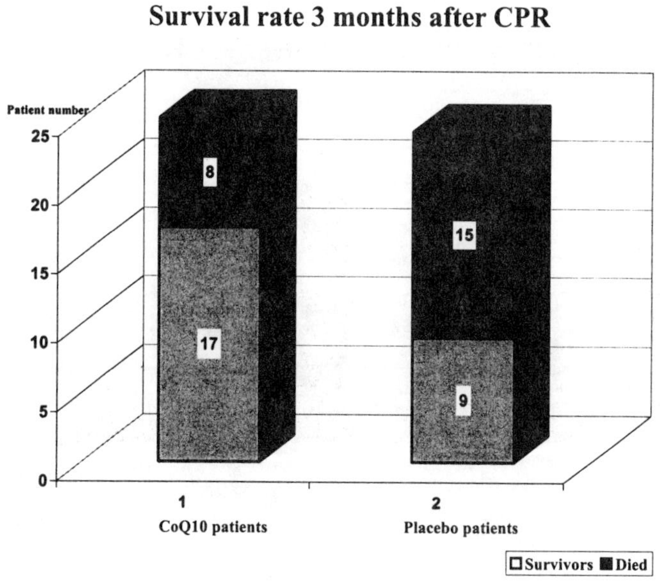

Survival rate 3 months after CPR

Figure 1. Three-month survival rate in CoQ10 vs. placebo patients.

inclusion did not differ. The creatine kinase was higher in the placebo group at inclusion and on day 1.

The 3-month survival rate was significantly higher in the CoQ10 group: 17 out of 25 (68%) in the CoQ10 group survived, but only 9 out of 24 (37.5%) in the placebo group (Fig. 1) ($p = 0.032$). 21 out of the 25 patients in the CoQ10 group (85%) survived until discharge from ICU, compared with 17 out of 24 (70.8%) patients in the placebo group (not significant). On 3-month follow-up examination, the GOS differed between the 2 groups, but sample sizes were too small for statistical comparison. Interestingly, in the CoQ10 group 9/25 patients (36%) had a good neurological outcome (GOS 4 or 5) vs. only 5/24 (20%) in the placebo group. Only 1 of the patients who survived 5 days, but died within 3 months, a patient in the placebo group, was neurologically intact and died of a recurrent cardiac disorder, whereas 9 of the 12 late fatalities (3 in the CoQ10 group and 6 in the placebo group) were patients who were discharged comatose and ventilator-dependent.

Serum S100 protein levels were significantly lower in the CoQ10 group 24 hours after CPR (Figure 2). The mean S100 in the CoQ10 group was 0.58 ng/ml vs. 1.4 ng/ml in the placebo group on day 0; on day 1 it was 0.47 ng/ml vs. 3.5 ng/ml, and on day 5 0.29 ng/ml vs. 0.71 ng/ml (normal: < 0.15 ng/ml).

There was no significant difference between overall SSEP amplitudes or N20/N13 amplitude ratios in the CoQ10 and placebo groups or in the bad outcome patients of both groups. However, in good outcome survivors the average N20/N13 amplitude ratio on day 5 was twice as high in the CoQ10 group ($p=0.019$).

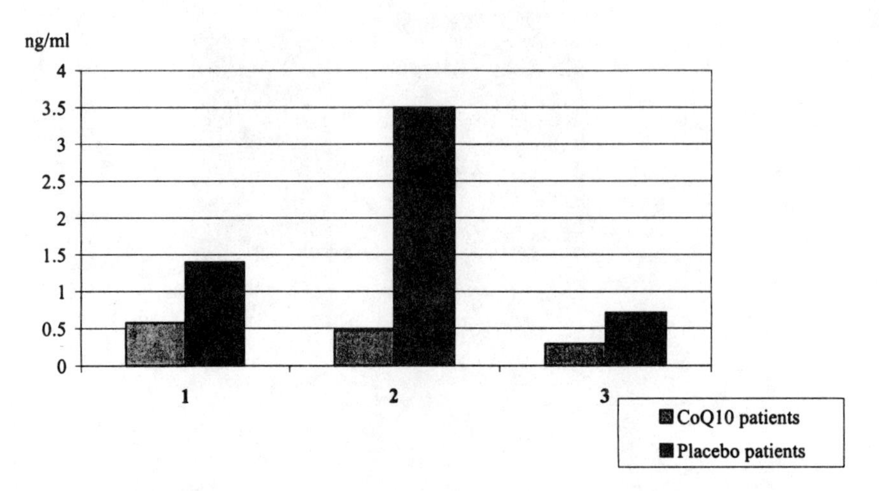

Figure 2. Mean serum S100 levels (1,2,3 = days 0, 1 and 5), showing a marked increase in the placebo group on day 1, but stable levels in the CoQ10 group (normal: < 0.15 ng/ml).

4.3. Discussion

These data show a significant increase in 3-month survival rate when patients were given high-dose CoQ10 after CPR in addition to mild hypothermia. The patient groups were well matched, and there was no difference in recognized outcome predictors. The more elevated CK in the placebo group during the first 24 hours suggest that either they suffered worse initial cardiac damage or the cardioprotective effect of CoQ10 postulated elsewhere[31] may also apply to CPR. S100 protein levels in serum are a robust indicator of neuronal damage, and peak 24 hours after brain injury. The attenuated release of S100 protein and the improved SSEP amplitude ratios in CoQ10 patients signify less severe global cerebral neuronal loss and a possible prolonged potential for improvement over several days. Neurological status largely determines long-term survival: late deaths (between days 5 and 90) occurred almost exclusively in patients with severe neurological deficits. We employed only mild hypothermia between 35° and 36°C, and allowed for an early transition to normothermia by the treating team. This may not have been necessary, and deeper hypothermia may be more effective. Early neuroprotection has consistently been more effective in laboratory studies, and therefore treatment initiated, for instance, at the site of CPR may be superior. Following data in neurodegenerative disease, we employed 450 mg CoQ10 daily and a loading dose of 250 mg. The optimal dose may be higher, given that Schults et al. have demonstrated that 1200 mg/d is more effective than 600 mg/d in Parkinson's disease.[26]

5. CONCLUSION

Today's understanding of complex pathophysiology is beginning to bring effective neuroprotective options to cerebral resuscitation after cardiac arrest. We now recognize that the mechanisms of hypoxic encephalopathy are multifactorial, and treatment needs to target several pathways. CoQ10 administration together with mild hypothermia appears to increase survival after cerebral hypoxia, and may improve neurological outcome in survivors. Neuroprotective aspects will play a major role in future CPR protocols.

6. ACKNOWLEDGEMENTS

The assistance of MSE Pharma, Bad Homburg, Germany, in providing the study medication and administrative support is gratefully acknowledged. Co-workers on the drug study reported were: Diana Ellenberg, Ramona Gildemeister, Jörg Lauermann, Gregor Simonis, Wolfgang Sauter, Christian Georgi; these data were presented in part at the 3rd International Conference of the Coenzyme Q10 Society, London, November 2002.

7. REFERENCES

1. de Vreede-Swagemakers JJ, Gorgels AP, Dubois-Arbouw WI, van Ree JW, Daemen MJ, Houben LG, Wellens HJ. Out-of-hospital cardiac arrest in the 1990's: a population-based study in the Maastricht area on incidence, characteristics and survival. *J Am Coll Cardiol* 1997;**30:**1500-5.
2. Safar P. Cerebral resuscitation after cardiac arrest: Research initiatives and future directions. *Ann Emerg Med* 1993;**22:**324-349.
3. Denton R, Thomas AN. Cardiopulmonary resuscitation: a retrospective review. *Anaesthesia* 1997;**52:**324-327.
4. Xiao F. Bench to Bedside: Brain edema and cerebral resuscitation. *Acad Emerg Med* 2002;**9:**933-946.
5. Jastremski M, Sutton-Tyrrell K, Vaagenes P, Abramson N, Heiselman D, Safar P. Glucocorticoid treatment does not improve neurological recovery following cardiac arrest. Brain Resuscitation Clinical Trial I Study Group. *JAMA* 1989;**262:**3427-3430.
6. Roine RO, Kaste M, Kinnunen A, Nikki P, Sarna S, Kajaste S. Nimodipine after resuscitation from out-of-hospital ventricular fibrillation. A placebo-controlled, double-blind, randomized trial. *JAMA* 1990;**264:** 3171-3117.
7. Brain Resuscitation Clinical Trial II Study Group. A randomized clinical study of a calcium-entry blocker (lidoflazine) in the treatment of comatose survivors of cardiac arrest. *N Engl J Med* 1991;**324:**1225-31.
8. Schroder R (FLUNA Study Group Berlin). Flunarizine i.v. after cardiac arrest (Fluna-study): study design and organisational aspects of a double-blind, placebo-controlled randomized study. *Resuscitation* 1989; **17(Suppl):**S121-7.
9. Monsalve F, Rucabado L, Ruano M, Cunat J, Lacueva V, Vinuales A. The neurologic effects of thiopental therapy after cardiac arrest. *Intensive Care Med* 1987;**13:**244-248.
10. Aldrete JA, Romo-Salas F, Mazzia VD, Tan SL. Phenytoin for brain resuscitation after cardiac arrest: an uncontrolled clinical trial. *Crit Care Med* 1981;**9:**474-477.
11. Maiese K, Vincent A, Lin SH, Shaw T. Group I and group III metabotropic glutamate receptor subtypes provide enhanced neuroprotection. *J Neurosci Res* 2000;**62:**257-272.
12. Maiese K. Hypoxic-ischemic encephalopathy. In: Medlink Neurology 2003 [online]. Available at: www.medlink.com. Accessed October 28th, 2003.
13. Green DR, Reed JC. Mitochondria and apoptosis. *Science* 1998;**281:**1309-1312.
14. Bernard SA, Gray TW, Buist MD, Jones BM, Silvester W, Gutteridge G, Smith K. Treatment of comatose survivors of out-of-hospital cardiac arrest with induced hypothermia. *N Engl J Med* 2002;**346:**557-563.
15. Hypothermia After Cardiac Arrest study group. Mild therapeutic hypothermia to improve the neurologic outcome after cardiac arrest. *N Engl J Med* 2002;**346:**549-56.

16. Vincent AM, Maiese K. Nitric oxide induction of neuronal endonuclease activity in programmed cell death. *Exp Cell Res* 1999;**246:**290-300.
17. Maiese K. The dynamics of cellular injury: transformation into neuronal and vascular protection. *Histol Histopathol* 2001;**16:**633-644.
18. Gandolfo C, Sandercock P, Conti M. Lubeluzole for acute ischaemic stroke. *Cochrane Database Syst Rev* 2002;**1:**CD001924.
19. Dubinsky J. Examination of the role of calcium in neuronal death. *Ann N Y Acad Sci* 1993;**679:**34-42.
20. Matthews RT, Yang L, Browne S, Baik M, Beal MF. Coenzyme Q10 administration increases brain mitochondrial concentrations and exerts neuroprotective effects. *Proc Natl Acad Sci USA* 1998;**95:**8892-8897.
21. Beal MF. Oxidative metabolism. *Ann N Y Acad Sci* 2000;**924:**164-9.
22. Beal MF. Energetics in the pathogenesis of neurodegenerative diseases. *Trends Neurosci* 2000;**23:**298-304.
23. Shibata M, Hattori H, Sasaki T, Gotoh J, Hamada J, Fukuuchi Y. Subcellular localization of a promoter and an inhibitor of apoptosis (Smac/DIABLO and XIAP) during brain ischemia/reperfusion. *Neuroreport* 2002;**13:**1985-1988.
24. Koroshetz WJ, Jenkins BG, Rosen BR, Beal MF. Energy metabolism defects in Huntington's disease and effects of coenzyme Q10. *Ann Neurol* 1997;**41:**160-165.
25. A randomized, placebo-controlled trial of coenzyme Q_{10} and remacemide in Huntington's disease. The Huntington Study Group. *Neurology* 2001;**57:**397-404.
26. Shults CW, Oakes D, Kieburtz K, Beal MF, Haas R, Plumb S, Juncos JL, Nutt J, Shoulson I, Carter J, Kompoliti K, Perlmutter JS, Reich S, Stern M, Watts RL, Kurlan R, Molho E, Harrison M, Lew M. Effects of coenzyme q10 in early Parkinson disease: evidence of slowing of the functional decline. *Arch Neurol* 2002;**59:**1541-1450.
27. Leonov Y, Sterz F, Safar P, Radovsky A, Oku K, Tisherman S, Stezoski SW. Mild cerebral hypothermia during and after cardiac arrest improves neurologic outcome in dogs. *J Cereb Blood Flow Metab* 1990; **10:**57-70.
28. Weinrauch V, Safar P, Tisherman S, Kuboyama K, Radovsky A. Beneficial effect of mild hypothermia and detrimental effect of deep hypothermia after cardiac arrest in dogs. *Stroke* 1992;**23:**1454-1462.
29. Kuboyama K, Safar P, Radovsky A, Tisherman SA, Stezoski SW, Alexander H. Delay in cooling negates the beneficial effect of mild resuscitative cerebral hypothermia after cardiac arrest in dogs: a prospective, randomized study. *Crit Care Med* 1993;**21:**1348-58.
30. Yanagawa Y, Ishihara S, Norio H, Takino M, Kawakami M, Takasu A, Okamoto K, Kaneko N, Terai C, Okada Y. Preliminary clinical outcome study of mild resuscitative hypothermia after out-of-hospital cardiopulmonary arrest. *Resuscitation* 1998;**39:**61-66.
31. Rosenfeldt FL, Pepe S, Linnane A, Nagley P, Rowland M, Ou R, Marasco S, Lyon W, Esmore D. Coenzyme Q10 protects the aging heart against stress: studies in rats, human tissues, and patients. *Ann N Y Acad Sci* 2002;**959:**355-359.

CONTROLLED HYPERTENSION FOR REFRACTORY HIGH INTRACRANIAL PRESSURE

Philippe Hantson*

1. INTRODUCTION

High intracranial pressure is a potential complication of severe traumatic brain injury (TBI). It has been shown from large studies that systemic hypotension could significantly increase the morbidity and mortality rates following TBI.[1] The monitoring of head injured patients in the intensive care unit (ICU) includes at least the determination of intracranial pressure (ICP) when there is clinical and radiological evidence of intracranial hypertension. This allows the calculation of cerebral perfusion (CPP) pressure, which is an important determinant of cerebral blood flow (CBF). CPP is the difference between mean arterial pressure (MAP) and mean ICP. The goal of intensive care is to lower ICP to values below 20 mmHg and to maintain CPP at least at 70 mmHg;[2,3] however, in some patients with severe injury, it could be difficult to reach this objective. The role of controlled hypertension in this setting is controversial and not well documented.[4] Some teams even advocate the use of anti-hypertensive agents in order to reduce interstitial fluid volume and limit brain edema (the Lund concept).[5,6]

Some aspects of induced hypertension therapy will be discussed with reference to four recent cases treated in our ICU.

2. CASE SERIES

Four patients (Pt) who were admitted to the ICU after severe head trauma were analyzed. Their mean age was 20 years. The patients' characteristics are presented in Table 1. They were all intubated and mechanically ventilated. A brain CT scan was obtained immediately after admission. All the patients exhibited evidence of high intracranial pressure and ICP monitoring was started immediately using an intraventricular or a fiberoptic catheter (Camino®).Early surgical intervention was

* Philippe Hantson, MD, PhD, Department of Intensive Care, Cliniques St-Luc, Université catholique de Louvain, Avenue Hippocrate, 10, 1200 Brussels, Belgium. Telephone : 32-2-7642755 Fax : 32-2-7648928 Email: hantson@rean.ucl.ac.be.

Table 1. Characteristics of the 4 patients admitted after severe head injury.

	Patient 1	Patient 2	Patient 3	Patient 4
Age (yr), gender	16, M	26, M	21, M	17, F
Initial GCS	< 8	7	7	3
Brain injuries	Left subdural hematoma, herniation, brain edema	Left extradural hematoma, brain edema	Right subdural hematoma, multiple frontal contusions, brain edema	Right subdural hematoma, left parietal contusions, brain edema
Early interventions	Drainage, craniectomy	Drainage	Drainage, craniectomy	-
Associated lesions	Isthmic aortic rupture	-	-	-
Days with ICP > 30 mmHg	14	15	25	17
Mean CPP (mmHg)	83.7±14.3	92.7±18.9	85.4±13	76.4±15.5
Autoregulation	-	+	+	+
Cumulative norepinephrine doses (mg)	1576	2868	1346	1532
Maximum norepinephrine administration rate (mg/day)	80	429	85	125
Cumulative phenylephrine doses (mg)	850	111	-	-
Maximum phenylephrine administration rate (mg/day)	88	40	-	-
Other vasopressors	No	Dopamine 6,887 mg	No	Dopamine 21,419 mg
Duration of vasopressive therapy (days)	35	17	31	25
Complications of vasopressors	No	Transient cardio-depression	No	No
ICU stay (days)	47	33	35	35

(continued next page)

Table 1. (continued)

Mechanical ventilation (days)	38	28	33	32
Outcome	Excellent	Excellent	Excellent	Excellent

intraventricular or a fiberoptic catheter (Camino®).Early surgical intervention was performed when required. Cerebrospinal fluid (CSF) was drained when possible. Decompressive craniectomy was done soon after admission in Pt 1 (Figure 1) and with delay in Pt 3. The first patient also had a traumatic injury of the ascending aorta, which was treated by an endovascular approach.

All the patients were also investigated by Transcranial Doppler (TCD) and by continuous monitoring of jugular bulb venous oxygen saturation ($SjvO_2$). It was also possible to obtain at the bedside repeated multimodality evoked potentials.

Despite sustained high ICP values, with peak values at 60 to 80 mmHg, the somatosensory and brainstem auditory evoked potentials never deteriorated significantly. This finding encouraged us to continue intensive care in all these patients.

The first objective of the treatment was to try to keep ICP below 20 mmHg and to maintain a CPP higher than 70 mmHg. Serial blood samples were drawn from the jugular vein and radial artery to determine the arteriovenous difference in oxygen. The objective was to maintain SjvO2 at least at 65%. When $SjvO_2$ fell below 50%, the cause of desaturation was systematically sought using a well accepted algorithm in order to

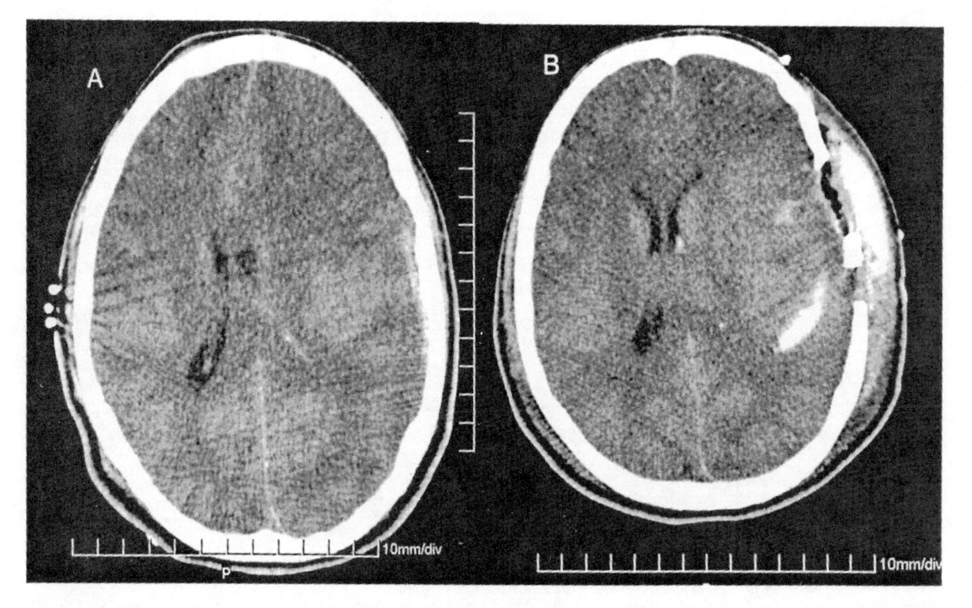

Figure 1. Brain CT scan in Pt 1 on admission (A) and 24 hours later, after decompressive craniectomy (B).

exclude technical problems, hypoxia, anemia, inadequate ventilation... The management of high ICP with low CPP was adjusted according to the arteriovenous difference (radial artery – jugular bulb) in oxygen content ($AVDO_2$). When $AVDO_2$ was low, it was assumed that hyperemia was present; hyperventilation was then applied and barbiturates were given for refractory high ICP. Ischemia was suspected with high $AVDO_2$; mannitol was given intravenously and CPP was increased by fluid therapy and vasopressors.

In some patients, ICP remained very high for several weeks. Their high intracranial pressure was unresponsive to usual treatment (mannitol, barbiturates,...). The therapeutic goal was then to increase CPP to values higher than 90 mmHg until $SjvO_2$ could be stabilized at values greater than 65%. This was achieved by the continuous infusion of vasopressive drugs. Norepinephrine and phenylephrine were the first line agents. The duration of vasopressive therapy ranged in the four patients from 17 to 35 days. The

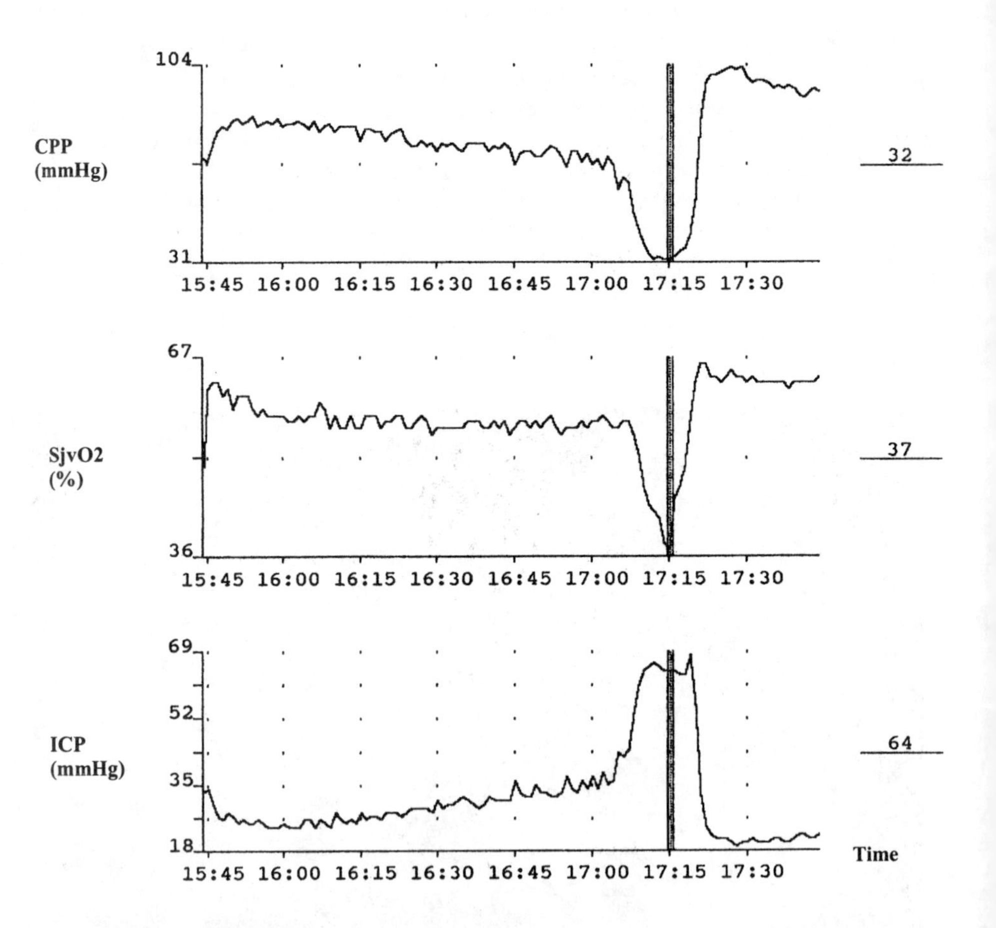

Figure 2. Patient 1 with abolished autoregulation. A progressive decrease in CPP (upper graph) was accompanied by a decrease in SjvO2 (middle graph) and an increase in ICP (lower graph). Numbers appearing in the right part refer to the values observed at the time point corresponding to the black line. The increase of the mean arterial pressure by vasopressors resulted in a marked fall in ICP (black line). When CPP was maintained at a higher level (> 90 mmHg) for the next hours, SjvO2 and ICP remained stable.

mean individual CPP value during treatment ranged from 76.4 to 92.7 mmHg. All the patients received norepinephrine with a cumulative dose during the ICU stay of 1830 ± 698 mg (SD). The maximal infusion rate in one patient was 429 mg/day. Phenylephrine was given in 2 patients, and dopamine in 2 patients also. The patient who received the highest doses of norepinephrine (combined with phenylephrine and dopamine) had a late complication of decreased left ventricle contractility on cardiac echography (ejection fraction 35%). This complication was completely reversible.

TCD examination was also repeated in order to detect vasospasm and to test autoregulation and vasoreactivity to changes in $PaCO_2$. The integrity of cerebral autoregulation was tested by the transient hyperemia test, obtained by the compression of the carotid artery. When autoregulation is intact, there is, after relaxation, a transient increase of mean blood flow velocity due to autoregulatory arteriole vasodilation as CPP

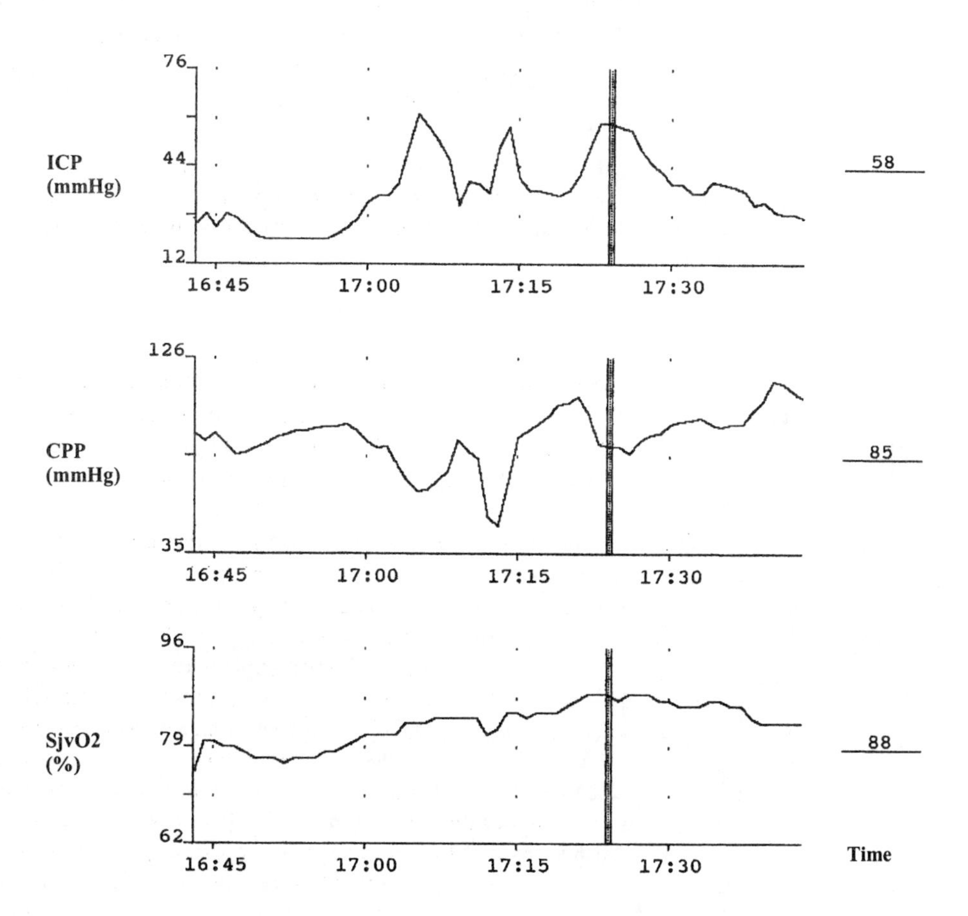

Figure 3. In Patient 2, episodes of high ICP (upper part) were initially associated with high SjvO2 values (lower part). Numbers appearing in the right part refer to the values observed at the time point corresponding to the black line. A carotid-venous fistula was causing increased venous oxygen saturation by the presence of arterial blood.

is decreased. All the patients except Pt 1 had normal autoregulation and vasoreactivity. We observed several patterns of mean blood flow velocity measured in each middle cerebral artery (MCA). Mean blood flow velocity was often asymmetric, with low values corresponding to reduced CBF in areas with high ICP. In Pt 4, TCD revealed on day 18 very high MCA blood flow velocities suggestive of bilateral vasospasm.

The jugular bulb venous oxygen saturation was monitored continuously by a fiberoptic catheter; recalibration was regularly performed. The continuous monitoring of $SjvO_2$ was extremely helpful in the four patients. Multiple episodes of low $SjvO_2$ with high ICP were recorded even with stable values of CPP at 60-70 mmHg before these events. The result of increasing CPP transiently to values sometimes higher than 100 mmHg, with a mean arterial pressure greater than 140 mmHg, was a dramatic decrease in ICP. Interestingly, this was also observed in Pt 1 who had theoretically an abolished autoregulation (Figure 2). Surprisingly, in Pt 2, episodes of high ICP were associated with $SjvO_2$ reaching 88% (Figure 3). As low mean blood flow velocities were noted at TCD examination, hyperemia seemed unlikely and further investigations were ordered. This patient was found to have a carotid-venous fistula explaining the increased venous oxygen content. After that the fistula closed spontaneously, and $SjvO_2$ returned to more common values.

The mean duration of ICU stay was 37 days, with a mean duration of mechanical ventilation of 32.7 days. All the 4 patients were discharged from the ICU with a Glasgow Outcome Scale (GOS) at 5. At 6-month follow, they had returned to usual activities.

3. DISCUSSION

Up to now, the literature concerning the benefit of increasing arterial blood pressure in patients with severe TBI and refractory high ICP has been inconclusive.[3] They are in the literature isolated case reports of patients who made a good recovery despite high ICP for sustained periods.[4] This illustrates that the prognosis is not related to the ICP *per se*, but to CBF and CPP.

The deleterious effect of hypotensive episodes is well documented, with resulting secondary ischemic injuries.[7,8] Global or regional ischemic lesions may appear in areas with extremely low blood flow. It seems also that ischemia may even be observed in late phases of TBI, as suggested by our observations in the second patient. The role of intensive care is to maintain CPP at a minimum of 70 mmHg. This can be achieved by fluid infusion or vasoactive drugs. The central venous pressure should be maintained between 5 and 10 mmHg, or the pulmonary capillary wedge pressure between 10 and 14 mmHg. Three vasopressors, norepinephrine, phenylephrine, and dopamine are employed most commonly in the critical care setting. For most young patients, norepinephrine is an excellent first line agent to increase arterial blood pressure. It acts as a peripheral vasoconstrictor and as an inotropic agent. The norepinephrine infusion generally should not exceed 0.20 µg/kg/min. Some patients require an additional drug to achieve adequate CPP. The recommended dosage for phenylpephrine should not exceed 5 µg/kg/min. It seems prudent to have an upper limit for systolic blood pressure of 180 to 200 mmHg, particularly when hemorrhagic lesions are found at the first CT investigations.

In our observations, the administered doses of norepinephrine, phenylephrine or dopamine largely exceeded the usual recommendations. Systolic blood pressure reached

240 mmHg for short periods of time in some patients. The risk related to the use of high-dose vasopressors is to induce ischemia in different organs. All the patients had cold extremities due to peripheral vasoconstriction but did not develop necrosis. Arterial blood gases were checked extremely frequently (every 4 hours); there were no episodes of metabolic acidosis with increased serum lactate. One patient (Pt 2) developed a transient decrease of cardiac contractility, revealed by echocardiography. There were no electrocardiographic changes. This side effect was likely related to the high doses of vasopressors administered as subendocardial necrosis has been documented in similar settings. This complication was fully reversible after discontinuation of norepinephrine and after a brief infusion of dobutamine. Finally, no complications related to high systolic pressure were observed. There was no episode of intraparenchymal bleeding and no evidence of worsening of brain edema.

In contrast to the strategy of controlled hypertension to maintain adequate CPP, authors from Lund University Hospital proposed a totally different approach known as the "Lund concept".[5,6] The two objectives of the Lund therapy are to prevent a significant rise in ICP and to improve a compromised microcirculation in areas surrounding contusions. This could be achieved by the preservation of a normal colloid osmotic pressure and by the reduction of the hydrostatic capillary pressure. Colloid osmotic pressure is maintained by albumin infusions. The goal is to maintain CPP around 60-70 mmHg, but transient episodes with CPP lower than 50 mmHg are tolerated. To reduce capillary hydrostatic pressure and sympathetic discharge, the patients are given a beta1-antagonist, an alpha2-agonist and a potential precapillary vasoconstrictor (dihydroergotamine). Infusion of low-dose prostacyclin is also proposed to improve microcirculation. The results obtained with this protocol are still a matter of debate.[9,10] Indeed, the results are usually compared to historical series and it is obvious that the general quality of care has improved over the last decade. In the literature, the group of patients treated by this strategy is heterogeneous. Craniectomy was performed in some patients, while others received thiopental or mannitol.

Our observations are not in fundamental disagreement with the Lund strategy, because the changes in CBF are not monophasic.[11-13] Successive phases of hyperemia and ischemia may be observed in a single patient. For exemple, in Pt 4, the initial TCD monitoring suggested hyperemia and elevated ICP was treated by optimizing hyperventilation. Later, when there was evidence of vasospasm, mean arterial pressure was increased in order to maximize CPP. It seems therefore essential that in any individual with TBI, CPP should continuously be adjusted according to the results of a comprehensive monitoring, including metabolic measurements. One component of this is the continuous monitoring of $SjvO_2$[14] There are obviously several limitations to $SjvO_2$ monitoring, the most important being anatomical restrictions. $SjvO_2$ reflects global cerebral oxygenation, not regional ischemia. There are several potential causes of jugular desaturation in patients with TBI: hypoxia, hypocarbia, anemia.... Nevertheless, $SjvO_2$ monitoring allows physicians to detect episodes of cerebral hypoxia which were not identifiable previously. It appears that there is a strong association between these episodes of desaturation and a poor outcome.[15] We observed in our 4 patients that the destabilization of intracranial pressure was triggered most of the time by a relative hypotension, even when CPP had been maintained well above 70 mm Hg for several hours. The only way to treat these episodes was to increase MAP. The consequence was a decrease in ICP and an increase in $SjvO_2$. It seemed surprising to observe this reaction also in the patient with abolished pressure autoregulation. However, it is likely that pressure autoregulation may be preserved in some regions of the brain, so that increasing CPP increases cerebral vascular resistance and reduces cerebral blood volume and, consequently, ICP. An increase in CBF may also improve perfusion of the ischemic areas, resulting in a lower ICP. $SjvO_2$ monitoring is also helpful, in combination with TCD monitoring, to detect cerebral vasospasm.[16] In this setting, $SjvO_2$ is usually low while TCD reveals high blood flow, indicating that ICP is not increased sufficiently to impair CPP. Here also, the first step is to increase MAP.

4. CONCLUSIONS

The management of severe TBI remains complex. Most patients require a CPP of 60 to 70 mmHg to maintain adequate CBF, but some will require a higher CPP sometimes for a long period of time. In some patients, considerable doses of catecholamines may be necessary to reach this goal. Experience with continuous $SjvO_2$ monitoring in head trauma patients suggests that ischemic episodes or vasospasm may occur at any time. The consequence is usually an increase in ICP that has to be treated by increasing MAP in order to optimize CPP at values higher than 70 mmHg. Rather than to aim for a fixed value of 60 to 70 mmHg for CPP, it seems preferable to adapt our treatment to the results of hemodynamic and metabolic monitoring.

5. REFERENCES

1. Chesnut RM. Avoidance of hypotension: conditio sine qua non of successful severe head-injury management. *J Trauma* 1997;**42:** S4-S9.
2. Changaris DG, McGraw CP, Richardson JD, Garreston HD, Arpin EJ, Shields CB. Correlation of cerebral perfusion pressure and Glasgow Coma Scale to outcome. *J Trauma* 1987;**27:**1007-1013.
3. Rosner MJ, Rosner SD, Johnson AH. Cerebral perfusion pressure: management protocol and clinical results. *J Neurosurg* 1995;**83:** 949-962.
4. Ferring M, Berré J, Vincent JL. Induced hypertension after head injury. *Intensive Care Med* 1999;**25:**1006-1009.
5. Grande PO, Asgeirsson B, Nordstrom CH. Physiologic principles for volume regulation of a tissue enclosed in a rigid shell with application for the injured brain. *J Trauma* 1997;**42:**S23-S31.
6. Eker C, Asgeirsson B, Grande PO, Schalen W, Nordstrom CH. Improved outcome after severe head injury treated with a new therapy based on principles for brain volume regulation and preserved microcirculation. *Crit Care Med* 1998;**26:**1181-1886.
7. Graham DI, Adams JH, Doyle D. Ischaemic brain damage in fatal non-missile head injuries. *J Neurol Sci* 1978;**39:**213-234.
8. Obrist WD, Langfitt TW, Jaggi JL, Cruz J, Gennarelli TA. Cerebral blood flow and metabolism in comatose patients with acute head injury. *J Neurosurg* 1984;**61:**241-253.
9. Gisvold SE. The Lund concept for treatment of head injuries – faith or science? *Acta Anaesthesiol Scand* 2001;**45:** 399-401.
10. Schneck MJ. Treating elevated intracranial pressure. Do we raise or lower the blood pressure? *Crit Care Med* 1998;**26:**1787-1788.
11. Martin NA, Patwardhan RV, Alexander MJ, Africk CZ, Lee JH, Shalmon E, Hovda DA, Becker DP. Characterization of cerebral hemodynamic phases following severe head trauma: hypoperfusion, hyperemia, and vasospasm. *J Neurosurg* 1997;**87:**9-19.

12. Lee JH, Martin NA, Alsina G, McArthur DL, Zaucha K, Hovda DA, Becker DP. Hemodynamically significant cerebral vasospasm and outcome after head injury: a prospective study. *J Neurosurg* 1997;**87:** 221-233.
13. Chan KH, Miller JD, Dearden NM, Andrews PJ, Midgley S. The effect of changes in cerebral perfusion pressure upon middle cerebral artery blood flow velocity and jugular bulb venous oxygen saturation after severe brain injury. *J Neurosurg* 1992;**77:**55-61.
14. Schoon P, Benito Mori L, Orlandi G, Larralde C, Radrizzani M. Incidence of intracranial hypertension related to jugular bulb oxygen saturation disturbances in severe traumatic brain injury patients. *Acta Neurochir Suppl* 2002;**81:**285-287.
15. Cruz J. On-line monitoring of global cerebral hypoxia in acute brain injury. Relationship to intracranial hypertension. *J Neurosurg* 1993;**79:** 228-233.
16. Chan KH, Dearden NM, Miller JD, Andrews PJ, Midgley S. Multimodality monitoring as a guide to treatment of intracranial hypertension after severe brain injury. *Neurosurgery* 1993;**32:**547-552.

ON IRREVERSIBILITY AS A PREREQUISITE FOR BRAIN DEATH DETERMINATION

James L. Bernat*

1. INTRODUCTION

Brain death is the colloquial term describing the determination of human death based upon showing the irreversible cessation of all clinical brain functions. Although the term *brain death* is potentially misleading because it suggests that there are different types of death, or that it is only the brain that is dead in such cases, it is the term that remains universally used and I will use it here.

Irreversibility has been a prerequisite for brain death determination in every set of brain death tests since the concept was first refined in the 1960s. For example, in Appendix F of the 1981 President's Commission report *Defining Death*, the medical consultants required that the cause of the brain damage be structural and known and that reversible metabolic and toxic factors be excluded.[1] And in my previously published analyses of the preconditions for brain death, I argued that irreversibility was essential.[2,3]

But making the claim of irreversibility and proving it are two different matters. Many authors have asserted that the diffuse neuronal damage in brain death can be proved to be irreversible by conducting sequential neurological examinations and demonstrating the persistent absence of all clinical brain functions. Despite the plausibility of this claim, it is not self-evidently true and its empirical basis is limited to a few studies.[4]

Moreover, since the earliest determinations of brain death, we have known that some examiners have been careless in performing, interpreting, or recording the clinical tests, particularly the apnea test.[5,6] Despite the presence of standardized, widely accepted, and highly publicized guidelines for brain death determination,[7,8] physicians perform it incorrectly in many settings. Even a recent study from a prestigious medical center demonstrated the inadequacies of the methods and recordings of routine brain death determinations.[9] It is probable that some of the alleged cases of "chronic brain death" described by Alan Shewmon,[10] in which the heartbeat and systemic circulation of purportedly brain dead patients were technologically maintained for many months or longer, represent cases that were improperly declared brain dead.

* James L. Bernat, M.D., Neurology Section, Dartmouth-Hitchcock Medical Center, Lebanon, NH 03756, USA.

In this paper, I propose that the prerequisite of irreversibility must not simply be assumed by repeated clinical examinations but must be proved before the determination of brain death can be made. The only reliable proof of irreversibility is demonstrating the complete absence of intracranial circulation. I therefore propose that a test demonstrating absent intracranial circulation should be required for brain death determination.

2. THE NATURE OF IRREVERSIBILITY

Here I use the common definition of the term "irreversible": that which is incapable of being undone. But this seemingly simple concept is subtle because it conveys at least two distinct definitions. The philosopher David Cole pointed out that "irreversible" could mean that which is incapable of being undone with any foreseeable future technology but also could mean that which is incapable of being undone by currently available technologies.[11] I follow Cole's second definition and use "irreversible" in this context to refer to any loss of function that cannot be restored using contemporary technologies. If future technologies were to permit reversal of these currently irreversible losses of functions, we would be forced to alter our concepts of life and death and to seek other irreversible functions to measure.

Physicians Joanne Lynn and Ronald Cranford applied these irreversibility concepts to the determination of brain death[12] and, using the analysis of L. P. Ivan,[13] identified four plausible times of death. T1 is the time of onset of coma and apnea; T2 is the time at which the neurological physical examination shows that the relevant brain functions have ceased; T3 is the time at which the cessation of brain functions becomes irreversible; and T4 is the time at which the cessation of brain functions can be proved to be irreversible. Although any one of these times constitutes a plausible time of death, I advocate for using time T4. As Lynn and Cranford point out, most or all of the brain damage leading to brain death occurs at time T1. Time T2 is when a physician first determines the loss of brain functions. Time T3 often may be inapparent at the time but can be determined in retrospect when the functions can be proved to be lost permanently at time T4.[12]

Despite the fact that the "moment" of brain death arguably occurs at time T3, a brain death determination must employ the time T4. Brain death is determined in retrospect in the same manner as cardiopulmonary death: by showing the irreversible cessation of the relevant vital functions. In declaring death, physicians note the time of their examination that they have certified that all vital functions have ceased, and formally declare death to have occurred at that time, despite the obvious fact that the vital functions had ceased earlier. Only in forensic cases is much attention paid to identifying the exact earlier moment when those vital functions ceased irreversibly.

3. ABSENT INTRACRANIAL BLOOD FLOW PROVES IRREVERSIBILITY

In a clinical determination of brain death, the standard for showing irreversibility requires satisfying three criteria: (1) the cause of the clinical brain dysfunction is known and structural; (2) contributions from potentially reversible metabolic and toxic causes have been excluded; and (3) the absence of clinical brain functions is documented in at least two sequential examinations separated by a time interval that varies as a function of age, etiology, and the use off confirmatory testing. If the absence of brain functions is from a structural cause with no reversible component and if the signs of absent brain functions persist between sequential examinations, they are presumed to be permanent.[1]

Although claims of permanence and totality are plausible if these criteria are satisfied, one can imagine cases in which they might be erroneous. Consider a patient with severe diffuse brain damage but who is not quite brain is dead. The superimposition of a seemingly small metabolic or toxic encephalopathy may produce a state that appears to be brain death but is not. Although some might argue that this problem is inconsequential because such a brain-damaged patient was "as good as dead." But that designation is insufficient for a determination of human death. Clinical certainty must be the standard.

The most confident way to demonstrate that the global loss of clinical brain functions is irreversible is to show the complete absence of intracranial blood flow. Brain neurons are damaged after just a few minutes of lack of blood flow and are globally destroyed when blood flow completely ceases for more than 20-30 minutes.[14] Thus, showing a total absence of intracranial blood flow that has persisted for more than 30 minutes proves the irreversibility as well as the totality of the loss of clinical brain functions.

Of course, the test proving the cessation of intracranial blood flow must be performed and reported accurately for it to be valid. Fortunately, a number of tests are available that can accurately and validly measure intracranial blood flow.

3.1. Intracranial Hemodynamics

Blood flow to the brain is tightly regulated by the homeostatic system of cerebral autoregulation that operates over a wide range of systemic blood pressures to assure adequate cerebral perfusion pressures. Normal cerebral autoregulatory mechanisms can be disturbed when systemic blood pressures become excessively high or low, or when intracranial pressure rises to very high levels.

Traumatic and vascular global brain lesions leading to brain death produce diffuse cerebral edema. The cerebral edema results in an increase in intracranial contents but intracranial volume remains fixed by the rigid skull. Consequently, intracranial pressure (ICP) rises. In most brain death cases, intracranial pressure rises until it exceeds mean arterial blood pressure. In many instances of massive head trauma and massive subarachnoid hemorrhage, ICP exceeds systolic blood pressure. When ICP exceeds systolic blood pressure, no blood can enter the cranial vault and the brain loses all circulation. When ICP is lower than systolic blood pressure but higher than diastolic pressure but exceeds mean arterial pressure, blood enters the cranium and brain during systole but is pushed back an equal amount during diastole. This phenomenon of so-called "reverberating flow" cannot result in perfusion of the brain because it produces no net forward circulation. Thus, whether there is no intracranial systolic blood flow or there is no net blood flow because of reverberating flow, the brain becomes diffusely and irreversibly destroyed within minutes.

The clinical examination evidence of absent intracranial blood flow is the presence of one of the syndromes of cerebral transtentorial herniation, as shown by Fred Plum and Jerome Posner (1980).[15] Central and uncal transtentorial herniation of the midbrain, results from intracranial tissue shifts caused by the development of lateralized intracranial pressure cones from an expanding mass lesion. The lateralized pressure cones induce a caudal shift of brain tissue that secondarily destroys brain stem neurons through a progressive pressure gradient-induced ischemia. All neurology residents are taught to seek the clinical evidence of these herniation syndromes because once the brain stem has been infarcted during transtentorial herniation, the loss of brain clinical functions has become irreversible.

An important added significance of requiring herniation syndromes resulting from raised ICP is that it provides proof that the destruction of brain neurons is widespread. The whole-brain criterion of death requires that all clinical functions of the brain cease irreversibly.[2] Once herniation has completed and all intracranial blood flow has stopped, examiners declaring brain death can be confident that neuronal damage is widespread and that the herniation has eliminated all clinical functions of the brain. Thus, requiring a demonstration of absence of intracranial blood flow at once confirms both irreversibility and totality of the cessation of neuronal function.

The natural history of cerebral circulatory arrest is known. If cardiopulmonary support is continued successfully despite the presence of brain death, as cerebral edema begins to subside, ICP begins to fall within hours to days. Once ICP has fallen to a level lower than mean arterial pressure, intracranial circulation restarts ("reflow") at least to a limited degree in the necrotic brain.[16] The so-called "respirator brain," described by Earl Walker and colleagues in the 1970s, is a result of neuronal and glial liquefactive necrosis in the setting of intracranial reflow once ICP has dropped.[17]

3.2. Benefits of Measuring Intracranial Blood Flow

Brain dead patients may have consumed or been treated with drugs that depress the central nervous system or that induce neuromuscular blockade. Brain death cannot be determined on clinical grounds in the presence of these agents because the drugs can induce a state resembling brain death yet one that is potentially reversible. Brain death determination in the presence of induced barbiturate coma or neuromuscular paralyisis requires the use of a confirmatory laboratory test. But because the EEG can be equally depressed in metabolic coma, most authorities recommend confirming brain death with a test showing the absence of intracranial blood flow.[7] Once intracranial blood flow can be proved to be absent, it is inconsequential if the patient also has intoxicating levels of barbiturates or neuromuscular blocking drugs that otherwise would confound the clinical determination of brain death.

Most importantly, showing the absence of intracranial blood flow proves that the loss of clinical brain functions is total and permanent. The regular use of this technique likely will decrease the number of erroneous brain death determinations and probably will reduce the number of reported instances of the syndrome of so-called "chronic brain death," many of whom were likely never brain dead in the first place.

It is essential that a total absence of intracranial blood flow not be confused with only a reduction of intracranial blood flow. The studies of Cicero Coimbra[18] on the ischemic penumbra surrounding lesions in purportedly brain dead patients demonstrated the importance of distinguishing between these two situations and emphasized the serious errors that can occur if they are confused.

4. TESTS OF INTRACRANIAL BLOOD FLOW IN BRAIN DEATH

Tests showing absent intracranial circulation are nearly as old as the subject of brain death. Contrast arteriography was first used in the 1970s to show absence of intracranial circulation distal to the intracranial portions of the internal carotid and vertebral arteries. It continues to be used by physicians in some settings that lack access to simpler alternative techniques. Its principal drawback is its invasiveness and that the patient must be transported to the radiology suite.[19]

Intravenous radionuclide angiography, perfected in the 1980s, was the next technique to be used to prove absent intracranial circulation. An intravenous infusion of the radioisotope pertechnetate is infused intravenously. The patient undergoes static and dynamic radionuclide brain scanning to measure entry of the radioisotope into the brain. Dynamic images show the istope stopping as the internal carotid and vertebral arteries enter the dura mater. Static images show only the presence of isotope in the scalp and face because of the patency of the external carotid artery and its branches.[20] A radiologist or nuclear medicine expert who is experienced in this technique can confidently interpret absence of blood flow in the brain. The patient must be transported to the nuclear medicine suite.

Transcranial Doppler ultrasound (TCD) was perfected in the 1990s and now is the test used by most medical centers now to document cessation of intracranial circulation in brain death.[21] TCD can be performed in the patient's bed in the ICU. Currents standards require three separate insonation sites.[22] Reproducible images of intracranial pulses usually can be obtained if they are present. Two principal patterns of TCD abnormalities have been documented in brain death: absent systolic spikes and reverberating flow. Systolic spikes are absent when ICP exceeds systolic blood pressure because no measurable systolic flow can be conducted to the intracranial arteries. When ICP exceeds mean arterial blood pressure but is lower than systolic blood pressure, reverberating flow is seen. Blood advances during systole but is pushed back an equal amount during diastole. Both patterns confirm the complete absence of intracranial blood flow.[21]

More recently, additional imaging techniques have been applied to this problem. There are several studies using single photon emission computed tomography (SPECT) scintigraphy with the radioisotope Tc-99 HMPAO that validate the complete absence of intracranial blood flow in brain death by this relatively simple technique.[23,24] Several case reports have been published demonstrating absent intracranial blood flow by magnetic resonance angiography (MRA), magnetic resonance (MRI) diffusion-weighted imaging, and computed tomography angiography (CTA).[25,26] An earlier technique, xenon-enhanced computed tomography,[27] has been replaced by the newer, simpler techniques of scintigraphy, MRI, MRA, and CTA

5. REFINEMENTS IN BRAIN DEATH DETERMINATION

I believe that a test showing the complete absence of intracranial circulation should become mandatory in the determination of brain death. Only such a test can plausibly prove the totality of neuronal damage and its complete irreversibility necessary for the tests of brain death to satisfy the whole-brain criterion of death.[28] Now that a variety of intracranial blood flow tests have been developed, the great majority of hospitals in which brain death might be determined should have access to at least one test.

I anticipate a few problems resulting from implementing this idea. If the intracranial blood flow test is not performed in a timely manner, namely during the period of raised intracranial pressure, it may be negative (i.e., reveal intracranial blood flow) despite actual brain death, because ICP later has fallen to more normal levels. This situation could create a false-negative determination in which the patient is truly brain dead but the test for death is negative.

Implementing a requirement for a blood flow test would exclude some patients currently determined dead under the United Kingdom program of "brain stem death." The brain stem death doctrine requires only that brain stem functions are irreversibly lost and does not require that cerebral hemispheric functions also must be lost.[29] Although most patients declared dead under a brain stem criterion also would be declared dead under the "whole brain" criterion prevailing in the United States and elsewhere in the world, one type of patient would not.[30] The patient with a primary brain stem catastrophe such as a massive brain stem hemorrhage, who did not develop raised intracranial pressure and transtentorial herniation, might be declared dead by the U.K. doctrine but not by the U.S.A. doctrine. Because of the absence of a markedly raised ICP in such a case, the patient would likely not have an absence of intracranial blood flow.[31,32]

It is likely that a few patients currently declared brain dead would not qualify under the new regulation. For those who truly were not brain dead, implementation of the intracranial blood flow test would be beneficial because it would increase the accuracy of brain death diagnosis. The price of obtaining slightly fewer organs for donation would be worth the benefit of greater accuracy and confidence in the diagnosis of brain death.

6. REFERENCES

1. Appendix F, in: President's Commission for the Study of Ethical Problems in Medicine and Biomedical and Behavioral Research. *Defining Death. Medical, Ethical, and Legal Issues in the Determination of Death.* Washington, DC: US Government Printing Office, 1981.
2. Bernat JL. A defense of the whole-brain definition of death. *Hastings Cent Rep* 1998;**28(2):**14-23.
3. Bernat JL. The biophilosophical basis of whole-brain death. *Soc Philos Policy* 2002;19(2):324-342.
4. Collaborative Study of Cerebral Survival, An appraisal of the criteria of cerebral death. *JAMA* 1977;**237:** 982-986.
5. Earnest MP, Beresford HR, McIntyre HB. Testing for apnea in brain death: methods used by 129 clinicians. *Neurology1986;* **36:**542-544.
6. Mejia RE, Pollack MM. Variability in brain death determination practices in children. *JAMA* 1995; **274:**550-553.
7. Wijdicks EFM. Determining brain death in adults. *Neurology* 1995;**45:**1003-1011.
8. The Quality Standards Subcommittee of the American Academy of Neurology, Practice parameters for determining brain death in adults [summary statement]. *Neurology* 1995;**45:**1012-1014.
9. Wang MY, Wallace P, Gruen JB. Brain death documentation: analyisis and issues. *Neurosurgery* 1998; **51:**731-735.
10. Shewmon DA. Chronic "brain death:" meta-analysis and conceptual consequences. *Neurology 1998;***51:** 1538-1545.
11. Cole DJ. The reversibility of death, *J Med Ethics* 1992;**18:** 26-30.
12. Lynn L Cranford R. The persisting perplexities in the determination of death, In: Youngner SJ, Arnold RM, Schapiro R, eds. *The Definition of Death: Contemporary Controversies.* Baltimore: Johns Hopkins University Press ,1999.
13. Ivan LP. Time sequence in brain death, In: Morley TP, ed. *Moral, Ethical, and Legal Issues in the Neurosciences.* Springfield, IL: Charles C. Thomas, 1981.
14. Miyamoto O, Auer RN. Hypoxia, hyperoxia, ischemia, and brain necrosis. *Neurology* 2000;**54:** 362-371.
15. Plum F, Posner JB. *The Diagnosis of Stupor and Coma.* 3rd ed. Philadelphia: F. A. Davis Co., 1980: 87-151.
16. Schroder R. Later changes in brain death: signs of partial recirculation. *Acta Neuropathol (Berl)* 1983;**62:** 15-23.
17. Walker AE, Diamond EL, Moseley J. The neuropathological findings in irreversible coma: a critique of the "respirator brain." *J Neuropathol Exp Neurol* 1975;**34:** 295-323.

18. Coimbra CG. Implications of ischemic penumbra for the diagnosis of brain death. *Brazil J Med Biol Res* 1999;**32:**1479-1487.
19. Bradac GB, Simon. RS. Angiography in brain death. *Neuroradiology* 1974;**7:**25-28.
20. Goodman JM, Heck LL, Moore B. Confirmation of brain death with portable isotope angiography: a review of 204 consecutive cases. *Neurosurgery* 1985;**16:**492-497.
21. Petty GW, Mohr JP, Pedley TA, et al. The role of transcranial Doppler in confirming brain death: sensitivity, specificity, and suggestions for performance and interpretation. *Neurology* 1990;**40:**300-303.
22. Ducrocq X, Braun M, Debouverie M, Junges C, Hummer M, Vespignani H. Brain death and transcranial Doppler: experience in 130 cases of brain dead patients. *J Neurol Sci* 1998;**160:**41-46.
23. Wilson K, Gordon L, Selby Jr. JB. The diagnosis of brain death with Tc-99m HMPAO. *Clin Nucl Med* 1993;**18:** 428-434.
24. Donohoe KJ, Frey KA, Gerbaudo VH, Mariani G, Nagel JS, Shulkin B. Procedural guidelines for brain death scintigraphy. *J Nucl Med* 2003;**44:**846-851.
25. Lovblad KO, Bassetti C. 'Diffusion-weighted magnetic resonance imaging in brain death. *Stroke* 2000;**31:**539-542.
26. Karantanas AH, Hadijigeorgion GM, Paterakis K, Sfiras D, Komnos A. Contributions of MRI and MR angiography in early diagnosis of brain death. *Eur Radiology* 2002;**12:** 2710-2716.
27. Darby JM,Yonas H, Gur D, Latchaw RE. Xenon-enhanced computed tomography in brain death. *Arch Neurol 1987;***44:** 551-554.
28. Bernat JL, Culver CM, Gert B. On the definition and criterion of death. *Ann Intern Med* 1981;**94:** 389-394.
29. Pallis C. *ABC of Brainstem Death.* 2nd ed. London: British Medical Journal Publishers, 1997.
30. Bernat JL. How much of the brain must die in brain death? *J Clin Ethics* 1992;**3**:21-26.
31. Kosteljanetz M, Ohrstrom JK, Skjodt S, Teglbjaerg PS. Clinical brain death with preserved cerebral arterial circulation. *Acta Neurol Scand* 1988;**78:**418-421.
32. Ferbert A, Buchner H, Ringelstein EB, et al, Isolated brainstem death. *Electroencephalogr Clin Neurophysiol* 1986;**65:**157-160.

HOW SHOULD TESTING FOR APNEA BE PERFORMED IN DIAGNOSING BRAIN DEATH?

Christoph J. G. Lang and Josef G. Heckmann*

1. INTRODUCTION

Apnea testing is a *conditio sine qua non* in determining brain or brain-stem death world-wide. It is an important sign of loss of brain stem functions and signifies that breath has vanished from man. It is, however, the most time-consuming, difficult and potentially harmful of all clinical assessments. It may induce not only hypotension but even cardiac arrhythmia or asystole.[1,2] Disagreement prevails as to which parameters to apply and how to proceed for best results. This paper reviews the preconditions and procedures for apnea testing and addresses some special problems and pitfalls to be overcome in order to obtain valid results.

2. METHOD

A literature review was conducted on apnea testing for diagnosing brain death world-wide, using computer-based scientific bibliographical data bases and monographs. After collecting data on requirements for apnea testing, the papers were scrutinized for special problems such as hypotension, excessive hypercarbia, hypoxia, and acidosis. Particular emphasis was laid on how to deal with patients who could not be well oxygenated or had pulmonary problems. The proposals and results were evaluated against the background of our own experience of more than 2000 apnea tests.

* Christoph J. G. Lang, M.D., Dipl-Psych, Josef G. Heckmann, M.D., Neurological Hospital with Outpatient Department, University of Erlangen-Nuremberg at Erlangen, Schwabachanlage 6, 91054 Erlangen, Germany.

Brain Death and Disorders of Consciousness, Edited by Machado and Shewmon
Kluwer Academic/Plenum Publishers, New York 2004

3. RESULTS

3.1. Preconditions

3.1.1. Partial Tension of Carbon Dioxide

The requirements for the final partial arterial tension of carbon dioxide ($PaCO_2$) differ according to national guidelines. Usually suggestions are made for obtaining a certain $PaCO_2$ (50 mmHg = 6.7 kPa [e.g. in the UK and Portugal][3,4] or 60 mmHg = 8.0 kPa [e.g. in the USA or Germany][5,6]).

3.1.2. Partial Tension of Oxygen

There are no recommendations for the partial arterial tension of oxygen (PaO_2) except preoxygenation with 100% O_2 for some time, mostly 10 minutes, and that hypoxia should be avoided. For purposes of lung transplantation a maximum PaO_2 of >500 mmHg is deemed desirable by transplant specialists. In practice, any PaO_2 level that is not below normal limits is acceptable.

3.1.3. Hydrogen Ion Concentration

There are no specific recommendations for the pH. Its drop is highly correlated inversely with the $PaCO_2$, and it is quickly restored with normoventilation or mild hyperventilation. It should not be corrected by metabolic means during the test. A normal pH or a value in the low basic range at the onset is useful.

3.1.4. Temperature

Testing with a body temperature below 32°C is discouraged by most authors. In these cases the body must be warmed. In other cases correction of blood gas values for actual body temperature is needed. Some authors recommend warming up to at least 36.5°C in every case[7] which may be a very time-consuming measure.

3.1.5. Duration of Apnea Testing

Fixed durations are nowadays not considered standard anymore. Some authors have recommended 10 min, even if blood gas levels cannot be determined. We would discourage such a procedure and insist on arterial blood gas determinations. In our experience the apnea test may last between one minute and more than one hour.

3.1.6. Duration of Observation of Apnea

The patient should be observed during the whole procedure for any respiratory movements and a sufficient time, about half a minute or so, after the recommended gas levels have been reached. Patients who fail the apnea test have been reported to start breathing at $PaCO_2$ values as low as 35 mmHg or as high as above 50 mmHg.

3.1.7. Proposed Increase over Baseline

Since a rapid increase in $PaCO_2$ to 20 mmHg above normal baseline is considered a maximal stimulus to the respiratory centers, such an increase is recommended when the baseline $PaCO_2$ is at or above 36 to 40 mmHg,[5,8,9] because this means that a $PaCO_2$ of about 60 mmHg will be reached. We recommend to determine the final blood gas in every case and not to rely on an anticipated increase in $PaCO_2$. The rate of increase may vary considerably but is usually biphasic being steeper during the first few minutes than thereafter.[10]

3.2. Problems and Pitfalls

3.2.1. Excessive Hypercarbia

CO_2 values over 120 mmHg cause narcosis and should be avoided, although general recommendations do not exist. There are no generally accepted $PaCO_2$ values for patients whose natural respiration is adapted to a $PaCO_2$ of more than 45 mmHg. In these cases confirmatory tests for loss of brainstem functions are recommended by the German guidelines.[6] If the respiratory center is being driven by hypoxia in patients known to have adapted to high $PaCO_2$ values, apnea testing may be used that includes cautious lowering of PaO_2. There are, however, no accepted criteria for this type of test.

3.2.2. Hypoxia

O_2 values less than 70 or 60 mmHg should be avoided. If pulse oxymetry is used, values should not drop below 80%.

3.2.3. Respiratory Acidosis

pH values less than 7.2 or 7.0 should be avoided.

3.2.4. Hypotension

Blood pressure values less than 70 or 60 mmHg systolic, 40 mmHg diastolic, and mean arterial pressure of 50 mmHg should be avoided. The presence of a blood pressure of 90 mmHg before testing is acceptable as a rule. Usually there is a mild increase of blood pressure with hyperoxygenation and a somewhat more marked decrease with hypercapnia.[1] Persistent hypotension may be corrected using intravenous fluids, 5% albumin, or an increase of intravenous dopamine or (nor)epinephrine.

3.2.5. Cardiac Arrhythmia and Arrest

Cardiac arrhythmias induced by apnea testing are rare and a cardiac arrest should be avoided at any rate. They are mostly due to excessive acidosis or to hypoxia and are mostly heralded by the onset of new cardiac arrhythmias or a marked drop in heart rate.

3.2.6. *Increase of Intracranial Pressure*

Intracranial (IC) pressure is usually not monitored during apnea testing except in some neurosurgical intensive care units. Since theoretically apnea testing may increase IC pressure via local hypercapnic vasodilation and ensuing increase of cerebral blood volume it should be the last of all clinical tests. For this reason the Japanese guidelines put the apnea test at the end of all tests, after a flat EEG has been demonstrated.

3.2.7. *False Readings*

We have repeatedly noted that sensible sensors in ventilators may be triggered by heartbeat-driven thorax excursions or extraneous movements. Wijdicks[7] also mentions this possibility when he remarks that with continuous positive airway pressure (CPAP) settings as low as 2, false readings ("spontaneous" respiratory rates of 20 to 30 per minute) may occur.

3.2.8. *Children vs. Adults*

Since it is assumed that the threshold for maximal stimulation is the same as with adults, children should be tested similarly. There is a single case report of a 3-year-old child with chronic severe neurological dysfunction who took a single breath after 8 minutes of apneic oxygenation at a $PaCO_2$ of 112 mmHg but ultimately died.[11] Another report is that of a 3-month-old infant who, after fulfilling the 1987 Task Force Criteria of pediatric brain death, developed two or three irregular breaths days later but also finally died.[12] In a 4-year-old child minimal respiratory effort began at 91 mmHg, raising the question whether children of a certain age have higher $PaCO_2$ thresholds, so that higher $PaCO_2$ levels should be required.

3.3. Special Situations

3.3.1. *Monitoring*

End-tidal capnometry, pulse oxymetry, transcutaneous blood determination, and intra-arterial blood gas analysis may all be used.[13,14] In vitro measurements, i.e. regular blood gas determinations, remain, however, the gold standard. Some authors recommend determining a basal value and then performing subsequent checks approximately every 5 minutes. Having obtained two or three values, the increase in $PaCO_2$ may be extrapolated rather precisely. Close monitoring – depending on the anticipated increase in CO_2 – is especially useful and necessary with artificial $PaCO_2$ augmentation. Final values should always be corrected for body temperature.

3.3.2. *Artificial CO_2 Insufflation*

When an especially long duration of observation is anticipated or if an increase in $PaCO_2$ is not easily achieved or if PaO_2 drops excessively with hypoventilation this method may be used.[15]

3.3.3. Replacement of Apnea Testing by Other Means

In cases where correct apnea testing is not possible, e.g., because of the impossibility of reaching the required $PaCO_2$ values or because a dangerous drop in PaO_2 is unavoidable, as may be the case in severe thoracic trauma or other pulmonary problems, instrumental brain stem testing, such as arteriography, perfusion single photon emission computed tomography (SPECT), ultrasound Doppler sonography, or short latency auditory or somatosensory evoked potentials, may replace this part of the examination.[6]

3.4. Recommendations and Proposals

Apnea testing should be the last of all clinical tests. Apneic oxygenation is the preferred means. If in doubt, an atropine test may be done beforehand. "The atropine test is performed by injecting 2 mg of atropine intravenously while observing the heart rate. If compared with the previous steady baseline it clearly increases by more than 10 percent within one minute it is considered positive. Utmost care has to be taken not to use a line that might be contaminated with sympathomimetics." If it is undoubtedly positive, apnea testing is not warranted. If atropine testing is negative, apnea testing must be done. First a body temperature of 32°C should be ascertained. After preoxygenation with 100% O_2 the highest possible PaO_2 should be achieved,[16] then ventilatory volume reduced to approximately 0.5 to 2 L/min (e.g. by synchronized intermittent mandatory volume ventilation, or by disconnection from the respirator and insufflation of O_2 into the trachea at a rate of about 6 L/min under close observation until requirements are met). The $PaCO_2$ should be checked after an appropriate period of time, depending on the initial blood gas values. An arterial line and the feasibility of rapid determination of blood gases are very helpful; as soon as the required $PaCO_2$ level is reached, the patient should be disconnected from the respirator (if not already) and observed for an appropriate period of time (usually about half a minute), then reconnected and mildly hyperventilated for some minutes. To detect respiratory activity we prefer to use a fine thread that is sensitive enough to show displacement by heartbeats, thereby proving that the system is capable of disclosing even the slightest respiratory effort. Other methods such as a moist thin paper membrane put over the endotracheal tube may serve as well.

4. CONCLUSIONS

Testing for apnea is considered indispensable for the determination of brain death world-wide and may be safely performed under almost any circumstances, although it sometimes poses a formidable task to the examiners. The better the patient is prepared for the test, the more easily the testing will be performed. In our experience it can be done *lege artis* in 99.9% of all cases; the remaining ones may be handled using subsidiary instrumental tests. From a pragmatic point of view it would be desirable to have identical guidelines world-wide, making questions such as "how can a patient be dead in one country and not dead in another?" unnecessary.

5. REFERENCES

1. Jeret JS, Benjamin J. Risk of hypotension during apnea testing. *Arch Neurol* 1994;**51**:595-599.
2. Marks SJ, Zisfein J. Apneic oxygenation in apnea tests for brain death: a controlled trial. *Arch Neurol* 1990;**47**:1066-1068.
3. Pallis C, Harley DH. *ABC of brainstem death.* 2nd ed. London: BMJ Publishing Group, 1996.
4. Garcia C, Ferro JM. European brain death codes: Portuguese guidelines. *J Neurol* 2000;**247**:140.
5. Wijdicks EFM. Clinical diagnosis and confirmatory testing of brain death in adults. In: Wijdicks EFM, ed. *Brain Death.* Philadelphia: Lippincott Williams & Wilkins, 2001:61-90.
6. Wissenschaftlicher Beirat der Bundesärztekammer. Richtlinien zur Feststellung des Hirntodes. Dritte Fortschreibung 1997 mit Ergänzungen gemäß Transplantationsgesetz (TPG). *Dtsch Ärztebl* 1998;**95**: B1509-B1516.
7. Wijdicks EFM, ed. *Brain Death.* Philadelphia: Lippincott Williams & Wilkins, 2001.
8. Belsh JM, Blatt R, Schiffman PL. Apnea testing in brain death. *Arch Intern Med* 1986;**146**:2385-2388.
9. Benzel EC, Mashburn JP, Conrad S, Modling D. Apnea testing for the determination of brain death: a modified protocol. *J Neurosurg* 1992;**76**:1029-1031.
10. Benzel EC, Gross CD, Hadden TA, Kesterson L, Landreneau MD. The apnea test for the determination of brain death. *J Neurosurg* 1989;**71**:191-194.
11. Brilli RJ, Bigos D. Altered apnea threshold in a child with suspected brain death. *J Child Neurol* 1995;**10**: 245-246.
12. Ashwal S. Clinical diagnosis and confirmatory testing of brain death in children. In: Wijdicks EFM, ed. *Brain Death.* Philadelphia: Lippincott Williams & Wilkins, 2001:91-114.
13. Lang CJG. Blood pressure and heart rate changes during apnoea testing with or without CO_2 insufflation. *Intensive Care Med* 1997;**23**:903-907.
14. Lang CJG, Heckmann JG, Erbguth F, Druschky A, Haslbeck M, Reinhardt F, Winterholler M. Transcutaneous and intra-arterial blood gas monitoring – a comparison during apnoea testing for the determination of brain death. *Eur J Emerg Med* 2002;**9**:51-56.
15. Lang CJ. Apnea testing by artificial CO_2 augmentation. *Neurology* 1995;**45**:966-969.
16. Goudreau JL, Wijdicks EF, Emery SF. Complications during apnea testing in the determination of brain death: predisposing factors. *Neurology* 2000;**55**:1045-1048.

EVOKED POTENTIALS IN THE DIAGNOSIS OF BRAIN DEATH

Enrico Facco and Calixto Machado*

1. INTRODUCTION

Brain death (BD) is a by-product of the modern intensive care: since the report on coma depassé by Mollaret and Goulon in 1959[1] and the new criterion of death on neurological grounds published by the Harvard Medical School Ad Hoc Committee in 1968,[2] both the definition and diagnostic of BD criteria have undergone a substantial evolution. Nowadays, most of the problems, dilemmas, and polemics raised in the past regarding the concept of BD, by terminological confusion, and by the grave responsibility of declaring dead a corpse with a still beating heart, are definitively overcome. However, despite the fact that BD is now widely accepted all over the world and regardless the certainty of its diagnosis when carefully formulated, its definition as well as diagnostic criteria are far from perfect and need further adjustments.

2. THE CONCEPT OF DEATH AND ITS RELATIONSHIP WITH DIAGNOSTIC CRITERIA

The concept of BD, as death of the individual, is the kernel of the whole topic: the diagnostic criteria spring from and closely depend on the definition of death, which includes biological as well as philosophical aspects. It is also important to emphasize that any complete standard of death should include three distinct elements: the definition of death, the criterion of death, and the tests to prove that the criterion has been satisfied. Undoubtedly, the term *criterion* for referring to the anatomical substratum introduces confusion in this discussion, because protocols of tests (clinical and instrumental) are also called *diagnostic criteria* or *sets of diagnostic criteria*. During the last decades, three main brain-oriented formulations of death have been proposed: whole brain, brain stem, and higher brain (neocortical) formulations.[3]

* Enrico Facco, MD. Department of Pharmacology and Anesthesiology, University of Padua, Italy. Calixto Machado, MD, Ph.D. Institute of Neurology and Neurosurgery, Havana, Cuba. Email: braind@infomed.sld.cu.

Table 1. Relationship between the criterion of brain death and diagnostic criteria.

CRITERION OF DEATH	KEY ASPECTS OF DIAGNOSTIC CRITERIA
WHOLE CENTRAL NERVOUS SYSTEM DEATH	EEG relevant Brain-stem reflexes relevant Spinal reflexes relevant
WHOLE BRAIN DEATH	EEG relevant Brain-stem reflexes relevant Spinal reflexes irrelevant
BRAIN STEM DEATH	EEG irrelevant Brain stem reflexes relevant Spinal reflexes irrelevant
NEOCORTICAL DEATH	EEG relevant Brain-stem reflexes irrelevant Spinal reflexes irrelevant

As a consequence, diagnostic criteria may be relevant or irrelevant according to the adopted standard of death: this may give raise to possible differences and even discrepancies in the diagnostic rules adopted by different countries. The main criterion-related differences are reported in Table 1.

Here, we do not discuss the value and limits of each possible concept of BD (a topic deeply discussed in other chapters of this book), but only emphasize the close dependence of criteria on definition: for example, the EEG, which is the most widely used test, is mandatory in Italy,[4,5] recommended in the USA[6] and irrelevant in the UK.[7] These differences reflect the concepts of "brain-stem death," adopted by UK, and of "whole brain death, adopted by USA and Italy. Whatever the definition of BD (apart from the unacceptable concept of neocortical death), the death of the brain stem is always the kernel and the *conditio sine qua non* for a proper diagnosis: therefore, the essential differences pertain to the need for checking hemispheric function also, when whole brain death is to be diagnosed. A few comments on the diagnosis of "whole brain death," which is the most widely adopted concept of brain death in the world (including USA and most European countries),[3,6,8-10] are necessary for a proper discussion on the role of EEG and evoked potentials.

2.1. Diagnosis of Whole Brain Death

2.1.1. The Term "Whole Brain Death"

The term "whole brain death" does not define whether death is "the loss of all neurons of the brain" or "the loss of the brain as a whole"; the latter seems more appropriate, since the persistence of small, isolated pools of neurons without integration is not a sign of life above the cellular level. On the other hand, the former is more precise and does not leave doubts about its meaning, while the latter requires an exact definition of all possible brain areas the persistence of which is compatible with death. Thus, the concept of whole brain death, when applied to criteria for diagnosis, necessarily implies the need to check for death of the whole brain. Of course, the persistence of brain-stem function excludes a diagnosis of death.

2.1.2. The Diagnostic Criteria

The diagnostic criteria commonly used for diagnosing the death of the whole brain seem to be conceptually rough. In fact, they cannot explore all brain functions, but to test consciousness, posture, some (not all!) explorable brain-stem structures apnea, and cortical electrical activity by EEG. Thus, the diagnosis is based on the assumption that all the brain is dead when the structures explored by the mentioned tests (essentially, parts of the brain stem and cortex) are dead. Curiously, the criteria usually applied for the diagnosis of whole brain death are intended to closely reflect the proposed concept of integrating critical system.[3,11,12] Even apart from considerations of the pathophysiology of intracranial hypertension and arrest of cerebral circulation in whole BD, the diagnosis is absolutely safe, although a strong discrepancy about the definition of death on neurological grounds does exist among authors.[3,12,13]

2.2. EEG

The EEG has been closely linked to BD since the publication of Harvard Criteria: at that time it was the only available technical investigation able to explore brain function and to give an "objective" confirmation of BD. However, in 1969 Beecher, the Chairman of the Ad Hoc Committee already asserted that the EEG, although valuable, was not essential.[14] Later on, several authors[15-20] have emphasized the technical pitfalls and limitations of the EEG, which can be summarized as follows: a) the EEG does not provide direct information on brain-stem function; b) the EEG may be flat in patients with preserved brain-stem function (e.g., comatose or vegetative patients following prolonged cardiac arrest); c) the EEG is unreliable in case of sedation, hypothermia, or presence of toxic or metabolic factors; d) there is a consistent risk of mistaking artifacts for residual cortical activity and vice versa.

When reversible causes of coma are excluded, the presence of artifacts may account for a non-negligible uncertainty about the presence or absence of electrocerebral activity.[21] Moreover, both inter- and intra-observer variability may account for diagnostic uncertainty in up to 20% of cases,[19] showing that EEG is much less "objective" than people believe.

Despite these limits, EEG is recommended by all countries adopting the "whole brain death" definition, and is mandatory in Italy.[4] The reasons for this may be found in the history of BD criteria, including the overestimation of EEG reliability, cultural factors (including psychological, social and political aspects) and the need for publication of official criteria to be shared by people. The need for an "objective" confirmation of BD sprang from the need to reassure patients' relatives and others about the certainty of diagnosis, as well as to protect the attending doctor from prosecution (essentially for murder). It is now time to redefine the role of EEG, taking into account the substantial progress in the definition of BD and the development of investigative techniques, in order to eliminate all the sources of over- and underestimation of their effectiveness: a conservative stance would only be manipulative (taking fallacies for facts) and, thus, be scientifically unacceptable (leading to avoidable errors).

2.3. Evoked Potentials

Auditory brainstem responses (ABRs) explore brainstem function; short latency somatosensory evoked potentials (SEPs) from upper limb stimulation evaluate lemniscal pathways in the brain stem and cerebral hemispheres; and visual evoked potentials (VEPs) evaluate visual pathways in the hemispheres. They can be considered an extension of the clinical examination, being able to check the function of specific nervous pathways.[22-24]

In order to differentiate between coma and BD, ABRs and SEPs allow the assessment of brain-stem structures which cannot be clinically explored; therefore, they are very good candidates for routine ancillary tests to confirm BD, in which the irreversible loss of brain-stem function is the kernel of the diagnosis (even more relevant than the assessment of cortical function). Moreover, ABRs and SEPs are relatively unaffected by sedatives and other reversible causes of coma and, thus retain all their value even when the EEG and clinical examination cannot provide reliable information. By including VEPs, the deep part of hemispheres from optic nerve to the calcarine cortex can be explored, adding further information relevant to whole brain death.[24-32]

Due to these features, a substantial interest has grown over the past two decades on multimodality evoked potentials (MEPs) in BD, and a wealth of data are now available in the literature.[22-42]

The use of each modality, singly and in combination, in the diagnosis of BD will now be briefly reviewed.

2.3.1. Role of VEPs

Visual evoked potentials may be a sensitive indicator of early impairment of cerebral function and may demonstrate changes sooner than electroencephalograms.[43-45] Therefore, several articles have appeared describing the use of VEPs to study brain-dead patients.[25,46] Nevertheless, there are still controversies among investigators about the possible contamination of VEP recordings by spread of the electroretinogram (ERG) to the occipital area, which could limit the use of this technique for the diagnosis of brain death.[43-44,46-49]

We found a characteristic pattern in all patients. When a cephalic reference was used for both VEPs and the ERG, the a- and b- waves of the ERG were recognized in all cases. The visual evoked responses consisted of waves with less amplitude but the same latency and morphologic features as in the ERG. When a noncephalic derivation was chosen for the ERG and VEPs, the ERG waves were the same in latency and morphologic characteristics, but the VEP channel showed no response.[50]

Trojaborg & Jørgensen[46] noted that when the summated ERG was superimposed with reverse polarity on the response recorded over the occipital region, it was cancelled, and they concluded that the VEP waves recorded in brain-dead patients are the spread of the ERG to the occipital area. These investigators were in disagreement with others,[43-44,47-49] who did not observe the spread of the ERG to the occipital region. Trojaborg and Jørgensen[46] explained this discrepancy by the fact that they applied up to 500-1000 stimuli and used amplification at least 10 times greater than those of other studies. We found that the amplitude of the waves recorded in the VEP channel decreased progressively when the reference electrode was more distant from the retina. Moreover, in our patients, the amplitudes of the ERG waves were greater in the ERG derivation with noncephalic reference, suggesting that the ERG was also picked up by the Fz electrode in the ERG channel with cephalic reference. Only in two cases was it possible to record waves in the VEP derivation with a noncephalic reference, indicating a spread of the ERG to the occipital area.

Trojaborg and Jørgensen[46] used A_1 or A_2 as references, and it is likely that the ERG was picked up mainly by the earlobe electrode and not by the occipital one. Because we used a noncephalic-referenced VEP derivation, the reference electrode was farther from the retina than the recording electrode placed in the occipital area. These results suggest that, although contamination of the VEP records by the spread of the ERG to the occipital area could occur, it is easy to confirm the absence of a true cortical visual response in brain-dead patients by means of a noncephalic reference.[50]

On the other hand, persistence of the ERG even long after the elimination of all respiratory support has led some investigators to assert that this technique is not useful for studying brain-dead patients.[44,47] We elicited and recorded VEPs and the ERG simultaneously, using a cephalic (Fz) and a noncephalic (C7) reference. In all cases, when a noncephalic reference was used, the ERG waves did not differ in either latency or morphologic features, while the VEP channel showed no response. This pattern clearly confirms that in the visual pathways of brain-dead patients, electrical activity is confined to the retina.[50] Our results further suggest the usefulness of combining VEPs and ERG in the same setting, using cephalic and noncephalic references to examine patients suspected to be brain dead. In other modalities of evoked potentials, the use of noncephalic references has also been suggested for a better elucidation of evoked potential component generation.[51] Moreover, as for median nerve SEPs, in which the Erb's potential serves as a stimulation input control,[52] by recording the ERG it is possible to ensure that adequate visual stimuli have been applied.

Although it is now more common to elicit the ERG with more sophisticated forms of stimuli, such as pattern reversal, color- and intensity- controlled light flashes, etc. [53], light-emitting diodes seem to be suitable to study unresponsive patients in intensive care units, where the conditions of illumination are difficult to manage. Moreover, goggles are easily attached to intubated and unconscious patients. When the visual pathways of suspected brain-dead patients are studied by VEP and ERG recordings, it is possible to assess mainly forebrain hemispheric function, while brain-stem evaluation is lacking. Therefore, it may be concluded that, although the setting we propose for VEPs and the ERG provides an objective electrophysiologic assessment of hemispheric function, which is essential for diagnosis of brain death, this technique is probably of limited value for this purpose when used in isolation.[54] Rather, we propose the use of a test battery, including electroencephalography, brainstem auditory evoked potentials, short-latency somatosensory evoked potentials, VEPs and ERG to study brain-dead patients.[55] Such a test battery would permit the assessment of several sensory pathways and the evaluation of both brainstem and cerebral hemispheric functions.[25,51,55-57] Thus, the reliability of the diagnosis of brain death could be considerably increased.

2.3.2. Role of ABRs

The detection of wave I only, which is usually delayed and with variable amplitude, is the most clear-cut ABR finding compatible with BD. The amplitude of wave I may be normal, low, or even increased, according to the timing of ABRs in relation to the onset of BD. These fluctuations eventually end in the disappearance of wave I, whereas the increased amplitude in some cases is related to the loss of descending inhibition of cochlear activity. The latency shift of wave I depends on intubation (likewise in comatose, ventilated patients) and the decrease of body temperature, which is a well known consequence of BD.[56]

Although the presence of wave I only is the paradigmatic ABR picture of BD, this pattern is present in only 25-30% of patients, while in the remaining ones ABRs are flat. The absence of wave I, paradoxically, cannot show the death of the brain stem, since a flat ABR indicates deafness only; however, it is compatible with BD, since it results from cochlear ischemia, following the arrest of cerebral circulation. This is the reason why ABRs were considered of limited value in the diagnosis of BD in the past. When ABRs are serially recorded during the process of dying, the disappearance of waves shows that the final picture (namely, flat ABRs) is the result of the patient's death. ABRs maintain all their value in the exclusion of false positives (that is, comatose patients looking brain dead), since a bilateral absence of response unrelated to BD is uncommon (deafness, ototoxic drug administration and bilateral temporal bone fractures must be taken into account). It may transiently occur in postanoxic coma (as may a flat EEG),[58,59] where some caution is to be used in their interpretation. Flat ABRs may be observed in basilar artery occlusion as well.[60] On the contrary, absent brain-stem reflexes and flat EEG with preserved evoked potentials have been found by us[42] and others,[61] illustrating their usefulness in identifying false-positive clinical diagnoses. Similar diagnostic problems have been reported in encephalitis.[34,62,63]

As far as generators are concerned, it is well known that waves III-V are generated in the brain stem, and, therefore, their presence *does* exclude BD. Wave II seems to be generated in the proximal part of the VIII nerve, probably at the CSF-brainstem interface, and in cochlear nuclei, with a mixed peripheral and central origin.[64] Wave I reflects the action potential of the VIII nerve, mainly generated in its caudal part. Thus, one can confirm the absence of brain-stem function when wave I only is preserved. The presence of wave II may give rise to some uncertainty in BD due to its dual origin, absence of the central component of which might perhaps be reflected in changes in latency and amplitude. Wave II may be still recordable in a few clinically brain-dead patients.[35,65-67] Machado et al. recorded a unilateral wave II in a terminal case with electrocerebral silence and no brain-stem reflexes, but not apneic.[68] However, the persistence of wave II seems to be of short duration in the progression towards BD, and at present it seems wiser to exclude its persistence from the certain patterns of BD.[35,42,66,68]

ABRs may show the presence of only wave I in deeply comatose, non-brain-dead patients following primary brain-stem lesions: here ABRs confirm the severity of pontine damage, but this picture may not be paradigmatic of BD, being the result of focal damage.[42,60]

2.3.3. Role of SEPs

A wealth of data regarding SEPs in BD are now available in the literature.[39,42,69-85] Unfortunately, several papers use a frontal reference, which prevents a proper recording of far-field potentials, an essential parameter in BD. This aspect calls for a short description of generators of cervical and far-field SEP components, which can be summarized as follows:[27,41,86-100] a) the brachial plexus for the cervical N9 and the far-field, scalp-recorded P9; b) the ascending volley in the dorsal columns of the cervical spinal cord for the cervical N11 and the far-field, scalp-recorded P11; c) the dorsal horn in the cervical spinal cord (with a dipole perpendicular to the spinal cord axis) for the cervical N13; d) the higher cervical segment or cervicomedullary junction for far-field scalp-recorded P13; e) the ascending volley in the caudal medial lemniscus for the P14. Concerning the N18, studies performed in the past two decades have allowed for considerable clarification of its generator. In the early eighties it was believed to originate in the thalamus or thalamic radiations; later Mauguière and Desmedt argued that its origin was more caudal, perhaps in the brain stem.[101] Thereafter its origin was identified in the pons or lower midbrain[102] and, finally, in the medulla oblongata, probably in the nucleus cuneatus.[100,103]

Since the assessment of brain-stem viability is the kernel of BD diagnosis, a proper SEP recording in BD cannot but use a non-cephalic reference, in order to detect the far-field components P13, P14, and N18. The use of a frontal reference, which had been widely used in the past, cancels out all far-field components and allows for the assessment of cortical function only, making SEPs similar to EEG in this context. One can only tell whether the activity of somatosensory cortex is preserved or not, a small thing in the diagnosis of BD. On the other hand, SEPs with a non-cephalic reference provide very useful information about the caudal part of the brain-stem. Thus, the combined use of ABRs and SEPs allows assessment of ponto-mesencephalic and bulbar levels, respectively, with a good capability of checking the level of rostrocaudal deterioration in preterminal states, down to brain-stem death.[42] A nasopharyngeal derivation may be added as well, to get a near-field recording of the caudal and rostral components of P14.[28,39,82]

SEPs demonstrate arrest of conduction at the cervico-medullary level in most patients, about 95% in our experience, while in the remaining cases the cervical components are not recordable; like flat ABRs, their absence depends on extracerebral damage, preventing the evaluation of central pathways. The main causes of flat SEPs are lesions in the brachial plexus and cervical spinal cord as well as polyradiculopathies.[63] Much caution is to be used in high cervical vertebral fractures, where a present cervical N13 with absent P14, resembling a BD pattern, only shows the arrest of conduction in the spinal cord.[89] Since SEPs allow checking the caudal brain stem, their far-field potentials may still be preserved when brain-stem reflexes and ABR waves II-V are already absent, and a close relationship between the onset of apnea and disappearance of N18 may be found.[42,104]

SEPs are a more reliable indicator of BD than ABRs, since in most patients they can show the arrest of conduction at the cervicomedullary level. The SEP patterns of BD are: a) absence of all components following the far-field, scalp-recorded P13; and b) the dissociation between N13 and P13, that is, the persistence of cervical N13 with absent P13 (in other words, absence of all components following P11). Furthermore, the medullary origin of N18 helps to avoid premature apnea testing; in fact, the apnea test may be dangerous in comatose, non-brain-dead patients, and SEP recording may allow postponing it until the disappearance of N18.[42]

2.3.4. Role of MEPs

Apart from the specific advantages and limitations of each modality, it is obvious that MEPs provide a better assessment of BD compared to any single modality, allowing assessment of different nervous pathways with different anatomic locations of their generators: VEPs explore fronto-occipital hemispheric structures, ABRs the pons and mesencephalon, and SEPs a long rostrocaudal path from parietal cortex to the cervical spinal cord (with relevant caudal generators in the medulla oblongata). Therefore, depending on the site and extent of primary and secondary lesions, a single modality might exclude BD. For example, VEPs might be preserved in brain-stem lesions that extinguish both ABRs and SEPs; conversely, patients with medullary lesions may retain ABRs, while in patients with hemispheric lesions and rostrocaudal evolution, SEPs alone may disclose a still viable brain stem with reserved P14 and/or N18.[25,42,51,55-57]

Furthermore, MEPs enable the exclusion of BD in sedated patients, easily and noninvasively, therefore helping to optimize the timing of contrast angiography; in a few patients with a clinical and EEG picture of BD and no reversible factors MEPs may even show residual function in the brain stem, illustrating how MEPs may also improve diagnostic safety.[42] These data once again emphasize that BD requires a careful diagnosis, regarding evaluation of the patient's history, clinical data, and appropriate use and interpretation of confirmatory tests, with no *a priori* choice of a single investigation. It seems odd nowadays that some published criteria still recommend EEG while underestimating, or not even mentioning, MEPs.[38,42,80]

3. SUMMARY

In conclusion, the wealth of data available on ancillary tests, the greater knowledge of the process of dying and the evolution of the concept of BD suggest a need to redefine the role of EEG and MEPs in the confirmation of BD. The main facts are the following:

3.1. The Concept of Death

The concept of death is the kernel of the whole matter: differences and apparent discrepancies in the adopted criteria by different countries cannot be overcome until a definitive, unique, and universally accepted definition of BD is adopted.

3.2. The Brain Stem is the Kernel

Whatever the adopted definition of BD, death of the brain stem is the kernel, and the *conditio sine qua non*, of the diagnosis. This is why the evaluation of all clinical signs of brain-stem function should be mandatory; since the death of the entire brain stem is to be diagnosed, the arbitrary exclusion of any of these signs from diagnostic criteria would be conceptually inconceivable.

3.3. Diagnostic Tools Capable of Exploring the Brain Stem

As a consequence, all diagnostic tools capable of exploring the brain stem should be deemed relevant for the diagnosis; there is no reason to regard brain-stem reflexes as essential, yet not even mention ABRs and SEPs. The absence of evoked potentials from published criteria reflects the historical approach of BD, since before the 1980's they were not widely available in hospitals; in the past two decades they have proved to be very relevant and able to provide objective data on brain-stem structures, which cannot be clinically explored in BD. Moreover, they remain effective even when both clinical examination and EEG are no longer able to provide information. It is now time to discuss and define their role in diagnostic criteria; any other stance would be manipulative, rather than scientific.

3.4. Visual Evoked Potentials

VEPs can provide evidence of absent activity in deep hemispheric structures and occipital cortex, thus improving the evaluation in comparison to the use of EEG in isolation.

3.5. The Electroencephalogram

The EEG, despite the overestimation of its sensitivity and specificity, remains a valuable tool in the diagnosis of whole brain death. In brain-dead patients following supratentorial primary lesions and severe intracranial hypertension, the viability of cortex seems very unlikely. Conversely, in patients with primary brain-stem insults and a clinical picture of BD, even short latency evoked potentials may be abolished, while the EEG may transiently show well preserved cortical activity. In these cases EEG and VEPs may be the most relevant tests, related to the concept of whole brain death.

3.5. Multimodality Evoked Potentials

MEPs may detect residual function in the brain stem in a few hopeless cases with the clinical and EEG picture of BD (despite the absence of any reversible cause of coma), showing that their use may improve diagnosis safety. In this regard, we must be aware that the essence of diagnosis is the determination that the process of dying has concluded. It must never be a prognosis, which could turn the diagnosis of death into euthanasia. The latter is a very relevant and ticklish but completely different philosophical debate. Hence, we propose the use of MEPs brain-dead patients to increase diagnostic reliability

4. REFERENCES

1. Mollaret P, Goulon M. Le coma depassé. *Rev Neurol* 1959;**101**:3-15.
2. Harvard Medical School ad Hoc Committee to examine the definition of brain death: a definition of irreversible coma. *JAMA* 1968;205:337-340.
3. Machado C. Consciousness as a definition of death: Its appeal and complexity. *Clin Electroencephalogr* 1999;**30**:156-164.
4. Legge 29 Dicembre 1993 n.578. Norme per l'accertamento e le certificazione di morte. *Gazzetta Ufficiale della Repubblica Italiana* 8/1/1994;**5**:4-5.
5. Ministero della Sanità. *Regolamento recante le modalità per l'accertamento e la certificazione di morte. Decreto* 1994;22/8/1994 n. 582.
7. Report of the Medical Consultants on the Diagnosis of Death to the President's Commission on the Study of Ethical Problems in Medicine and Biomedical and Behavioral Research. Guidelines for the determination of death. *JAMA* 1981;**246**;2184-2186.
8. Royal Colleges and Faculties of the United Kingdom. Diagnosis of brain death. *Br Med J* 1976;**2**:1187-1188.
9. Le Ministre Delegue charge de la Sante. *Decret n° 90-844 du 24 septembre 1990*. 1990.
10. Wissenschaftlicher Beirat der Bundesärtzekammer. Kriterien des Hirntodes. *Dtsch Artzebl Mitteilg* 1991;**87**:2855-2860.
11. Korein J. The problem of brain death: development and history. *Ann N Y Acad Sci* 1978;**315**:19-38.
12. Machado C. Is the concept of brain death secure? In: Zeman A, Emanuel L, eds. *Ethical Dilemmas in Neurology*. London: W. B. Saunders Company, 2000; **Vol 36**:193-212.

13. Machado C. Una nueva formulación de la muerte: Definición-criterio-pruebas confirmatorias. *Rev Neurol* 1998;**26(154):**1040-1047.
14. Beecher HK. After the definition of irreversible coma. *N Engl J Med* 1969;**281:**1070-1071.
15. Pallis C. Prognostic value of brain stem lesion. *Lancet* 1981;**i:**379.
16. Pallis C. ABC of brain stem death. The arguments about the EEG. *Br Med J* 1983;**286:**284-287.
17. Egol AB, Guntupalli KK. Intravenous infusion device artifacts in the EEG-confusion in the diagnosis of electrocerebral silence. *Intensive Care Med* 1983;**9:**29-32.
18. Spudis EV, Penry JK, Link SL. Paradoxical contribution of EEG during protracted dying. *Arch Neurol* 1984;**41:**153-156.
19. Buchner H, Schuchardt V. Reliability of electroencephalogram in the diagnosis of brain death. *Eur Neurol* 1990;**30(3):**138-141.
20. Nau R, Prange HW, Klingelhofer J, Kukowski B, Sander D, Tchorsch R, et al. Results of four technical investigations in fifty clinically brain dead patients. *Intensive Care Med* 1992;**18(2):**82-88.
21. Bennett DR. The EEG in the determination of brain death. *Ann N Y Acad Sci* 1978;**315:**110-119.
22. Goldie WD, Chiappa KH. Brainstem auditory and short latency somatosensory evoked responses in brain death. *Neurology* 1981;**31:**248-256.
23. Mauguière F, Grand C, Fischer C, Courjon J. Aspects of early somatosensory and auditory evoked potentials in neurologic comas and brain death. *Rev Electroencephalogr Neurophysiol Clin* 1982;**12(3):** 280-285.
24. Facco E, Caputo P, Casartelli LM, Munari M, Toffoletto F, Fabiani F, et al. Auditory and somatosensory evoked potentials in brain-dead patients. *Riv Neurol* 1988;**58:**140-145.
25. Ganes T, Lundar T. EEG and evoked potentials in comatose patients with severe brain damage. *Electroencephalogr Clin Neurophysiol* 1988;**69(1):**6-13.
26. Facco E, Casartelli LM, Munari M, Toffoletto F, Baratto F, Giron GP. Short latency evoked potentials: new criteria for brain death? *J Neurol Neurosurg Psychiatry* 1990;**53(4):**351-353.
27. Wagner W. SEP testing in deeply comatose and brain dead patients: the role of nasopharyngeal, scalp and earlobe derivations in recording the P14 potential. *Electroencephalogr Clin Neurophysiol* 1991;**80(5):** 352-363.
28. Wagner W. SEP testing in deeply comatose and brain dead patients: the role of nasopharyngeal, scalp and earlobe derivations in recording the P14 potential. *Electroencephalogr Clin Neurophysiol* 1991;**80(5):** 352-363.
29. Firsching R, Frowein RA, Wilhelms S, Buchholz F. Brain death: practicability of evoked potentials. *Neurosurg Rev* 1992;**15(4):**249-254.
30. Guérit JM. Evoked potentials: a safe brain-death confirmatory tool? *Eur J Med* 1992;**1(4):**233-243.
31. Machado C, Valdés P, García O, Coutin P, Miranda J, Román J. Short latency somatosensory evoked potentials in brain-dead patients using restricted low cut filter setting. *J Neurosurg Sci* 1993;**37(3):**133-140.
32. Machado C. Multimodality evoked potentials and electroretinography in a test battery for an early diagnosis of brain death. *J Neurosurg Sci* 1993;**37(3):**125-131.
33. Facco E, Munari M, Baratto F, Giron GP. Role of the somatosensory evoked potentials in the diagnosis of brain death. In: Machado C, ed. *Brain Death*. Amsterdam: Elsevier Science Publishers B.V., 1995:135-139.
34. Guérit JM, de Tourtchaninoff M, Hantson P, Mahieu P. Multimodality evoked potentials in the differential diagnosis of brain death. In: Machado C, ed. *Brain Death*. Amsterdam: Elsevier Science Publishers B.V., 1995:119-126.
35. Shiogai T, Takeuchi K. Diagnostic reliability in loss of brainstem function evaluated by brainstem auditory evoked potentials and somatosensory evoked potentials in impending brain death. In: Machado C, ed. *Brain Death*. Amsterdam: Elsevier Science Publishers B.V., 1995:111-118.
36. Wagner W. Scalp, earlobe and nasopharyngeal recordings of the median nerve somatosensory evoked P14 potential in coma and brain death. Detailed latency and amplitude analysis in 181 patients. *Brain* 1996; **119(Pt 5):**1507-1521.
37. Facco E, Baratto F, Munari M, Behr AU, Volpin SM, Gallo F, et al. Short latency evoked potentials in the diagnosis of brain death. *Electroencephalogr Clin Neurophysiol* 1997;**103(1):**106.
38. Machado-Curbelo C, Roman-Murga JM. [Usefulness of multimodal evoked potentials and the electroretinogram in the early diagnosis of brain death]. *Rev Neurol* 1998;**27(159):**809-817.
39. Roncucci P, Lepori P, Mok MS, Bayat A, Logi F, Marino A. Nasopharyngeal electrode recording of somatosensory evoked potentials as an indicator in brain death. *Anaesth Intensive Care* 1999;**27(1):**20-25.
40. Ruiz-Lopez MJ, Martinez dA, Serrano A, Casado-Flores J. Brain death and evoked potentials in pediatric patients [see comments]. *Crit Care Med* 1999;**27(2):**412-416.

41. Sonoo M, Tsai-Shozawa Y, Aoki M, Nakatani T, Hatanaka Y, Mochizuki A, et al. N18 in median somatosensory evoked potentials: a new indicator of medullary function useful for the diagnosis of brain death. *J Neurol Neurosurg Psychiatry* 1999;**67(3)**:374-378.
42. Facco E, Munari M, Gallo F, Volpin SM, Behr AU, Baratto F, et al. Role of short latency evoked potentials in the diagnosis of brain death. *Clin Neurophysiol* 2002;**113(11)**:1855-1866.
43. Arfel G. Visual stimuli and cerebral silence. *Electroencephalogr Clin Neurophysiol* 1967;**23(2)**:172-175.
44. Arfel G. *Problemes electroencéphalographiques de la mort.* Paris: Masson, 1970.
45. Reilly EL, Kondo C, Brunberg JA, Doty DB. Visual evoked potentials during hypothermia and prolonged circulatory arrest. *Electroencephalogr Clin Neurophysiol* 1978;**45(1)**:100-106.
46. Trojaborg W, Jørgensen EO. Evoked cortical potentials in patients with "isoelectric" EEGs. *Electroencephalogr Clin Neurophysiol* 1973;**35(3)**:301-309.
47. Wilkus RJ, Chatrian GE, Lettich E. The electroretinogram during terminal anoxia in humans. *Electroencephalogr Clin Neurophysiol* 1971;**31(6)**:537-546.
48. Walter S, Arfel G. Responses to visual stimuli in acute and chronic coma states. *Electroencephalogr Clin Neurophysiol* 1972;**32(1)**:27-41.
49. Cobb WA, Dawson. The latency and form in man of the occipital potentials evoked by bright flashes. *J Physiol* 1960;**152**:108-121.
50. Machado C, et al. Visual evoked potentials and electroretinography in brain-dead patients. *Doc Ophthalmol* 1993;**84**:89-96.
51. Ferbert A, Buchner H, Bruckmann H, Zeumer H, Hacke W. Evoked potentials in basilar artery thrombosis: correlation with clinical and angiographic findings. *Electroencephalogr Clin Neurophysiol* 1988;**69(2)**: 136-147.
52. Anziska BJ, Cracco RQ. Short latency SEPs to median nerve stimulation: comparison of recording methods and origin of components. *Electroencephalogr Clin Neurophysiol* 1981;**52(6)**:531-539.
53. Chiappa KH. *Evoked potentials in clinical medicine.* New York: Raven Press, 1989.
54. Guérit J, Mathieu P. Are evoked potentials a valuable tool for the diagnosis of brain death? *Transplant Proc* 1986;**18**:38-67.
55. Machado C. A multimodal evoked potential and electroretinography test battery for the early diagnosis of brain death. *Int J Neurosci* 1989;**49**:241-242.
56. Machado C, Valdes P, Garcia-Tigera J, Virues T, Biscay R, Miranda J, et al. Brain-stem auditory evoked potentials and brain death. *Electroencephalogr Clin Neurophysiol* 1991;**80(5)**:392-398.
57. Machado C. Muerte encefálica. Criterios diagnósticos. *Rev Cubana Med* 1991;**30**:181-206.
58. Brunko E, Delecluse F, Herbaut AG, Levivier M, Zegers dB. Unusual pattern of somatosensory and brain-stem auditory evoked potentials after cardio-respiratory arrest. *Electroencephalogr Clin Neurophysiol* 1985;**62(5)**:338-342.
59. Fauvage B, Combes P. Isoelectric electroencephalogram and loss of evoked potentials in a patient who survived cardiac arrest. *Crit Care Med* 1993;**21(3)**:472-475.
60. Ferbert A, Buchner H, Zeumer H, Hacke W. Evoked potentials in basilar artery thrombosis: correlation with clinical and angiographic findings. *Electroencephalogr Clin Neurophysiol* 1988;**69(136)**:147.
61. Barelli A, Della CF, Calimici R, Sandroni C, Proietti R, Magalini SI. Do brainstem auditory evoked potentials detect the actual cessation of cerebral functions in brain dead patients? *Crit Care Med* 1990; **18(3)**:322-323.
62. Chandler JM, Brilli RJ. Brainstem encephalitis imitating brain death. *Crit Care Med* 1991;**19(7)**:977-979.
63. Hantson P, Guerit JM, de Tourtchaninoff M, Deconinck B, Mahieu P, Dooms G et al. Rabies encephalitis mimicking the electrophysiological pattern of brain death. A case report . *Eur Neurol* 1993; **33(3)**:212-217.
64. Martin WH, Pratt H, Schwegler JW. The origin of the human auditory brain-stem response wave II. *Electroencephalogr Clin Neurophysiol* 1995;**96(4)**:357-370.
65. Facco E, Munari M, Gallo F, Volpin SM, Behr AU, Baratto F, et al. Role of short latency evoked potentials in the diagnosis of brain death. *Clin Neurophysiol* 2002;**113(11)**:1855-1866.
66. Hall JW, Mackey-Hargadine JR, Kim EE. Auditory brain-stem response in determination of brain death. *Arch Otolaryngol* 1985;**111(9)**:613-620.
67. Paniagua-Soto M, PiñeiroBenítez M, GarcíaJiménez MA, RodríguezElvira M. Evoked potentials in brain death. *Electroencephalogr Clin Neurophysiol* 1987;**67(42P)**:43P.
68. Machado C, Valdes P, Garcia-Tigera J, Virues T, Biscay R, Miranda J, et al. Brain-stem auditory evoked potentials and brain death. *Electroencephalogr Clin Neurophysiol* 1991;**80(5)**:392-398.
69. de la Torre JC. Evaluation of brain death using somatosensory evoked potentials. Biol Psychiatry 1981; **16(10)**:931-935.

70. Mauguière F, Grand C, Fischer C, Courjon J. [Aspects of early somatosensory and auditory evoked potentials in neurologic comas and brain death]. *Rev Electroencephalogr Neurophysiol Clin* 1982;**12(3):** 280-285.
71. Fotiou F, Tsitsopoulos P, Sitzoglou K, Fekas L, Tavridis G. Evaluation of the somatosensory evoked potentials in brain death. *Electromyogr Clin Neurophysiol* 1987;**27(1):**55-60.
72. Stohr M, Riffel B, Trost E, Ullrich A. Short-latency somatosensory evoked potentials in brain death. *J Neurol* 1987;**234(4):**211-214.
73. Besser R, Dillmann U, Henn M. Somatosensory evoked potentials aiding the diagnosis of brain death. *Neurosurg Rev* 1988;**11(2):**171-175.
74. Buchner H, Ferbert A, Hacke W. Serial recording of median nerve stimulated subcortical somatosensory evoked potentials (SEPs) in developing brain death. *Electroencephalogr Clin Neurophysiol* 1988;**69(1):**14-23.
75. Facco E, Caputo P, Casartelli LM, Munari M, Toffoletto F, Fabiani F, et al. Auditory and somatosensory evoked potentials in brain-dead patients. *Riv Neurol* 1988;**58(4):**140-145.
76. Chancellor AM, Frith RW, Shaw NA. Somatosensory evoked potentials following severe head injury: loss of the thalamic potential with brain death. *J Neurol Sci* 1988;**87(2-3):**255-263.
77. Buchner H, Ferbert A, Hacke W. Serial recording of median nerve stimulated somatosensory evoked potentials in brain death. *Neurosurg Rev* 1989;**12(Suppl 1):**349-352.
78. Haupt WF. Evoked potentials in non-traumatic brain death. *Neurosurg Rev* 1989;**12(Suppl 1):**369-374.
79. Facco E, Casartelli LM, Munari M, Toffoletto F, Baratto F, Giron GP. Short latency evoked potentials: new criteria for brain death? *J Neurol Neurosurg Psychiatry* 1990;**53(4):**351-353.
80. Machado C. Multimodality evoked potentials and electroretinography in a test battery for an early diagnosis of brain death. *J Neurosurg Sci* 1993;**37(3):**125-131.
81. Facco E, Giron GP. Multimodality evoked potentials in coma and brain death. *Minerva Anestesiol* 1994; **60(10):**593-599.
82. Wagner W. Scalp, earlobe and nasopharyngeal recordings of the median nerve somatosensory evoked P14 potential in coma and brain death. Detailed latency and amplitude analysis in 181 patients. *Brain* 1996; **119(Pt 5):**1507-1521.
83. Tomita Y, Fukuda C, Maegaki Y, Hanaki K, Kitagawa K, Sanpei M. Re-evaluation of short latency somatosensory evoked potentials (P13, P14 and N18) for brainstem function in children who once suffered from deep coma. *Brain Dev* 2003;**25(5):**352-356.
83. Sonoo M, Tsai-Shozawa Y, Aoki M, Nakatani T, Hatanaka Y, Mochizuki A, et al. N18 in median somatosensory evoked potentials: a new indicator of medullary function useful for the diagnosis of brain death. *J Neurol Neurosurg Psychiatry* 1999;**67(3):**374-378.
84. Facco E. Review: the diagnosis of brain death - role of short latency evoked potentials. *J Audiol Med* 2001; **10:**1-19.
85. Desmedt JE, Cheron G. Prevertebral (oesophageal) recording of subcortical somatosensory evoked potentials in man: the spinal P13 component and the dual nature of the spinal generators. *Electroencephalogr Clin Neurophysiol* 1981;**52(4):**257-275.
86. Mauguière F. Cervical somatosensory evoked potentials in the healthy subject: analysis of the effect of the location of the reference electrode on aspects of the responses. *Rev Electroencephalogr Neurophysiol Clin* 1983;**13(3):**259-272.
87. Desmedt JE. Critical neuromonitoring at spinal and brainstem levels by somatosensory evoked potentials. *Cent Nerv Syst Trauma* 1985;**2(3):**169-186.
88. Mauguière F, Ibanez V. The dissociation of early SEP components in lesions of the cervico-medullary junction: a cue for routine interpretation of abnormal cervical responses to median nerve stimulation. *Electroencephalogr Clin Neurophysiol* 1985;**62(6):**406-420.
89. Möller AR, Jannetta PJ, Burgess JE. Neural generators of the somatosensory evoked potentials: recording from the cuneate nucleus in man and monkeys. *Electroencephalogr Clin Neurophysiol* 1986; **65(4):**241-248.
90. Morioka T, Tobimatsu S, Fujii K, Fukui M, Kato M, Matsubara T. Origin and distribution of brain-stem somatosensory evoked potentials in humans. *Electroencephalogr Clin Neurophysiol* 1991;**80(3):**221-227.
91. Sonoo M, Shimpo T, Takeda K, Genba K, Nakano I, Mannen T. SEPs in two patients with localized lesions of the postcentral gyrus. *Electroencephalogr Clin Neurophysiol* 1991;**80(6):**536-546.
92. Sonoo M, Genba K, Zai W, Iwata M, Mannen T, Kanazawa I. Origin of the widespread N18 in median nerve SEP. *Electroencephalogr Clin Neurophysiol* 1992;**84(5):**418-425.
93. Mavroudakis N, Brunko E, Delberghe X, Zegers dB. Dissociation of P13-P14 far-field potentials: clinical and MRI correlation. *Electroencephalogr Clin Neurophysiol* 1993;**88(3):**240-242.
94. Raroque HG Jr, Batjer H, White C, Bell WL, Bowman G, Greenlee R Jr. Lower brain-stem origin of the median nerve N18 potential. *Electroencephalogr Clin Neurophysiol* 1994;**90(2):**170-172.

95. Hashimoto T, Miyasaka M, Yanagisawa N. Short-latency somatosensory evoked potentials following median nerve stimulation in four patients with medial medullary infarction. *Eur Neurol* 1995;**35(4)**:220-225.
96. Restuccia D, Di Lazzaro V, Valeriani M, Conti G, Tonali P, Mauguière F. Origin and distribution of P13 and P14 far-field potentials after median nerve stimulation. Scalp, nasopharyngeal and neck recording in healthy subjects and in patients with cervical and cervico-medullary lesions. *Electroencephalogr Clin Neurophysiol* 1995;**96(5)**:371-384.
97. Noel P, Ozaki I, Desmedt JE. Origin of N18 and P14 far-fields of median nerve somatosensory evoked potentials studied in patients with a brain-stem lesion. *Electroencephalogr Clin Neurophysiol* 1996;**98(2)**: 167-170.
98. Sonoo M, Hagiwara H, Motoyoshi Y, Shimizu T. Preserved widespread N18 and progressive loss of P13/14 of median nerve SEPs in a patient with unilateral medial medullary syndrome. *Electroencephalogr Clin Neurophysiol* 1996;**100(6)**:488-492.
99. Sonoo M. Anatomic origin and clinical application of the widespread N18 potential in median nerve somatosensory evoked potentials. *J Clin Neurophysiol* 2000;**17(3)**:258-268.
100. Mauguière F, Desmedt JE, Courjon J. Neural generators of N18 and P14 far-field somatosensory evoked potentials studied in patients with lesion of thalamus or thalamo-cortical radiations. *Electroencephalogr Clin Neurophysiol* 1983;**56(4)**:283-292.
101. Urasaki E, Wada S, Kadoya C, Tokimura T, Yokota A, Yamamoto S, et al. Amplitude abnormalities in the scalp far-field N18 of SSEPs to median nerve stimulation in patients with midbrain-pontine lesion. *Electroencephalogr Clin Neurophysiol* 1992;**84(3)**:232-242.
102. Sonoo M, Tsai-Shozawa Y, Aoki M, Nakatani T, Hatanaka Y, Mochizuki A, et al. N18 in median somatosensory evoked potentials: a new indicator of medullary function useful for the diagnosis of brain death. *J Neurol Neurosurg Psychiatry* 1999;**67(3)**:374-378.
103. Sonoo M, Tsai-Shozawa Y, Aoki M, Nakatani T, Hatanaka Y, Mochizuki A, et al. N18 in median somatosensory evoked potentials: a new indicator of medullary function useful for the diagnosis of brain death. *J Neurol Neurosurg Psychiatry* 1999;**67(3)**:374-378.

RECOVERY FROM NEAR DEATH FOLLOWING CEREBRAL ANOXIA

A case report demonstrating superiority of median somatosensory evoked potentials over EEG in predicting a favorable outcome after cardiopulmonary resuscitation

Ted L. Rothstein*

1. INTRODUCTION

Prognostication of comatose patients after cardiac arrest presents many challenges. Most clinicians rely upon clinical neurological examination findings to predict outcome. The outcome is usually considered unfavorable when a patient experiences persistent brainstem dysfunction. However, prognostic scales, which rely upon brainstem dysfunction alone, are flawed. Some patients with brainstem dysfunction recover while patients with preserved brainstem function often die without awakening as a result of irreversible destruction of the cerebral cortex.

Electrophysiologic studies can assist in the evaluation and prediction of outcome in comatose patients following cardiorespiratory arrest (CPA). Median Somatosensory evoked potentials (SSEP) and electroencphalography (EEG) have been used to prognosticate based on patterns associated with poor outcome.[1-11] The early absence of an N20/P23 cortical response on SSEP is always predictive of death without awakening.[11] Malignant changes on EEG such as electrocerebral silence (ECS) and burst suppression also predict little chance for recovery.[12] However, a normal EEG or SSEP can be misleading as a prognostic indicator as some patients with normal studies die without awakening.[4,11,13,14]

I report the case of a patient deeply comatose after CPA with spontaneous generalized myoclonus and apnea with an admitting Glasgow Coma Score (GCS) of 3 (no verbal response, no spontaneous eye opening, and no motor response), and whose initial EEG performed 5 hours after his resuscitation was isoelectric. These findings would place him in the category of patients with the poorest chance of survival. However, SSEP performed immediately following the EEG and on 3 successive days were normal. The SSEP proved more reliable as a predictor of outcome than either clinical examination

* Ted Rothstein, MD, Department of Neurology, George Washington University Medical Center, Washington, DC; e-mail: trothstein@mfa.gwu.edu.

Brain Death and Disorders of Consciousness, Edited by Machado and Shewmon
Kluwer Academic/Plenum Publishers, New York 2004

or EEG, as the patient regained awareness on the 4th hospital day and eventually made a complete recovery.

2. CASE REPORT

A 50 year old man was admitted to the hospital in coma. He had complained of indigestion, collapsed and became unresponsive. A bystander performed immediate CPR until the arrival of emergency medical technicians. He was found to be in pulseless ventricular fibrillation. He received intravenous lidocaine and epinephrine and underwent repeated electronic defibrillations. After 20 minutes there was return of sinus rhythm.

The patient had enjoyed an active healthy life. He had been hospitalized twelve years previously with a "mild" myocardial infarction. He had hypertension for several years and was treated with a beta-blocker and diuretic. The family history was non-contributory.

On physical examination he was deeply comatose with occasional generalized myoclonic movements. The temperature was 98 degrees F, the pulse 116 and blood pressure was 115 systolic and 65 diastolic. The neck veins were flat and the lungs were clear. The heart sounds were distant and the point of maximum impulse was not felt. The abdomen was normal and there was no peripheral edema.

On neurologic examination there was no spontaneous eye opening. The optic fundi were normal and the pupils measured 4mm in diameter and reacted promptly to light. The oculocephalic eye movements were absent. Ice water calorics produced no response. Corneal reflexes were absent bilaterally. Muscle tone was reduced and when picked up the extremities fell limply. There was no motor response to noxious stimulation. The deep tendon reflexes were equal and normal in the arms and knees and absent in the ankles. The plantar response was extensor on the right and mute on the left. There were no spontaneous respirations and he required full ventilatory support.

The urine was normal. The hematocrit was 41.7 per cent, and the white-cell count was 11,400. The urea nitrogen was 26 mg./dl, serum glucose 223 mg/dl, SG0T 127 IU, SGPT 141 IU, CPK 127 IU and troponin 0.1 ng. A toxicology screen was negative. An electrocardiogram revealed well developed Q waves and diffuse ST depression consistent with acute inferior myocardial injury. A left ventriculogram disclosed inferobasal akinesis. Coronary artery angiography showed a shelf-like thrombus in the left anterior descending artery and extensive narrowing in the right coronary artery and circumflex arteries.

Within one hour after admission the patient experienced a generalized tonic-clonic seizure and was treated with 1 gram intravenous fosphenytoin and 1 mg intravenous lorazepam.

A digital computerized EEG performed 5 hours following the arrest was isoelectric using gains of 7 uV per millimeter (Figure 1). At maximal gains of 2 μV per millimeter there was extensive EKG, respirator and muscle artifact but no demonstrable cerebral electrical activity. A median nerve somatosensory evoked potential study (SSEP) revealed a normal N20/P23 cortical response (Figure 2).

On the second hospital day, the patient remained deeply comatose with small reactive pupils, no spontaneous eye movements, and oculocephalic maneuver produced 10 degrees of horizontal excursion. The corneal reflexes were weakly reactive. He was observed to spontaneously cough and gag. The limbs were flaccid. He had a flexion

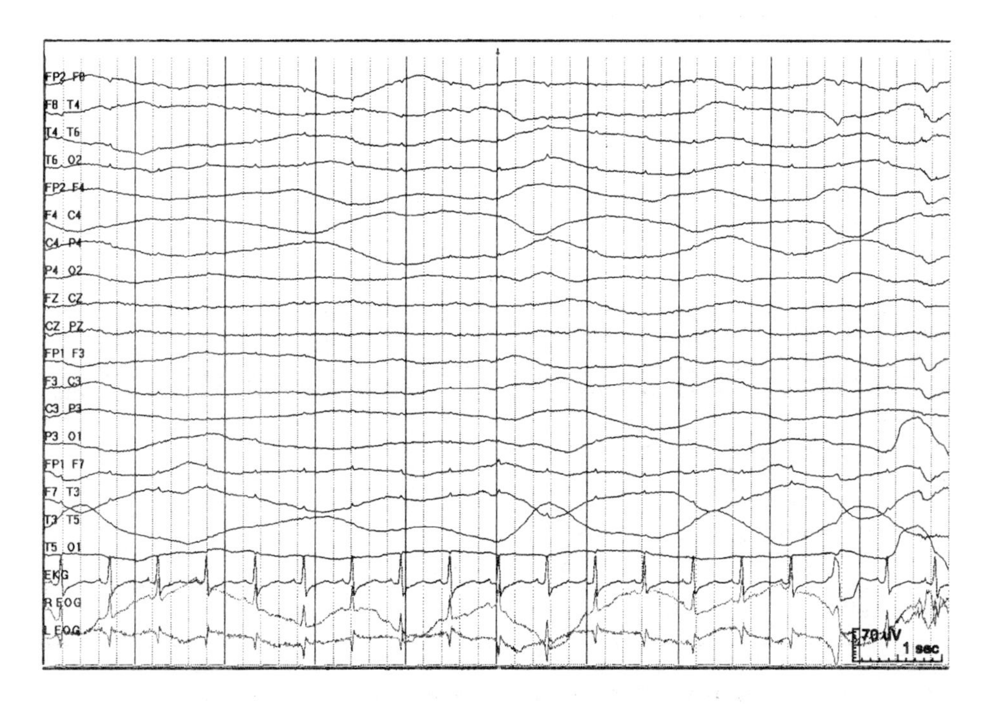

Figure1. Generalized suppression (isoelectric EEG) at 7 μV per millimeter taken 5 hours after cardiac arrest.

response in the upper extremities to stimuli. The right plantar response was extensor and the left was flexor. EEG disclosed alpha coma without reactivity (Figure 3). A median SSEP was normal.

On the third hospital day, he remained unresponsive but opened his eyes to painful stimulation. Oculocephalic movements were present but dysconjugate and corneal reflexes had returned. He would withdraw to pain. He required ventilatory support. An

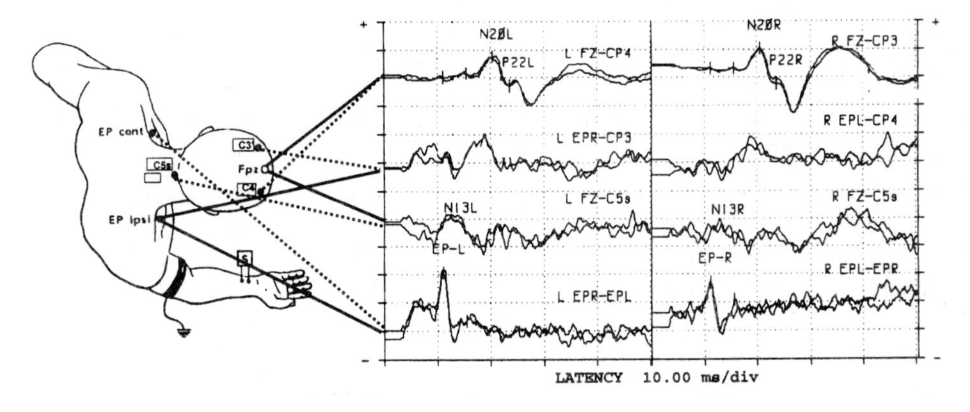

Figure 2. Median SSEP reveals normal central conduction times and N20/P23 cortical response bilaterally.

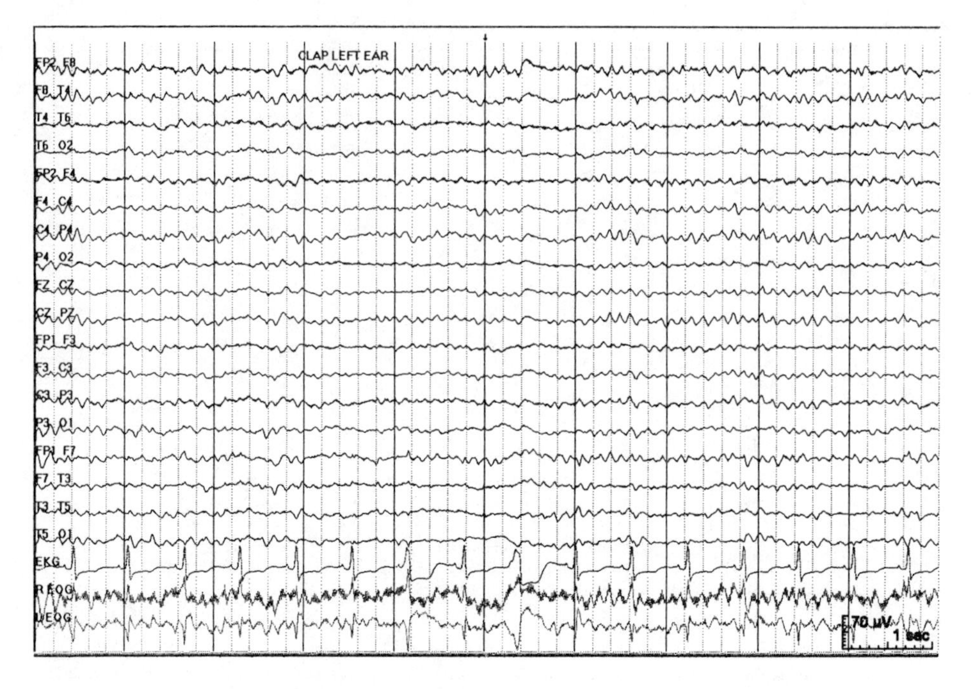

Figure 3. Alpha rhythms without reactivity to noise stimulation.

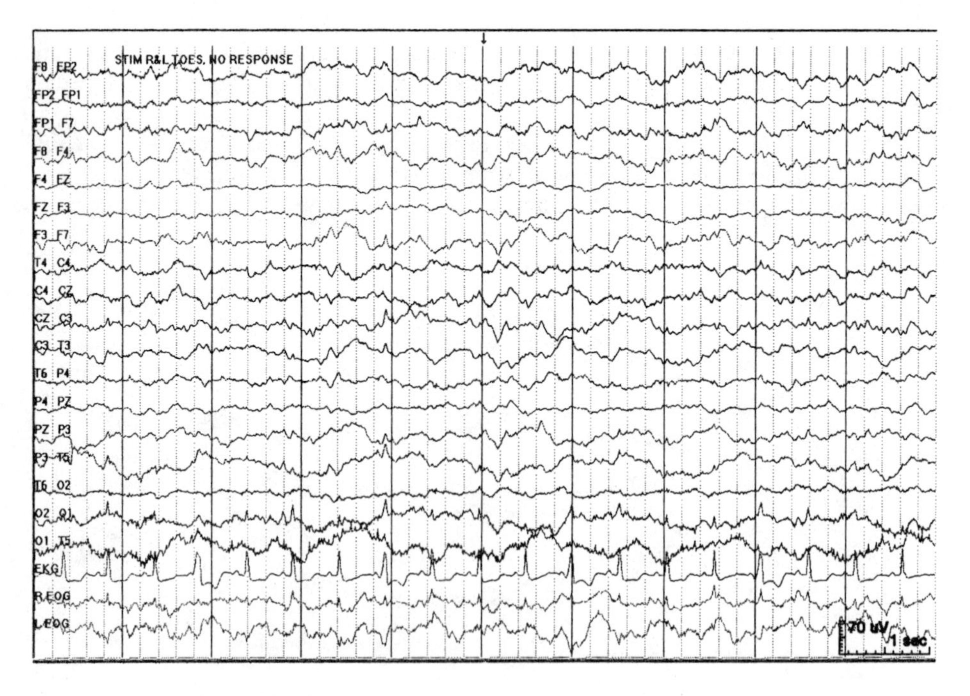

Figure 4. Alpha rhythms with admixed theta and delta activity.

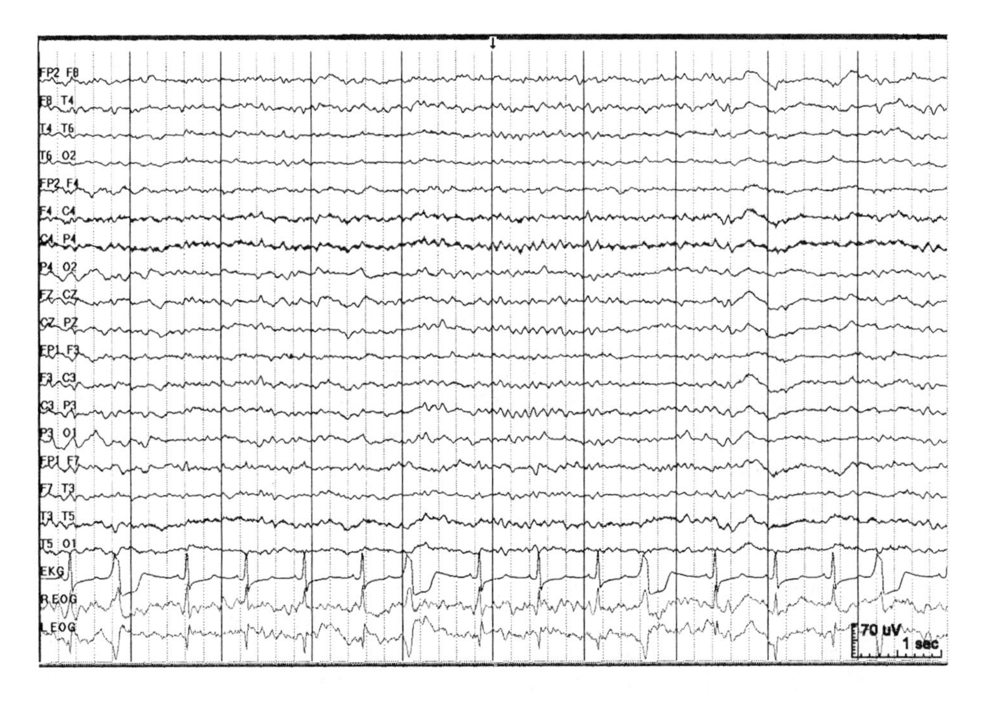

Figure 5. Normal EEG with mixed alpha and beta activity.

EEG revealed alpha activity with intermixed theta and delta rhythms (Figure 4), and a median SSEP was normal.

On the fourth hospital day he could be roused and followed simple commands. Upon request the patient would turn his eyes right or left, squeeze with his right hand and wiggle his toes. His breathing had become spontaneous and he was removed from assisted ventilation. He made no verbal response. An EEG disclosed mainly theta activity with admixed delta and some alpha rhythms (Figure 5). A CT brain scan was normal. Upon the family's request he was transferred to an out of state hospital in his hometown that evening.

The patient ultimately made a full recovery and returned to work one month after his cardiac arrest. (Ping Chow M.D., personal communication).

3. DISCUSSION

Most clinicians rely upon the neurologic examination as an early predictor of outcome following CPA. There is a close correlation between the number of brainstem reflex abnormalities and the possibility of survival.[15] Levy *et al*[16] found that pupillary reflexes, motor response and spontaneous eye movements were the most reliable markers. The loss of pupillary responses was recognized as an early predictor of unfavorable prognosis.[16,17] Similarly, the absence of oculocephalic reflexes correlated with severe disability or death.[18]

Brainstem structures are relatively resistant to the effects of anoxia[19,20] and when the brainstem is damaged by anoxic insult, it is reasonable to assume the cerebral cortex has sustained an even worse injury.[20,21] However, the preservation of brainstem reflexes does not necessarily predict a favorable outcome because some patients sustain a critical measure of anoxia sufficient to destroy the cerebral cortex while preserving some or all brainstem functions.[4] Seizures and myoclonus do not correlate with outcome unless they are continuous.[22]

Two of three criteria evaluated on the GCS are linked to brainstem function. However, predictive scales, which are dependent on brainstem function, can be misleading. Some patients with normal brainstem function can attain moderately high GCS scores but die without awakening due to irreversible destruction of the cerebral cortex. Conversely, patients who present with the minimal GCS of 3, can make a full recovery as documented in this case report.

The value of electrophysiologic study in predicting outcome in anoxic coma is well established. Certain EEG patterns have strong association with poor outcome.[1-11] Several systems for grading EEG have been devised to aid in prognostication after cardiac arrest. EEG with normal or nearly normal frequency, voltage, and reactivity are assigned low grades while those with diffuse delta activity, burst suppression, or ECS are assigned high grades.[12,23-6] Synek,[12] identified diffuse delta, burst suppression and very flat or ECS patterns on EEG as "malignant," and these were associated with an unfavorable outcome. Other studies supported a relationship between generalized voltage suppression, burst suppression and generalized epileptiforme activity with a hopeless prognosis.[4,11,23-27]

Some authors contend that ECS predicts a uniformly poor prognosis (excluding sedative drug effect or hypothermia).[12,28] and it remains a confirmatory test for validation of brain death in many countries.[29] In the American Collaborative Study[30] only two of 187 patients presenting with ECS regained consciousness and both were resuscitated after drug overdoses.

In past years, the Ad Hoc Committee of the American EEG Society concluded that except for drug overdose and hypothermia, ECS when taken together with "unreactive coma" was strong presumptive evidence for irreversible coma and cerebral death.[31] The EEG has been used in deciding to discontinue resuscitation in some cases.[32]

However, physiological experiments have shown that a severe anoxic insult to the cerebral cortex, leading to ECS may be reversible.[25] Jørgenson *et al*,[17] in a series of 231 patients resuscitated from circulatory arrest studied 125 patients with ECS immediately after resuscitation, and found that 32 patients followed an orderly sequence of recovery of EEG and neurologic function. They found an inverse relationship between the duration of ECS and extent of recovery.

Møller[32] studied 185 patients with acute myocardial infarction and cardiac arrest and concluded that it was not possible to predict the extent of anoxic brain damage on the basis of a single EEG.

SSEP provide valuable information regarding the functional state of the cerebral cortex during coma but have not been widely utilized as a predictor of outcome.[28] SSEP assess the functional integrity of the posterior columns of the spinal cord, brainstem medial lemniscus, thalamus, and frontoparietal sensorimotor cortex. However, normal SSEP do not assure a favorable prognosis. Those comatose patients with an anoxic-ischemic brain injury and normal SSEP who fail to awaken most likely have damaged or dying neurons which are still capable of generating normal or near normal cortical responses.[4] SSEP tend to flatten and attenuate over time in these patients.[4]

Chen *et al*[10] performed a prospective study involving 34 patients in anoxic-ischemic coma, who had "malignant" changes on EEG. Patients with ECS were excluded from their analysis. Two patients with burst suppression patterns on EEG recovered, with minimal or no neurological residual. They could be distinguished by having some brainstem reflexes and normal median SSEP. The authors concluded that patients with an initial malignant EEG pattern and normal SSEP have potential for recovery and should be supported until their condition had changed to a more prognostically definitive category.

My patient sustained an anoxic insult sufficiently severe to produce absence of some brainstem reflexes, apnea, seizure activity, myoclonus, a GCS of 3 on admission, an isoelectric EEG 5 hours after onset, and coma lasting more than 72 hours, none of which precluded a full recovery. The normal N20/P23 obtained on SSEP in the same time frame as an isoelectric EEG would indicate that cortical neurons are capable of responding to afferent stimuli at a time they are incapable of generating spontaneous electrical activity.

The attempt to establish neurologic prognosis in comatose patients immediately following CPA has major ethical and socioeconomic implications but remains uncertain. Caution must be applied when relying on the neurologic examination and the presence or absence of brainstem function. Electrophysiological study can provide valuable information in these patients. When there is inconsistency between SSEP and EEG results the clinician must use caution in predicting outcome. While normal SSEP do not invariably anticipate neurologic recovery, some comatose patients with malignant EEG and normal SSEP can and do recover. Such patients warrant full support until a more definitive prognosis can be ascertained over time.

4. ACKNOWLEDGMENTS

I thank Elizabeth M. Thomas and Grady Woodward who performed the EEG and SSEP studies. I also thank David Trees who provided technical support. Julie Heyn, M.D. performed the cardiac evaluation.

5. REFERENCES

1. Hockaday JM, Potts F, Epstein E, et al. Electroencephalographic changes in acute cerebral anoxia from cardiac or respiratory arrest: prognostic value of early electroencephalographic findings. *Electroencephalogr Clin Neurophysiol* 1965;**18:**575-86.
2. Binnie CD, Prior PF, Lloyd, DSL Scott DF, Margerison JH. Electroencphalographic prediction of fatal anoxic brain damage after resuscitation from cardiac arrest. *Br Med J* 1970;**4:**265-8.
3. Goldie WD, Chiappa, KH, Young RR, Brooks EB. Brainstem auditory and short-latency somatosensory evoked responses in brain death. *Neurology* 1981;**31:**248-56.
4. Rothstein TL, Thomas EM, Sumi SM. Predicting outcome in hypoxic-ischemic coma. A prospective and electrophysiologic study. *Electroencephalogr Clin Neurophysiol* 1991;**79:**101-107.
5. Chiappa KH, Hoch DB. Electrophysiologic monitoring. In: Roper AH, ed. *Neurological and neurosurgical intensive care.* New York: Raven press, 1993;147-83.
6. Madl C, Grimm G, Kramer L, et al. Early prediction of individual outcome after cardiopulmonary resuscitation. *Lancet* 1993;**341:**855-858.
7. Young GB, Blume WT, Campbell VM, et al. Alpha, theta, and alpha-theta coma; a clinical outcome study using serial recordings. *Electroencephalogr Clin Neurophysiol* 1994;**91:**93-99.
8. Berek K, Lechleitner P, Leuf G, et al. Early determination of neurological outcome after prehospital cardiopulmonary resuscitation. *Stroke* 1995;**26:**543-549.

9. Madl C, Kramer L, Yegaehfar W, et al. Detection of nontraumatic comatose patients with no benefit from intensive care treatment by recording sensory evoked potentials. *Arch Neurol* 1996;**53**:512-16.
10. Chen R, Bolton CF, Young GB. Prediction of outcome in patients with anoxic coma: A clinical and electrophysiologic study. *Crit Care Med* 1996:**24**:672-675.
11. Rothstein TL. The role of evoked potentials in anoxic-ischemic coma and severe brain trauma. *J Clin Neurophysiol* 2000;**17**:486-497.
12. Synek VM. Value of a revised EEG coma scale for predicting survival in anoxic-encephalopathy. *Clin Exp Neurol* 1989;**26**:119-127.
13. Bassetti C, Bomio F, Mathis J, Hess CW. Early prognosis in coma after cardiac arrest: a prospective clinical, electrophysiological, and biochemical study of 60 patients. *J Neurol Neurosurg Psychiatry* 1996; **61**:610-615.
14. Pohlmann-Eden B, Dingethal K, Bender H-J, Koelfen W. How reliable is the predictive value of SEP (somatosensory evoked potentials) patterns in severe brain damage with special regard to the bilateral loss of cortical responses? *Intensive Care Med* 1997;23:301-308.
15. Snyder BD, Gumnit, RJ, Leppik IE, Hauser MD, Loewenson RB, Ramirez-Lassepas M. Neurologic prognosis after cardiopulmonary arrest: 1V. Brainstem reflexes. *Neurology* 1981;**31**:1092-1099.
16. Levy DE, Caronna JJ, Singer BH, Lapinski RH, Frydman H, Plum F. Predicting outcome from hypoxic-ischemic coma. *JAMA* 1985;**253**:1420-1426.
17. Jørgenson EO, Malchow-Møller A. Natural history of global and critical brain ischemia. *Resuscitation* 1981;**9**:133-138.
18. Mueller-Jensen A, Neunzig HP, Emskotter T. Outcome prediction in comatose patients: significance of reflex eye movement analysis. *J Neurol Neurosurg Psychiatry* 1987;**50**:389-392.
19. Brierly, JB, Graham DI, Adams JH, Simpson JA. Neocortical death after cardiac arrest: a clinical, neurophysiological and neuropathological report of two cases. *Lancet* 1971;**ii**:560-566.
20. Mullie A, Buylaert W, Michem N, et al. Predictive value of Glasgow coma score for awakening after out-of-hospital cardiac arrest. *Lancet* 1988;**i**:137-40.
21. Longstreth WT Jr. Neurological Complications of Cardiac Arrest. In Aminoff MJ, ed. *Neurology and General Medicine*. New York: Churchill Livingstone, 1994:166.
22. Krumholz A, Stern BJ, Weiss HD. Outcome from coma after cardiopulmonary resuscitation: relation to seizures and myoclonus. *Neurology* 1988;**38**:401-405.
23. Scollo-Lavizzari G, Bassetti C. Prognostic value of EEG in post-anoxic coma after cardiac arrest. *Eur Neurol* **26**:161-170, 1987
24. Pampiglione G, Hardan A. Resuscitation after circulatory arrest. Prognostic evaluation of early electroencephalographic findings. *Lancet* 1968;**1**:1261-1265.
25. Prior P. *The EEG in acute cerebral anoxia*. Assessment of cerebral function and prognosis in patients resuscitated after cardiorespiratory arrest. Amsterdam: Excerpta Medica, 1973.
26. Bassetti C, Karbowski K. Prognostic value of electroencephalography in non-traumatic comas. *Schweiz Med Wochenschr* 1990;**120**:1425-1434.
27. Young GB, Kreeft JH, McLachlan RS, Demelo J. EEG and clinical associations with mortality in comatose patients in a general ICU. *J Clin Neurophsyiol* **16**:354-360, 1999.
28. Attia J, Cook DJ. Prognosis in anoxic and traumatic coma. *Crit Care Clin* 1998;**14**:497-511.
29. Wijdicks EFM. The diagnosis of brain death. *N Engl J Med* 2001;344:1215-1221.
30. Walker, AE. *Cerebral death*. Dallas: Professional Information Library, 1977
31. Silverman D, Saunders MG, Schwab RS, Masland RL. Cerebral death and the electroenceophalogram. Report of the ad hoc committee of the American Electroencephalographic Society on EEG criteria for determination of cerebral death. *JAMA* 1969;**209**:1505-1510.
32. Møller M, Holm B, Sindrup E, Nielson BL. Electroencephalographic prediction of anoxic brain damage after resuscitation from cardiac arrest in patients with acute myocardial infarction. *Acta Med Scand* 1978; **203**:33-37.

HUMAN BRAIN-DEAD DONORS AND ^{31}P MRS STUDIES ON FELINE MYOCARDIAL ENERGY METABOLISM

George J. Brandon Bravo Bruinsma and Cees J. A. van Echteld*

1. INTRODUCTION

In clinical heart transplantation, the heart of a brain dead donor is used. The quality of the donor heart is one of the key factors for a succesful transplantation. A hemodynamically unstable brain dead donor is often rejected because survival of the recipient is reduced.[1-3] The precise mechanisms of brain death-related hemodynamic instability remain unknown. One of the reported contributing factors is myocardial injury resulting from the acutely increased discharge of endogenous catecholamines during the onset of brain death.[4,5] This injury has been supposed to change aerobic to anaerobic energy metabolism, causing depletion of myocardial high-energy phosphates and contractile dysfunction.[6-8] Energy depletion appeared to be even more pronounced when high dosages of inotropic agents were used to treat contractile dysfunction of the canine donor heart.[9] Notwithstanding these results, the presence of anaerobic metabolism is disputed by others.[10-13]

Explantation, preservation, and storage of the heart are essential aspects of transplantation. The harmful impact of these aspects on the integrity of the myocardium has been related to the method of cardioplegic arrest and conditions of storage.[14-16] Up to now, static information on the myocardial energy status was derived principally from small biopsies of the left ventricle. In addition, biopsies are liable to phosphocreatine (PCr) and adenosine triphosphate (ATP) breakdown during the assessment procedure. These factors may, in part, explain the contradicting results. Phosphorus Magnetic Resonance Spectroscopy (MRS) provides continuous information of a large area of the left ventricle or the whole isolated heart. It has contributed important insights, mainly derived from animal experiments, to current controversies in human donor issues.[17] Up to now, assessment of myocardial energy metabolism by applying ^{31}P MRS on the human donor is too complicated. The ultimate goal presently possible is to correlate myocardial energy status of the isolated donor heart with post transplantation heart performance.[18]

* G.J. Brandon Bravo Bruinsma, M.D., Ph.D., Cardiothoracic Surgeon, Department of Cardiothoracic Surgery, Isala Clinics, Groot Wezenland 20, 8011 JW Zwolle, The Netherlands, Telephone: + 31384242866, Fax: + 31384243163, Email: g.j.bbb@planet.nl. C.J.A. van Echteld, Ph.D., F.E.S.C., Heart Lung Center Utrecht, CardioNMR Laboratory, University Medical Center Utrecht, The Netherlands.

In this review we explain the principles of ^{31}P MRS and its application in the research of brain death-related issues. Furthermore, we have combined the results of a series of experiments to address the impact of brain death on *in vivo* and *ex vivo* high-energy phosphate metabolism of the feline donor heart during the onset of brain death, on the hemodynamic instability which follows up to 6 h afterwards, on the treatment of hemodynamic instability with an inotropic agent, and during perservation and reperfusion after explantation.

2. PHOSPHORUS MAGNETIC RESONANCE SPECTROSCOPY

Magnetic resonance spectroscopy is a non-destructive technique, depending on the emission of radio frequencies (rf) of resonating atomic nuclei with an odd number of protons and/or neutrons after absorption of energy from rf pulses. These rf pulses are transmitted by a surface coil, applied locally on the left ventricle, or when the heart is explanted, by a volume (Helmholtz) surrounding coil. As the atomic nuclei, *e.g.* ^{1}H and ^{31}P, exhibit a rotational motion (spin) their behavior resembles the action of magnets. When they are brought into the magnetic field of a cryomagnet they align in a disproportional but balanced parallel or antiparallel orientation.[19,20] Absorption of rf energy, affects the distribution of the P nuclei over the parallel and antiparallel orientations. After cessation of the rf pulse, the equilibrium situation will be regained as the resonance frequency signal of the nuclei is emitted. The return to the initial situation is determined by longitudinal and transversal relaxation time constants, T1 and T2, respectively. The emitted signal, detected by the same coil, can be described as free induction decay, FID, and is visualized as a spectrum of frequencies following Fourier transformation.

^{31}P MRS is an ideally suited technique to study *in vivo* energy metabolism of the heart and was introduced as such by Gadian *et al.*[21] in 1976. A phosphorus MR spectrum contains several peaks, which correspond to phosphorus nuclei in different chemical environments, *e.g.* phosphomonoesters (PME), phosphodiesters (PDE), inorganic phosphate (P*i*), PCr and adenosine triphosphate (γ, α, and β-ATP). The peaks are positioned along a horizontal axis which is divided in parts per million (ppm), a magnetic field independent, dimensionless parameter of frequency differences. The integrated area under each peak is proportional to the quantity of the ^{31}P nuclei detected by the coil. Some peaks, like PME and PDE, represent more than one kind of phosphorus containing molecules. Others, like Pi, may be obscured by resonances of phosphate coming from another origin, *e.g.* 2,3-diphosphoglycerate in blood or P*i* in muscle tissue adjacent to the coil. This is especially the case during *in vivo* experiments and quantification of those metabolites cannot be achieved reliably.

Adenosine, when phosphorylated, appears in three configurations: mono-(AMP), di-(ADP), and triphosphate (ATP). AMP is part of the PME peak. ADP appears at two frequencies representing phosphate in the α and β-configuration in the molecule. ATP appears at frequencies representing phosphate in the configurations α, γ, and β. However, the α and β resonances of ADP occur at the same frequency as the resonances of α and γ-ATP, respectively. Therefore, resonances from the β-ATP configuration are not influenced by ADP and may serve as the most accurate estimation of the amount of free ATP.

PCr content is a very sensitive indicator of myocardial ischemia. Within seconds after the onset of ischemia PCr decreases, and P*i* increases, in order to maintain ATP content.

Anaerobic glycolysis cannot produce sufficient ATP to cope with the rate of ATP hydrolysis during myocardial ischemia.[22,23] Therefore, ATP will progressively decrease as ischemia continues. This reduction in energy supply is accompanied by a reduction of contractile capacity of the heart. Under aerobic conditions, the P*i*/PCr ratio is a very sensitive index for oxidative phosphorylation.[24] But, as mentioned earlier, accurate assessment of *in vivo* P*i* is very difficult. Therefore, the PCr/ATP ratio is often applied and related to changes in myocardial workload.[25-27]

3. BRAIN DEATH AND ^{31}P MRS RESEARCH OF THE MYOCARDIUM

Although many *in vivo* and *ex vivo* MRS studies have been performed on the myocardial energy metabolism since 1976, only quite recently, Kitai *et al.*[28] were the first to apply this technique to study the impact of brain death on the energy status of the canine donor heart. Despite the challenging technical difficulties of phosphorus spectroscopy and the experimental brain death model, they reported no change in myocardial PCr/P*i* ratio in the hypotensive brain dead dog. Pinelli *et al.*,[29] have applied *ex vivo* MRS on the isolated heart of pigs during 6 h storage, subsequent to 3 h brain death. The investigators observed a significant decrease in β-ATP/S (= total phosphosus pool) in the brain death group vs control group, already present at the onset of preservation. Carteaux *et al.*[30] have shown in an *ex vivo* MRS study of isolated and subsequently blood perfused hearts, explanted from not-brain dead pigs, that PCr/S and pHi, measured at the onset and end of preservation, correlated with left ventricular contractility, expressed by LV dP/dt_{max}, during reperfusion. Van Dobbenburgh *et al.*[18] were able to relate the functional recovery after heart transplantation to the metabolic condition of the hypothermic human donor heart. Their most important conclusion was that the cardiac index of heart transplant recipients, 1 week after transplantation, correlated significantly with the extrapolated PCr/ATP, PME/ATP, and PCr/P*i* ratios at the time of reperfusion.

4. MATERIALS AND METHODS

In our four brain death studies, cats were sedated and placed in supine position. The first study was meant to investigate the hemodynamic response to brain death.[31] Cats were assigned alternately to the brain death group (n=15) or the sham operated control group (n=12). In the following three MRS studies, the cats were assigned alternately to three brain death groups (each n=6) or three control groups (each n=6). The animals were intubated and mechanically ventilated.

4.1. Assessment of Hemodynamic Alterations

Catheters served for continuous recordings of central venous pressure (CVP) and arterial blood pressure (and mean arterial blood pressure (MAP)). The latter catheter was also used for arterial blood sampling. Continuous intravenous administration of anesthetics and Geloplasma was applied. Heart rate (HR) was measured from three-lead surface ECG electrodes. CVP was kept constant. Myocardial workload was expressed as rate-pressure product (RPP = HR x systolic arterial blood pressure). Brain death-related hemodynamic instability was characterized by a significantly lower MAP vs the control group.

4.2. Registration and Induction of Brain Death

In prone position, four silver electrodes were inserted into the scalp of the cat. Recordings (sensitivity 50 μV/cm, band width 0.53 - 70 Hz (-3dB), paperspeed 1 cm/sec) of electrical cerebral activity were performed prior and during the induction of brain death, and for 2 min every hour after the induction of brain death.

A trephine hole, 2 mm in diameter and 1 cm lateral from the sutura sagittalis, was drilled in the right dorsoparietal cranial area. A 6 Fr balloon catheter was inserted epidurally and parallel to the sutura sagittalis. At t = 0 min, brain death was induced by inflation of the balloon with a fixed volume of 3.5 ml saline at a rate of 0.6 ± 0.1 ml/s, 120 min after premedication.[31] The ballon was kept inflated during the experimental time. In the control group (n=12) of the first study, the intracranial balloon was inserted but not inflated. In the control groups (n=6) of the MRS experiments, the balloon was not inserted. The criterion for brain death was development of complete electrocerebral inactivity.

4.3. Assessment of Myocardial Energy Metabolism

4.3.1. *In vivo* ^{31}*P MRS assessment of Myocardial Energy Metabolism*

The heart was exposed by a midsternal sternotomy. A circular, single turn RF surface coil was sutured to the free wall of the left ventricle. The cat was positioned in an 4.7 T horizontal magnet, interfaced to a spectrometer. The surface coil was pretuned and field homogeneity was optimized. Data acquisition was gated both to the midexpiration phase of the ventilation cycle (frequency set at approximately 25/min) and to the upstroke of the systolic arterial blood pressure curve. After excitation, ^{31}P MR spectra were obtained. Two *in vivo* studies were performed.

4.3.1a. Onset of Brain Death and Hemodynamic Instability

In this first protocol, 240 min lasting, *in vivo* ^{31}P MRS spectra were obtained from 12 or 24 FIDs with a repetition time (T_R) of 2.4 s. Between 2.5 min before and 15 after the induction of brain death, 30 sec spectra (12 FID's) were aquired. For assessment of the basal condition of the myocardial energy metabolism and during the period of hemodynamic instability, 60 s (24 FID's) spectra were acquired 5 min before and from 15 min till 240 min after the induction of brain death.

4.3.1b. Inotropic Treatment of Hemodynamic Instability

In a second protocol, 360 min lasting, *in vivo* ^{31}P MRS spectra were obtained from 60 FID's (5 min) with a T_R of 4.8 s during 5 min before the induction of brain death until 360 min afterwards. At 210 min dopamine was infused intravenously at 5μg/kg/min for 30 min. The dosage was consecutively increased to 10, 20, and 40 μg/kg/min every 30 min. Dopamine infusion was discontinued during the last 30 min of the 360 min protocol.

4.3.2. Ex vivo ^31P MRS Assessment of Myocardial Energy Metabolism

In the third protocol, 24 h lasting, the heart was explanted, as in standard clinical practice, in a state of progressive hemodynamic deterioration at 360 min after the induction of brain death. It was mounted to a Langendorff perfusion system, placed in a glass MR tube and lowered into a vertical 4.7 T magnet, interfaced to a spectrometer. *Ex vivo* ^{31}P MRS spectra were obtained from 128 FID's with a T_R of 2.35 s at 4^{0}C. The first 5 min spectra were obtained 1 h after explantation. A spectrum was collected every hour during 17 h unperfused storage at 4^{0}C. During the following 60 min reperfusion at 38^{0}C, 5 min spectra were collected continuously. In both *in vivo* studies and in the *ex vivo* study, peak areas in all spectra were determined. PCr/ATP ratios were calculated. PCr and ATP were also expressed as a percentage of the initial value at t = 0 min (*4.3.1b.*), or in arbitrary units per g dry heart weight (*4.3.2.*). The changing conditions in all protocols did not allow proper measurement of T1. Therefore, no correction for partial saturation was applied.

4.4. Statistical Analysis

All results were expressed as mean ± SEM. The unpaired, two-tailed Student's *t*-test was used to determine the significance of differences between the means of the brain death group and the control group. Analysis of variance (ANOVA) for repeated measurements was applied when the values obtained within each group at various times were compared to the basal value. If significant differences were present the Bonferroni test was applied. A difference was considered significant when $p < 0.05$.

5. RESULTS

5.1. Brain Death

Electrocerebral activity disappeared in all animals within 30 s after inflation of the epidural balloon. In the sham operated group, insertion of the catheter without inflation was followed by only slight amplitude reduction of the right parietally recorded cortical electrical activity. Brain stem death was confirmed histologically by the presence of intraparenchymal haemorrhages at the junction of the pons and the mesencephalon.[31]

5.2. Ventilatory and Metabolic Variables

No significant differences in metabolic or ventilatory variables were observed between the two groups at t = -5 min and at t = 360 min after induction of brain death.

5.3. Hemodynamic Response

Heart rate increased from a basal value of 143 ± 7 to a maximum value of 241 ± 10 beats/min (p < 0.0001 vs control group) at 2 min, returned to normal levels at 20 min, and remained so for the rest of the experimental period. MAP initially increased from 135 ± 5 to a maximum value of 260 ± 8 mmHg (p < 0.0001 vs control group) at 2 min, and returned to basal values at 20 min. Subsequently, MAP deteriorated progressively, became significantly different from the control group at 150 min (93 ± 11 vs 128 ± 7 mmHg ($p < 0.05$) and was 53 $\pm$ 8 mmHg ($p < 0.001$ vs control group) at 360 min. The response in RPP was similar to the reponse in MAP.

5.4. Myocardial Energetic Response (MER) to Brain Death

5.4.1. MER During Onset of Brain Death and Hemodynamic Instability

Although HR, MAP, and RRP were significantly higher than in controls during the hyperdynamic phase (2 to 15 min after $t = 0$ min) and significantly lower after 150 min (MAP) to 180 min (RPP), no significant variation in the PCr/ATP ratios was seen during the hyperdynamic phase or during the hemodynamic deterioration which followed. The average PCr/ATP ratio was 1.61 ± 0.12 in the brain death group and 1.61 ± 0.08 in the control group.[32]

5.4.2. MER During Inotropic Treatment of Hemodynamic Instability

In a state of progressive hemodynamic instability, at 240 min after the induction of brain death, dopamine was infused at 5 μg/kg/min for 30 min. MAP (and RPP) increased in both groups and the differences disappeared between the groups. During the stepwise increase of dopamine to 10, 20, and 40 μg/kg/min a further increase in MAP (and RPP) was seen without significant differences between the two groups. However, when dopamine infusion was discontinued at $t = 330$ min, MAP deteriorated rapidly to 35 ± 3 mmHg in the brain death group and to 74 ± 7 mmHg in the control group ($p < 0.01$) at $t = 360$ min. RPP decreased to $9.5 \pm 0.3 \times 10^3$ $mmHg.min^{-1}$ in the brain death group and to $14.8 \pm 2.3 \times 10^3$ $mmHg.min^{-1}$ in the control group ($p < 0.05$). Again, no significant changes of the PCr/ATP ratio or the PCr or ATP content were observed during the experimental period. The average PCr/ATP ratio was 2.06 ± 0.10 in the brain death group and 2.14 ± 0.08 in the control group.[33]

5.4.3. MER During Preservation and Reperfusion

^{31}P MR spectra of explanted feline hearts were studied during hypothermic unperfused storage (17 h) and normothermic reperfusion (60 min). During storage, a progressive decrease in PCr was observed which was not only paralleled by an increase of inorganic phosphate (P*i*) but also by an increase of phosphomonoesters (PME), while during reperfusion only a partial recovery occurred. The first PCr/ATP ratio (at 1 h storage) was 1.08 ± 0.17 in the brain death group and 0.56 ± 0.07 in the controls. A subsequent decrease of the PCr/ATP ratio was observed, which was significantly sharper in controls than in the brain death group, until PCr/ATP ratio was practically zero in both groups at 6 h storage. However, this observation does not indicate a better metabolic condition of the donor hearts of the brain death group, but is the result of a sharper decrease of ATP in this group which already occurred before the first spectrum was acquired. During reperfusion PCr/ATP increased to approximately 1.4 in both groups.

PCr levels at 1 h storage were similar in both groups and decreased to practically zero at 6 h storage. During reperfusion PCr in controls recovered to a level similar to the level of 1 h storage but in the brain death group recovery was only approximately 50% of that level. Pi and PME showed an increase during storage, which was much sharper in controls than in the brain death group. Upon reperfusion Pi and PME normalized in both groups. No difference in intracellular pH (pHi) between the groups was observed during storage. The pH of the applied cardioplegic solution was 7.8. This resulted in a pHi of approximately 7.6 at 1 h storage which decreased to 6.2 at 17 h storage. During reperfusion, using a solution with a pH of 7.35 ± 0.05, Pi-peaks had a broadened appearance, which indicated an average pHi of 6.7. Only one heart, from the brain death group, showed restoration of sinus rhythm and contractile recovery upon reperfusion.[34]

6. DISCUSSION

The reviewed *in vivo* studies in the cat showed that neither the short hyper-hemodynamic state following the acute induction of brain death, nor the subsequent hemodynamic deterioration in the hours afterwards, nor the treatment of this deteriorated hemodynamic state with excessive high dose dopamine affect the heart energetically. These findings are in contrast with reported results of myocardial biopsies.[4-9] However, the *ex vivo* studies have shown a detrimental impact of brain death to the myocardial energy metabolism which became apparent only during storage and subsequent reperfusion.

As PCr/ATP ratios were not corrected for partial saturation, an overestimation of the ratios may have been present during the hyperdynamic phases at the onset of brain death and during the period of dopamine infusion. The differences in PCr/ATP ratios between the paragraphs *5.4.1.* and *5.4.2.* can be attributed to differences in partial saturation too, caused by different repetition times. The results described in paragraph *5.4.3.* emphasize the importance of correct interpretation of PCr/ATP ratios. The higher PCr/ATP ratio in the brain death group at 1 h storage compared with controls did not represent a better energetic status of these hearts but was merely due to the fact that myocardial ATP content in the brain death group was already lower than in the controls. This is supported by the lesser extent of recovery of PCr and ATP in the brain death group compared with controls during reperfusion, whereas the PCr/ATP ratios showed equal partial recovery in both groups.

The contribution of our results to clinical heart transplantation is as follows: when hemodynamic deterioration of a properly managed donor is encountered, acceptance of the heart may be considered on the basis of the myocardial energy status which is unaffected in the altered milieu of brain death. However, it cannot be excluded that brain death increases the susceptibility of the heart to ischemic injury during the explantation procedure. This stresses the importance of improvement of the explantation conditions by applying an myocardial energy preserving cardioplegic solution, rapid cardiac arrest, and adequate topical cooling, thus shortening the harmful period of warm ischemia of the heart.

The importance of the explantation procedure may be underestimated in clinical conditions. In addition, the energy profile of the only heart (from a brain dead cat) that regained contractile function upon reperfusion showed that its initial ATP content during storage and its content by the end of reperfusion were close to the respective means of the brain death group. As reported by others,[30] this observation demonstrates that the myocardial ATP content during storage and reperfusion is not the sole determinant of the return of contractile function. Loss op contractile function may have resulted from defects at the subcellular level of the excitation-contraction coupling system. This includes disturbances of calcium uptake and myofibrillar ATPase activity of the sarcoplasmic reticulum due to calcium overload when the ischemic heart is reperfued. Many other factors, for instance brain death-induced elevated plasma levels of neurotransmitters,[35] may add to the progression of hemodynamic deterioration or to myocardial contractile function. The precise impact of these substances is complicated and needs to be evaluated.

7. CONCLUSION

In conclusion, by applying *in vivo* ^{31}P MRS, it was demonstrated that the onset of brain death and the associated hemodynamic instability in the cat are not related to significant changes of myocardial energy metabolism. However, *ex vivo* ^{31}P MRS of the explanted feline donor heart, showed that brain death does affect the myocardial energy metabolism. The application of ^{31}P MRS in the research of brain death-related phenomena signifies a new dimension that will have important implications for clinical heart transplantation. We recommend further experimental and clinical ^{31}P MRS studies to unravel the influence of brain death on the high-energy phosphate metabolism of the donor heart before procurement, during storage, and after implantation.

8. REFERENCES

1. Kormos RL, Donato W, Hardesty RL, et al. The influence of donor organ stability and ischemia time on subsequent cardiac recipient survival. *Transplant Proc* 1988;**20:**980-983.
2. O'Connell B, Bourge RC, Costanzo-Nordin MR, et al. Cardiac transplantation: recipient selection, donor procurement, and medical follow-up. *Circulation* 1992;**86:**1061-1079.
3. Baldwin JC, Anderson JL, Boucek MM, et al. Task force 2: donor guidelines. *J Am Coll Cardiol* 1993;**22:**15-20.
4. Novitzky D, Wicomb WN, Cooper DKC, et al. Electrocardiographic, hemodynamic and endocrine changes occurring during experimental brain death in the Chacma baboon. *Heart Transplant* 1984;**4:**63-69.
5. Cooper DKC, Novitzky D, Wicomb WN. The pathological effects of brain death on potential donor organs, with particular reference to the heart. *Ann R Coll Surg Engl* 1989;**71:**261-266.
6. Novitzky D, Cooper DKC, Morrell D, Isaacs S. Change from aerobic to anaerobic metabolism after brain death, and reversal following triiodothyronine therapy. *Transplantation* 1988;**45:**32-36.
7. Szark F, Thicoïpé M, Lassié P, Dabadie P. Modification of mitochondrial energy metabolism in brain dead organ donors. *Transplant Proc* 1996;**28:**52-55.
8. Szark F, Erny P. Energy metabolism in brain dead organ donors. *Nutrition* 1997;**13:**691-692.
9. Tixier D, Matheis G, Buckberg GD, Young HH. Donor hearts with impaired hemodynamics. *J Thorac Cardiovasc Surg* 1991;**102:**207-214.
10. Mertes PM, Burtin P, Carteaux JP, et al. Changes in hemodynamic performance and oxygen consumption during brain death in the pig. *Transplant Proc* 1994;**26:**229-230.
11. Bittner HB, Chen EP, Milano CA, et al. Myocardial β-adrenergic receptor function and high-energy phosphates in brain death-related cardiac dysfunction. *Circulation* 1995;**92[suppl II]:** II-472-II-478.
12. Galiñanes M, Hearse DJ. Brain death-induced impairment of cardiac contractile performance can be reversed by explantation and may not preclude the use of hearts for transplantation. *Circ Res* 1992;**71:**1213-1219.
13. Szabó G, Sebening C, Hachert T, et al. Effects of brain death on myocardial function and ischemic tolerance of potential donor hearts. *J Heart Lung Tranplant* 1998;**17:**921-930.
14. Flameng W, Dyszkiewics W, Minten J. Energy state of the myocardium during long-term cold storage and subsequent reperfusion. *Eur J Cardiothorac Surg* 1988;**2:**244-255.
15. Pratschke J, Wilhelm MJ, Kusaka M, et al. Brain death and its influence on donor organ quality and outcome after transplantation. *Transplantation* 1999;**67:**343-348.
16. Bittner HB, Kendall SWH, Chen EP, van Trigt P. The combined effects of brain death and cardiac graft preservation on cardiopulmonary hemodynamics and function before and after subsequent heart transplantation. *J Heart Lung Transplant* 1996;**15:**764-777.
17. McLean AD, Rosengard BR. Aggressive donor management. *Curr Opin Org Transplant* 1999;**4:**130-133.

18. Van Dobbenburgh O, Lahpor JR, Woolley SR, et al. Functional recovery after human heart transplantation is related to the metabolic condition of the hypothermic donor heart. *Circulation* 1996;**94:**2831-2836.
19. Bloch F. Nuclear induction. *Phys Rev* 1946;**70:**460-466.
20. Purcell EM, Torrey HC, Pound CV. Resonance absorption by nuclear magnetic moments in a solid. *Phys Rev* 1946;**69:**37-46.
21. Gadian DG, Hoult DI, Radda GK, et al. Phosphorus nuclear magnetic resonance studies on normoxic and ischemic cardiac tissue. *Proc Natl Acad Sci USA* 1976;**73:**4446-4448.
22. Murphy DA, O'Blenes S, Nassar BA, Armour JA. Effects of acutely raising intracranial pressure on cardiac sympathetic efferent neuron function. *Cardiovasc Res* 1995;**30:**716-722.
23. Opie LH. *The Heart. Physiology and Metabolism.* 2nd ed. New York: Raven Press, 1991.
24. Chance B, Leigh Jr JS, Kent J, et al. Multiple controls of oxidative metabolism in living tissues as studied by phosphorus magnetic resonance. *Proc Natl Acad Sci USA* 1986;**83:**9458-9463.
25. Saeed M, Wendland MF, Wagner S, Derugin N, Higgins CB. Preservation of high-energy phosphate reserves in a cat model of post-ischaemic myocardial dysfunction. *Invest Radiol* 1992;**27:**145-149.
26. Wendland MF, Saeed M, Kondo C, Derugin N, Higgins CB. Effect of lidocaine on acute regional myocardial ischemia and reperfusion in the cat. *Invest Radiol* 1993;**28:**619-623.
27. Osbakken M, Young M, Huddell J, et al. Acute volume loading studied in cat myocardium with ^{31}P nuclear magnetic resonance. *Magn Reson Med* 1988;7:143-149.
28. Kitai T, Tanaka A, Terasaki M, et al. Energy metabolism of the heart and the liver in brain-dead dogs as assessed by ^{31}P NMR spectroscopy. *J Surg Res* 1993;**55:**599-605.
29. Pinelli G, Mertes PM, Carteaux JP, et al. Myocardial effects of experimental acute brain death: evaluation by hemodynamic and biological studies. *Ann Thorac Surg* 1995;**60:**1729-1734.
30. Carteaux JP, Mertes PM, Pinelli G. et al. Left ventricular contractility after hypothermic preservation: predictive value of phosphorus 31-nuclear magnetic resonance spectroscopy. *J Heart Lung Transplant* 1994;**13:**661-668.
31. Brandon Bravo Bruinsma GJ, Nederhoff, Geertman HJ, et al. Acute increase of myocardial workload, hemodynamic instability, and myocardial histological changes induced by brain death in the cat. *J Surg Res* 1997;**68:**7-15.
32. Brandon Bravo Bruinsma GJ, Nederhoff MGJ, te Boekhorst BCM, et al. Brain-death induced alterations in myocardial workload and high-energy phosphates: a phosphorus-31 magnetic resonance spectroscopy study in the cat. *J Heart Lung Transplant* 1998;**17:**894-990.
33. Brandon Bravo Bruinsma GJ, Nederhoff MGJ, van de Kolk CWA, et al. Myocardial bio-energetic response to dopamine after brain death-induced reduced workload: a phosphorus-31 magnetic resonance spectroscopy study in the cat. *J Heart Lung Transplant* 1999;**18:**1189-1197.
34. Brandon Bravo Bruinsma GJ, van de Kolk CWA, Nederhoff MGJ. Brain death-related energetic failure of the donor heart becomes apparent only during storage and reperfusion: an ex vivo phosphorus-31 magnetic resonance spectroscopy study on the feline heart. *J Heart Lung Transplant* 2001;**20:**996-1004.
35. Brandon Bravo Bruinsma GJ, Bredée JJ, Ruigrok TJC, et al. No evidence for participation of non-adrenergic non-cholinergic substances in brain death-related hemodynamicdeterioration of the feline potential heart donor. *Ann Transplant* 2001;**4:**43-47.

ORGAN DONATION AFTER FATAL POISONING

An update with recent literature data

Philippe Hantson*

1. INTRODUCTION

In Belgium, as in many countries, the number of individuals awaiting allograft organ transplantation exceeds the number of available grafts. Patients dying from head trauma or massive intracranial bleeding represent the majority of donors. In contrast, it can be estimated that very few allograft organs are obtained from patients who died from acute poisoning. In the US and UK, a survey of organ donor characteristics shows that poisoned donors currently represent usually less than 1% of all organ donors. The data from the Organ Procurement and Transplantation Network (OPTN) indicate that among the 6081 patients considered as potential organ donors in 2001, poisoning was the cause of death in 83 cases (1.4%).[1] This percentage slightly increased in comparison to the data obtained in 1994 (0.6%).

Several factors may explain that organ donation from poisoned donors remains marginal. First of all, brain death after acute poisoning is a rare condition because the majority of patients admitted to the Emergency Department with acute poisoning will survive with supportive care. A second reason is likely the reluctance of some transplantation centers to consider these patients as suitable organ donors because scientific data are lacking. There is always a concern of a potential transmission of the poisoning to the recipient. Finally, the diagnosis of brain death after acute poisoning could be difficult according to the methods used for the electrophysiological assessment of cessation of brain function.[2]

* Philippe Hantson, MD, PhD, Department of Intensive Care, Cliniques St-Luc, Université catholique de Louvain, Avenue Hippocrate, 10, 1200 Brussels, Belgium. Telephone : 32-2-7642755 Fax : 32-2-7648928. Email: hantson@rean.ucl.ac.be.

2. SURVEY OF PHYSICIANS' OPINION

In 1996, a questionnaire was sent to heart transplant surgeons in the UK.[3] Sixteen respondents gave their opinion on elective or emergency heart donation following paracetamol or barbiturate overdose and carbon monoxide (CO) poisoning. Most considered the hearts of CO victims unsuitable for transplantation. In contrast, there was a consensus in favor of hearts from paracetamol or barbiturate overdoses. We performed a limited survey in 1999 by sending questionnaires to European and US physicians involved in transplantation centers or in clinical toxicology (physicians working in a poison control center or in an intensive care unit). Although the response rate was relatively low, it was interesting to note that there was a discrepancy between the opinion of these two categories when some specific toxins (cyanide, carbon monoxide, methanol...) were considered, the transplant physicians being more reluctant to accept organs from patients who had been poisoned by these substances. This trend is confirmed by other surveys that have appeared in the literature. Wood et al. in the UK recently published the results of a postal survey of transplant centers and intensive care units.[4] The questionnaire consisted of four different scenarios involving brain death resulting from acute poisoning: deliberate methanol ingestion, cocaine use, accidental carbon monoxide inhalation, and accidental cyanide inhalation. Most of the intensive care directors would offer poisoned patients as potential organ donors after having obtained information from toxicology services. As for transplantation centers, more than 70% of the physicians would consider or accept organs from patients poisoned by methanol, cyanide or carbon monoxide. The percentage felt to 50 for cocaine exposure. Only 40% of the respondents could give reasons for refusal in the different situations. In some cases, the reason was directly related to the potential toxic effect on the graft. Physicians also cited risks of viral transmission in patients using illicit drugs. Most of the physicians agreed that standard criteria could be applied in the decision whether to remove organs. In particular, only a small number of respondents suggested performing additional toxicological investigations.

3. CASE SERIES

A first series was published in 1994 by Leikin et al., who identified 17 poisoned patients as organ donors.[5] Poisoning was not the direct cause of death in 4 patients who had also head trauma. The toxins involved were: ethanol (8), cocaine (5), carbon monoxide (5), barbiturates (2), cannabis (1), lead (1), phenylpropanolamine (1), multiple drugs (6). There were 41 recipients: 32 kidney transplantations and 9 liver transplantations. Two liver transplanted patients died during the procedure (death definitely not related to the toxic origin of the graft). In the kidney group, renal function immediately after the procedure was good in 28/32 patients (one graft loss on day 5 due to thrombosis, 3 with transient dysfunction) and remained satisfactory at 8-month follow-up in 31/32 cases.

Our personal experience in a series of 21 poisoned donors has been presented in previous papers and is summarized in Table 1.[6-11] From1989 to 1997, 864 organs were procured from 293 donors.. Of the 293 donors, 21 patients (7%) had developed brain death after acute poisoning. The toxins involved were : benzodiazepines (2), tricyclic

Table 1. Toxic substances and organs procured from poisoned donors. Legend: X: organ transplanted; BZD: benzodiazepine, TCA: tricyclic antidepressant; DPH: phenytoin; n: number of transplanted organs; X: number of transplantation procedures.

Toxic substance	Right kidney (n=20)	Left kidney (n=19)	Heart (n=6)	Lung (n=2)	Liver (n=9)	Pancreas (n=2)
Methaqualone					X	
BZD	X	X				
BZD	X	X	X		X	
BZD + TCA	X	X	Valves			
TCA	X	X				
Barbiturates + DPH	X	X			X	
Barbiturates	X	X	X		X	
Insulin	X	X	X			Islet cells
Insulin	X	X	Valves		X	
Carbon monoxide	X	X				
Carbon monoxide	X	X				
Carbon monoxide	X	X	X		X	X
Cyanide	X	X	Valves			X
Paracetamol	X	X	X			Islet cells
Methanol	X	X	Valves			
Methanol	X	X	X			
Methanol	X	X	Valves			
Methanol	X	X	Valves		X	Islet cells
Methanol	X	X	Valves		X	Islet cells
Methanol	X	X	Valves	X	X	Islet cells
Methanol	X		Valves			

antidepressants (1), benzodiazepines and tricyclic antidepressants (1), barbiturates (2), insulin (2), carbon monoxide (3), cyanide (1), methaqualone (1), paracetamol (1),methanol (7). On the whole, 58 grafts were obtained: 39 kidneys, 6 hearts, 2 lungs, 9 livers and 2 pancreases. Overall recipient survival rates appear in Table 2. One liver graft and one kidney graft did not function after one month (with also the delayed loss of two additional kidney grafts and one liver graft). An analysis of the causes of mortality or morbidity (graft loss) did not reveal any correlation with the toxic origin of the grafts.

In addition to these case series, other case reports or small series have recently appeared in the literature. We would like to focus the discussion on some substances that remain controversial.

Table 2. Recipient survival rates.

	1 year	5 years
Kidney	100%	88%
Pancreas	100%	100%
Liver	67%	67%
Heart	50%	33%
Lung	100%	100%

3.1. Methanol

The experience with organ procurement from patients poisoned by methanol is growing, as illustrated by some recent publications.[12-22]

Methanol poisoning could still be considered by some physicians as an absolute or relative contraindication to organ donation. Indeed, methanol poisoning is usually complicated by a severe metabolic acidosis due to the metabolization of methanol to formate by alcohol dehydrogenase. Patients poisoned by methanol could either survive when they are treated very early or die due to the consequences of cerebral edema. In the survivors, the retina is clearly a target organ of methanol poisoning, resulting in permanent visual impairment. As for the other organs (heart, liver, kidney, lung), the experience obtained from non fatal cases of methanol poisoning indicates that even some dysfunction may occur, a complete recovery is observed most of the time. The only exception is the pancreas. It seems that acute methanol poisoning is associated with a high incidence of pancreatic injury (increase of pancreatic enzymes or clinically severe acute necrotizing pancreatitis).[23] However, methanol poisoned patients are often also chronic ethanol abusers, and ethanol is also used to treat methanol poisoning. Although the differential involvement of methanol or ethanol is not precisely known, success in pancreatic transplantation may in any case be difficult.

Successful kidney or liver transplantations have now been published by several transplant teams.[12-17,19,22] The largest series comes from Spain with a multicenter team approach.[19] A total of 38 organs were obtained from methanol poisoned donors: 29 kidneys, 4 hearts and 5 livers. Two recipients died (one after heart transplantation and one after liver transplantation); the other recipients had favorable outcomes. Actuarial recipient survival rate in the kidney group after 1, 3, and 5 years was respectively 100%, 89% and 83% ; during the same interval, the graft survival rate was 93%, 78% and 75%. After 1 year, the 3 liver and heart-transplanted patients had normal graft function. Experience with lung or heart transplantation is still limited.[18-20] The lung is not a target organ of methanol poisoning and transplantation is acceptable. As for the heart, it seems also that methanol poisoning is not an absolute contraindication to organ donation. Cardiovascular collapse may be observed in fatal cases of methanol poisoning. Cardiac dysfunction may then be related to the severity of metabolic acidosis. However, there is no clear evidence that methanol or its toxic metabolites may provoke direct injury to the myocardium. Some transplant centers have accepted hearts from methanol poisoned donors without complications. These hearts may, however, require longer inotropic support postoperatively in the recipient.

As expected, skin can also be procured from a methanol-poisoned victim for banking and grafting.[21]

The diagnosis of methanol poisoning relies on the determination of serum methanol concentration. However, there is no correlation between serum methanol level and the severity of poisoning. Metabolic acidosis due to the accumulation of formate best reflects the severity, but the determination of formate in the blood is not possible on a routine basis. So, the clinician can only rely on the correction of metabolic acidosis before considering organ donation. It is safe to wait until methanol has completely disappeared from blood (it can last sometimes more than 100 hours under antidotal therapy) and metabolic acidosis is fully corrected before starting any harvesting procedure. There is no case in the literature of transmission of methanol toxicity to the recipient.

3.2. Carbon monoxide

Fatalities due to CO accidental exposure are still quite common even in the UK or US. It is, however, surprising that only a few of these fatalities result in organ procurement.[24] Hypoxia is the direct consequence of CO exposure, and the brain and the heart are particularly susceptible to this type of injury. There are many case reports of successful kidney, pancreas, lung or liver transplantation from CO poisoned donors.[25-34] There is still some debate concerning the possibility of heart donation,[26-33] and no consensus among transplant teams. To date, of the 17 reported cases of heart transplantation after CO poisoning, 10 had good long-term outcome. But it can be suspected that negative experiences are less frequently reported than positive ones. A direct comparison between the reported cases is difficult because not all patients had the same investigations and postoperative death can be caused by factors independent from CO toxicity (technical failure, sepsis). There are, however, reports of impaired cardiac contractility during the postoperative phase, and the role of CO toxicity can be at least suspected.[31] The determination of CO blood levels is not helpful for the decision whether to remove the heart. To minimize risks for the recipient, it seems safe to exclude CO poisoned donors who had a prolonged cardiac arrest with difficult resuscitation. Also, patients with an abnormal electrocardiogram or echocardiography, increased cardiac enzymes, or requiring significant inotropic support should not be considered candidates for organ donation.

3.3. Cyanide

Cyanide poisoning also exemplifies the possibility of organ donation in selected cases. Cyanide inhibits mitochondrial cytochrome oxidases, resulting in cellular hypoxia. Brain and heart are the organs most severely and rapidly damaged. At least two cases of successful heart transplantation have been reported in the literature.[35-37] In addition, there are numerous reports of cornea, kidney, pancreas, and liver transplantations.[38-43] The severity of cyanide poisoning is partially reflected by the degree of lactic acidosis. The duration of cyanide poisoning is usually short. Treatment includes supportive care with oxygen administration and antidotes to promote cyanide elimination.

The decision to remove organs should be based on standard criteria. It is safest to wait until cyanide blood concentration is below 7 μmol/l. Heart donation is possible when the donor is hemodynamically stable without inotropic support.

3.4. Illicit Substances

Drugs abusers are usually not considered donor candidates because of the risks of viral transmission.[24] However, the abuse of illicit substances by the intravenous route is decreasing, while the use of drugs by ingestion or inhalation is increasing. Cocaine and amphetamine derivatives are now frequently used by young people. Cocaine has been associated with an increased incidence of intracranial bleeding. This explains how brain death can occur in some cocaine users. Fatalities due to hyperthermia and cerebral edema have also been reported following ecstasy (3,4-methylenedioxymethamphetamine) exposure. There is a limited experience with organ donation in this setting. Cocaine may directly affect the cardiovascular system.[44] Acute events include hypertensive crisis, chest pain or even myocardial infarction. It seems also that chronic use of cocaine may cause premature atherosclerotic disease. To date, there are no published case reports of heart transplantation following cocaine-related deaths. By contrast, 2 successful liver transplantations and 12 kidney transplantations were reported by Leikin et al.[5] A series of 8 organ transplantations (one heart, one bilateral lung, 3 kidneys, one kidney-pancreas, and 2 livers) from ecstasy poisoned donors appeared recently.[45] There were two deaths unrelated to ecstasy toxicity.

4. CONCLUSIONS

The literature documents an increasing number of successful organ transplantations from brain dead donors following acute poisoning. It may be possible to increase the donor pool by a careful selection of such patients. The criteria for brain death determination have to be very strictly applied to avoid any misleading interpretation of the clinical findings. Finally, clinical toxicologists have to be consulted to discuss the toxicological data.

5. REFERENCES

1. Based on OPTN data as of January 1, 2002. http://www.optn.org/data/annualReport.asp
2. Hantson P, Mahieu P, de Tourtchaninoff M, Guérit JM. The problem of "brain death" and organ donation in poisoned patients. In: Machado C, ed. *Recent developments in Neurology. "Brain Death."* Amsterdam: Elsevier Science BV,1995:119-126.
3. el Oakley RM, Yonan NA, Simpson BM, Deiraniya AK. Extended criteria for cardiac allograft donors: a consensus study. *J Heart Lung Transplant* 1996;**15:**255-259.
4. Wood DM, Dargan PI, Jones AL. Poisoned patients as potential organ donors: postal survey of transplant centres and intensive care units. *Crit Care* 2003;**7:**147-154.
5. Leikin JB, Heyn-Lamb R, Erickson T, Snyder J. The toxic patient as a potential organ donor. *Am J Emerg Med* 1994;**12:**151-154.
6. Hantson P, Vekemans MC, Squifflet JP, Mahieu P. Outcome following organ removal from poisoned donors: experience with 12 cases and a review of the literature. *Transpl Int* 1995;**8:**185-189.
7. Hantson P, Mahieu P, Hassoun A, Otte JB. Outcome following organ removal from poisoned donors in brain death status: a report of 12 cases and review of the literature. *J Toxicol Clin Toxicol* 1995;**33:**709-712.
8. Hantson P, Vekemans MC, Vanormelingen P, De Meester J, Persijn G, Mahieu P. Organ procurement after evidence of brain death in victims of acute poisoning. *Transplant Proc* 1997;**29:**3341-3342.
9. Hantson P, Mahieu P. Organ donation after fatal poisoning. *Q J Med* 1999;**92:**415-418.
10. Hantson P. The poisoned donors. *Curr Opin Organ Transpl* 1999;**4:**125-129.
11. Hantson P, de Tourtchaninoff M, Mahieu P, Guérit JM. Prélèvements d'organes consécutifs aux décès par intoxication: expérience et problèmes diagnostiques. *Réan Urg* 2000;**9:**197-209.
12. Chari RS, Hemming AW, Cattral M. Successful kidney pancreas transplantation from donor with methanol intoxication. *Transplantation* 1998;**66:**674-675.
13. Friedlaender MM, Rosenmann E, Rubinger D, Silver J, Moskovici A, Drantizki-Elhalel M, Popovtzer MM, Berlatzky Y, Eid A. Successful renal transplantation from two donors with methanol intoxication. *Transplantation* 1996;**61:**1549-1552.
14. Hantson P, Kremer Y, Lerut J, Squifflet JP, Mahieu P. Successful liver transplantation with a graft coming from a methanol-poisoned donor. *Transpl Int* 1996;**9:**437.
15. Zavala A, Nogue S. Methanol poisoning and renal transplant. *Rev Esp Anestesiol Reanim* 1986;**33:**373.
16. Hantson P, Vanormelingen P, Lecomte C, Dumont V, Squifflet JP, Otte JB, Mahieu P. Fatal methanol poisoning and organ donation: Experience with seven cases in a single center. *Transplant Proc* 2000;**32:**491-492.
17. Hantson P, Vanormelingen P, Squifflet JP, Lerut J, Mahieu P. Methanol poisoning and organ transplantation. *Transplantation* 1999;**68:**165-166.

18. Bentley MJ, Mullen JC, Lopushinsky SR, Modry DL. Successful cardiac transplantation with methanol or carbon monoxide-poisoned donors. *Ann Thorac Surg* 2001;**71:**1194-1197.
19. Lopez-Navidad A, Caballero F, Gonzalez-Segura C, Cabrer C, Frutos MA. Short- and long-term success of organs transplanted from acute methanol poisoned donors. *Clin Transplant* 2002;**16:**151-162.
20. Evrard P, Hantson P, Ferrant E, Vanormelingen P, Mahieu P. Successful double lung transplantation with a graft obtained from a methanol poisoned donor. *Chest* 1999;**115:**1458-1459.
21. Bogdanov-Berezovsky A, Glesinger R, Kachko L, Arbel E, Rosenberg L, Grossman N. Accreditation of skin from a methanol-poisoned victim for banking and grafting. *Transplantation* 2002;**73:**1913-1917.
22. Zota V, Popescu I, Ciurea S, Copaciu E, Predescu O, Costandache F, Turcu R, Herlea V, Tulbure D. Successful use of the liver of a methanol-poisoned, brain dead donor. *Transplant Int* 2003;16:444-446.
23. Hantson P, Mahieu P. Pancreatic injury following acute methanol poisoning. *J Toxicol Clin Toxicol* 2000; **38:**297-303.
24. Jones AL, Simpson KJ. Drug abusers and poisoned patients: a potential source of organs for transplantation. *Q J Med* 1998;**91:**589-592.
25. Verran D, Chui A, Painter D, Shun A, Dorney S, McCaughan G, Sheil R. Use of liver allografts from carbon monoxide poisoned cadaveric donors. *Transplantation* 1996;**62:**1514-1515.
26. Hantson P, Vekemans MC, Squifflet JP, Mahieu P. Organ transplantation from victims of carbon monoxide poisoning. *Ann Emerg Med* 1996;**27:**673-674.
27. Smith JA, Bergin PJ, Williams TJ, Esmore DS. Successful heart transplantation with cardiac allografts exposed to carbon monoxide poisoning. *J Heart Lung Transplant* 1992;**11:**698-700.
28. Iberer F, Königsrainer A, Wasler A, Petutschnigg B, Auer T, Tscheliessnigg K. Cardiac allograft harvesting after carbon monoxide poisoning. Report of a successful orthotopic heart transplantation. *J Heart Lung Transplant* 1993;**12:**499-500.
29. Roberts JR, Bain M, Klachko MN, Seigel EG, Wason S. Successful heart transplantation from a victim of carbon monoxide poisoning. *Ann Emerg Med* 1995;**26:**652-655.
30. Koerner MM, Tenderich G, Minami K, Morshuis M, Mirow N, Arusoglu L, Gromzik H, Wlost S, Koerfer R. Extended donor criteria : use of cardiac allografts after carbon monoxide poisoning. *Transplantation* 1997;**63:**1358-1360.
31. Rodrigus IE, Conraads V, Amsel BJ, Moulijn AC. Primary cardiac allograft failure after donor carbon monoxide poisoning treated with biventricular assist device. *J Heart Lung Transplant* 2001;**20:**1345-1348.
32. Luckraz H, Tsui SS, Parameshwar J, Wallwork J, Large SR. Improved outcome with organs from carbon monoxide poisoned donors for intrathoracic transplantation. *Ann Thorac Surg* 2001;**72:**709-713.
33. Karwande SV, Hopfenbeck JA, Renlund DG, Burton NA, Gay WA. The Utah transplantation program. An avoidable pitfall in donor selection for heart transplantation. *J Heart Transplant* 1989;**8:**422-424.
34. Shennib H, Adoumie R, Fraser R Successful transplantation of a lung allograft from a carbon monoxide-poisoning victim. *J Heart Lung Transplant* 1992;**11:**68-71.
35. Snyder JW, Unkle DW, Nathan HM, Yang S-L. Successful donation and transplantation of multiple organs from a victim of cyanide poisoning. *Transplantation* 1993;**55:**425-427.
36. Swanson-Biearman B, Krenzelok EP, Snyder JW, Unkle DW, Nathan HM, Yang S-L. Successful donation and transplantation of multiple organs from a victim of cyanide poisoning. *J Toxicol Clin Toxicol* 1993;**31:**95-99.
37. Barkoukis TJ, Sarbak CA, Lewis D, Whittier FC. Multiorgan procurement from a victim of cyanide poisoning. A case report and review of the literature. *Transplantation* 1993;**55:**1434-1436.
38. Puig JM, Lloveras J, Knobel H, Nogues X, Aubia J, Masramon J. Victims of cyanide poisoning make suitable organ donors. *Transpl Int* 1996;**9:**87-88.
39. Ravishankar DK, Kashi SH. Organ transplantation from donor who died of cyanide poisoning : A case report. *Clin Transplant* 1998;**12:**142-143.
40. Hantson P, Squifflet JP, Vanormelingen P, Mahieu P. Organ transplantation after fatal cyanide poisoning. *Clin Transplant* 1999;**13:**72-73.
41. Brown P, Buckels J, Jain A, McMaster P. Successful cadaveric renal transplantation from a donor who died from cyanide poisoning. *Br Med J* 1987;**294:**1325.
42. Lindquist TD, Oiland D, Weber K. Cyanide poisoning victims as corneal transplant donors. *Am J Ophthalmol* 1988;**106:**354-355.
43. Hantson P, Vekemans MC, Squifflet JP, Evrard P, Mahieu P. Successful pancreas-renal transplantation after cyanide poisoning. *Clin Transplant* 1991;**5:**419-421.
44. Virmani R, Robinowitz M, Smialek JE, Smyth DF. Cardiovascular effects of cocaine: An autopsy study of 40 patients. *Am Heart J* 1988;**115:**1068-76.
45. Caballero F, Lopez-Navidad A, Cotorruelo J, Txoperena G. Ecstasy-induced brain death and acute hepatocellular failure: multiorgan donor and liver transplantation. *Transplantation* 2002;**74:**532-537.

THE ABC OF PVS
Problems of definition

D. Alan Shewmon*

1. INTRODUCTION

A prerequisite for any scientific discussion or study is a clear definition of the matter under consideration. With few topics is this both more pertinent and less heeded than with persistent vegetative state (PVS). The more one studies the literature on PVS, the less clear an entity it becomes. Multiple quasi-official definitions have been proposed, not all consistent with each other, and the literature is full of examples of "PVS" being formally defined in one way but used in other ways, even within the same article, resulting in invalid conclusions and inability to relate one study to another.

The goal of this chapter will not be to provide a simple, standard, or "correct" definition of vegetative state (VS) and PVS, but rather to review the complexity of the definitional task and the variety of definitions that have been proposed, with their respective pros and cons.

To begin with, it is useful to distinguish two main definitional axes, the "V" and the "P" so to speak, corresponding respectively to the state of a patient at a given moment and the evolution of that state over time. Let us call them, for want of better terms, phenomenal and temporal axes, or alternatively, diagnostic and prognostic axes. This chapter will focus primarily on the phenomenal, or diagnostic, V-axis (what is "vegetative"?), and conclude with a brief consideration of the temporal, or prognostic, P-axis (what makes a VS "persistent" or "permanent"?).

2. DIAGNOSTIC AXIS – THE VEGETATIVE "V"

Within the diagnostic axis, the various explicit and implicit definitions of VS encountered in the literature segregate into three conceptual domains, which I like to designate mnemonically as the "ABC of PVS": "A" for Anatomy, "B" for Behavior, and

* D. Alan Shewmon, MD, Professor of Neurology and Pediatrics, David Geffen School of Medicine at UCLA, Los Angeles, CA 90095. Email: ashewmon@mednet.ucla.edu.

Brain Death and Disorders of Consciousness, Edited by Machado and Shewmon
Kluwer Academic/Plenum Publishers, New York 2004

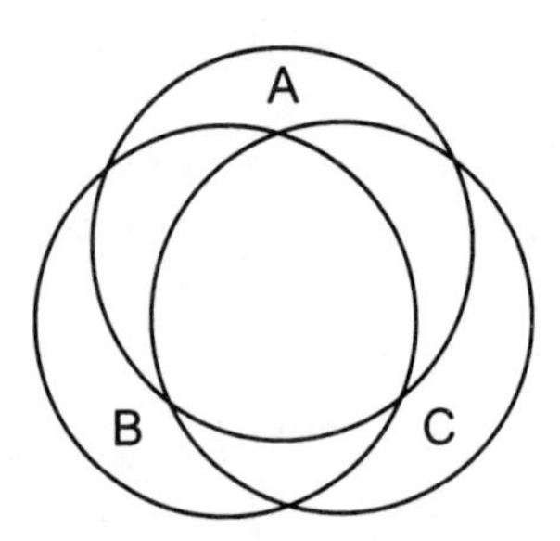

Figure 1. Hypothetical relationships between Anatomical, Behavioral, and Consciousness definitions of VS. Ideally they should superimpose; next most desirable would be for the central area of triple intersection to be relatively as large as possible. The actual relationships turn out to be quite different.

"C" for Consciousness. "A" is from the neuropathologist's or neurophysiologist's perspective, "B" from the clinical observer's perspective, and "C" from the patient's perspective. It would be nice if the three domains perfectly corresponded, but reality is seldom so simple. Rather, the three definitional approaches overlap as in a Venn diagram (Fig. 1). As we proceed, we shall refine the relative size and pattern of overlap among the subsets of patients corresponding to A-, B- and C-type definitions.

2.1. "A" for Anatomy – "Apallic Syndrome," "Neocortical Death"

Anatomical definitions of VS can be structural or functional, respectively specifying massive destruction or nonfunction of the cerebral cortex, with relative preservation of brain-stem structure and function. The structural-anatomical definition was promoted in the 1970s by Swedish neurologist David Ingvar.[1] The anatomical concept of VS is especially well reflected in the German term, "apallisches Syndrom"[2,3] (which coincidentally also begins conveniently with "A"), indicating loss of the pallium, or gray-matter mantle of the brain. If the cortical destruction is extreme (e.g., as manifested by an isoelectric electroencephalogram), the condition has sometimes been referred to as "neocortical death,"[4-6] an unfortunately ambiguous term in my opinion.[7]

Structural-anatomical VS, or "apallic syndrome," has the definitional advantage of objectivity and relative clarity. I emphasize "relative," because the brain insults and their resultant neuropathology obviously occur across continua both of degree and spatial extent, so there is necessarily a fuzzy, gray zone between pathology that definitely qualifies as "apallic syndrome" and pathology that definitely does not (however one may care to define it operationally).

The price for clarity in anatomical definitions is that they apply to relatively few cases that one would like to define as "VS" on behavioral grounds. The cases that inspired Ingvar's "apallic" designation were all due to severe hypoxia-ischemia. By contrast, clinically indistinguishable VS due to severe head trauma stems primarily from widespread axonal injury, involving white matter more prominently than gray. Moreover, the autopsy findings in the quintessential case of Karen Ann Quinlan greatly surprised the neurological world with the very reverse of the expected pattern of damage: primarily thalamic with relative preservation of cortex.[8] Others have also reported cases with primarily thalamic damage.[9] It is probably fair to say that, of all the cases that one would intuitively want to label "VS," only a minority would fulfill any strict anatomical-domain definition. Although most cases of apallic syndrome are clinically in a VS, most cases of clinical VS do not have apallic syndrome, understood in its literal, etymological sense.

The incomplete domain-correspondence also works in the other direction. Note that I wrote that *most*, not *all*, cases of apallic syndrome are clinically in a VS. For some decades the assumption has prevailed that cortical function is absolutely necessary for consciousness, in terms of both adaptive interaction with the environment (B domain) and inner awareness of self and environment (C domain). But this is more a dogma of neurological faith than a scientifically established fact. In cases of apallia with apparent VS, there is no empirical way to distinguish whether the lack of adaptive interaction with the environment is due to a true lack of subjective consciousness or merely to a supracritical impairment of the sensorimotor systems necessary for mediating purposeful interaction between a conscious self and the environment. Therefore, even if the A-domain implied "B," it wouldn't necessarily imply "C" (unless prior, independent proof existed that cortical function were necessary for any and all forms of subjective consciousness, which intrinsically cannot be proved and in at least one context has been disproved).

There have been children with congenital apallia who were clearly not in a "behavioral VS" and presumably therefore not in a "subjective-consciousness VS" either. Many neurologists have heard anecdotally from colleagues about such iconoclastic cases. I and two colleagues had the rare opportunity to study several such children in detail in their own homes.[10] Three will be briefly described here.

The first was a girl with hydranencephaly, examined at age 13 years. The diagnosis was reliably established by the combination of classical findings on computed tomography (CT), an isoelectric electroencephalogram (EEG), and even a "brain" biopsy revealing only meninges and a thin layer of gliotic tissue devoid of neurons. A more dramatic case of "apallia" would be hard to imagine. Nevertheless, she manifested discriminative awareness of the environment, for example, consistently distinguishing close family members from others. (At the approach or touch of strangers she assumed a fearful affect, became tense and withdrew, but relaxed to the touch and voice of mother.) She had favorite pieces and types of music, to which she would consistently smile and vocalize, in contrast to other music, to which she consistently remained indifferent. On "good days" she manifested visual tracking.

Another girl had a severe brain malformation documented on MRI scan, largely resembling hydranencephaly with an element of holoprosencephaly in the form of a thin sliver of frontal lobe without midline fissure. I examined her at age 5 and was amazed how well she could see without any visual cortex. Though spastic, she had considerable voluntary movement in the absence of any motor cortex. She had obvious tactile sensation without somatosensory cortex, excellent hearing without auditory cortex, and even a receptive vocabulary of a few words without language cortex.

The last example is a boy with a variation on hydranencephaly. CT scan revealed the supratentorial space to be full of fluid, in the center of which was a cyst with fluid of a different density; to one side of the cyst was a small amount of abnormal brain tissue. His EEG showed no activity except for some epileptic discharges emanating from that bit of tissue. Some mesial-basal temporal lobe tissue was present in the left middle fossa. As with the other cases, the thalamus, brain stem and cerebellum were intact. He clearly had visual orientation, fixation and tracking, and could scoot around the house on his back by pushing with his legs, avoiding collisions through visual guidance. Upon listening to Prokofiev's "Peter and the Wolf," his affect changed appropriately with the mood of the music. He clearly enjoyed being played with by his mother. When I called to him from a distance, he oriented with head and eye turning in my direction. He was fascinated by his own reflection in a mirror: despite attempts to get his attention, he kept turning back to the reflection.

Cases like these seriously undermine the concept of "apallic syndrome" or "neocortical death," understood as anything beyond neuropathology, because they unequivocally prove that the absence of cortex does not necessarily result in what is generally understood as VS.

Can such rare, extraordinary cases teach us anything about VS in general? Since they are congenital, one must be careful about extrapolating to older children and adults. Clearly they demonstrate something important about the plasticity of the fetal and early infant brain – especially what could be called "vertical" plasticity of the brain stem and diencephalon for supposedly "cortical" functions, a fascinating possibility that has so far received little investigative attention.

Such cases also suggest a plausible reason for their rarity. In each instance doctors told parents at birth that the children would certainly remain forever vegetative. If the parents or guardians had accepted this prediction and treated the children accordingly, institutionalizing them with relatively little stimulation, they surely would have remained vegetative like so many children with similar pathology. Instead, they were taken home and given much affection and stimulation, no doubt permitting development of the limited cognitive functions described. If a perfectly normal baby is treated impersonally like a thing, it will develop very poorly. So much the more so for a severely disabled infant. The remarkable parenting in these extraordinary cases reinforces the suspicion that "developmental vegetative state" may be, at least in some (perhaps many?) cases, a self-fulfilling prophecy.

Do such congenital cases carry any implications for acquired apallia? One difference between older patients and children is that, for the same cortical lesion, the former suffer greater motor impairment than the latter (e.g., the degree of disability in adults with large strokes contrasted with the remarkable contralateral sensorimotor function in children treated for epilepsy by hemispherectomy). If someone has severe spastic quadriplegia and pseudobulbar palsy, plus apraxia of any residual voluntary movement, of course they will exhibit no apparently purposeful movement. We should be careful, therefore, not to conclude that the essential difference between the acquired and congenital cases lies entirely in degree of brain-stem plasticity for consciousness; it could also have to do with plasticity for voluntary motor function. What the congenital cases imply about acquired VS in older patients is to suggest a plausible alternative to the cortex-consciousness dogma. It has never been scientifically ruled out (nor can it be), that some (unknowable number of) acquired-apallia patients have a limited form of consciousness but simply cannot manifest it due to extreme motoric disability.

It is well known that the neuroanatomical pathways of pain sensation involve mainly subcortical structures.[11-16] In the intact brain secondary sensory and limbic cortical areas are also involved in pain processing,[17-20] but lesional data indicate that they play a modulatory role, mediate the affective component of pain, and are not necessary for experiencing the raw sensation. There can be perfectly intact pain perception in the absence of sensory cortex, as exemplified in stroke[21(pp.89-90,113-114)] and hemispherectomy cases. Patients with limbic lesional surgery for the treatment of pain report (mysteriously but consistently) that they still feel the pain "just as before" but it "no longer bothers" them.[22,23] Since the necessary, and apparently also sufficient, pathways for pain (at least the raw sensation abstracted from the affective component) are intact in both congenital and postnatally acquired apallia, there is no reason to assume that apallia precludes all experience of pain and discomfort.[24] The hydranencephalic children described above clearly experienced pain, as do newborns and fetuses with relatively non-functioning cortices.[25] Therefore, when adults or older children with similar acquired lesions withdraw limbs, grimace and cry to noxious stimuli, on what grounds can anyone assert that these responses are "*merely* primitive reflexes," even if the motor reaction is simple and stereotyped?

In contrast to the structural approach, a *functional*-anatomical definition (absence of all cortical functions) is advocated by the UK's Institute of Medical Ethics Working Party (IMEWP)[26] and the Royal College of Physicians Working Group.[27] Such a definition should be carefully distinguished from a (putative) conclusion or corollary from an essentially behavioral or consciousness-type definition. The main problem with the functional-anatomical approach is the very notion of "cortical function." Does it imply that if some area of cortex is present it is *involved* in the function, or rather that that area is *required* for the function? If some other brain structure (e.g., thalamus) is also involved or required, is the function still a "cortical" one? What if the function is ordinarily "cortical," but in a particular case it is mediated by subcortical structures (e.g., functional vision in some of the decorticate children mentioned above)? How can one determine scientifically whether consciousness of self and environment (as distinguished from adaptive interaction with the environment) is a "cortical function" (whether understood as either mediated exclusively by the cortex or requiring some minimal amount of cortex)?

In summary, anatomical definitions of VS as "apallia" (whether structural or functional) incompletely specify the behavioral and consciousness domains, and a probably relatively small proportion of clinical VS is due to pure "apallia." The "A" domain of the Venn diagram should therefore be considerably shrunken relative to the other two domains, as illustrated in Figure 2.

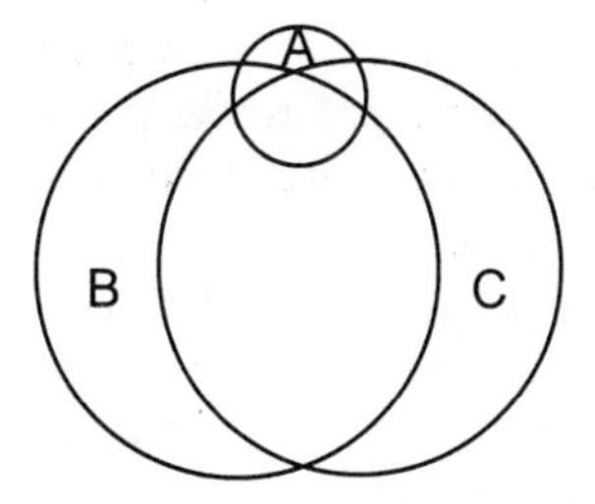

Figure 2. Adjustment of Anatomical-VS (apallia) relative to Behavioral and Consciousness definitions.

2.2. "B" for Behavior – Absence of Evidence vs. Evidence of Absence

According to the behavioral approach, VS is a state of alert appearance with sleep/wake cycles but no *evidence* of consciousness. This is a clinical definition, from the third-person perspective of doctor, nurse, family member or other observer. It corresponds to the German term "Wachkoma" ("awake coma"), which much more aptly captures the essence of the condition at issue than "apallisches Syndrom." Jennett and Plum, who in 1972 coined the term "vegetative state," understood it as essentially a behavioral syndrome.[28,29(p.4)]

Unfortunately, behavioral-VS does not necessary imply anything about either anatomy or subjective consciousness. As discussed above and illustrated in Figure 2, most cases of VS, behaviorally defined, do not feature total destruction of the cerebral cortex as the main neuropathology. Less generally acknowledged is that behavioral-VS implies nothing about inner consciousness either. What if the brain lesion caused merely an impediment to the manifestation of consciousness without erasing consciousness itself – something like a "super locked-in syndrome"? Indeed, there is such an impediment. These patients have not only severe spastic quadriplegia and pseudobulbar palsy, but also most likely severe apraxia, agnosia, global aphasia, and short-term memory impairment. If any have some form of inner consciousness (reflective self-awareness, dream-like memories, experiences of pleasure, pain, frustration, depression...) they could not possibly communicate that fact to others. Although everyone knows that "absence of evidence" does not *per se* constitute "evidence of absence," a perusal of the PVS literature suggests that not a few neurologists employ principles of logic in practice that they would never endorse in theory.[24,30-32]

Furthermore, how hard must one search for evidence before concluding that there is none? What often happens in these cases is that a neurologist will spend ten or fifteen minutes examining the patient, elicit some brain-stem and spinal cord reflexes, not observe any evidence of consciousness, and declare the patient to be in a VS. But nurses or family members, who spend all day with the patient, may notice subtle signs of adaptive interaction with the environment, perhaps only intermittently. Too often their observations are dismissed as "subjective," "denial," or "projection." Sometimes that is the case, but other times the one in denial is a proud physician who refuses to be diagnostically contradicted by non-physicians. Whose "evidence" counts?

Often behavioral definitions include phrases like "no evidence of meaningful or purposeful motor activity." "Meaningful" or "purposeful" to whom? Maybe a movement doesn't look meaningful to you or me, but the patient is trying to do something (unsuccessfully) and the attempt is meaningful and purposeful to the patient. A major rehabilitation hospital in London found that as many as 43% of patients referred with a diagnosis of chronic VS turned out, upon careful examination by specialists in this field, to have evidence of consciousness after all.[33]

Even patients in frank coma (eyes-closed, unarousable unresponsiveness) can sometimes be inwardly aware, as proved by subsequent accurate recall of events and discussions around them while supposedly "unconscious."[34,35] Who knows what proportion of behaviorally comatose patients have inner conscious experiences but are simply amnestic for them upon recovery (or die before recovering enough to report them)? All the more should we not rule out such a possibility in unresponsive patients

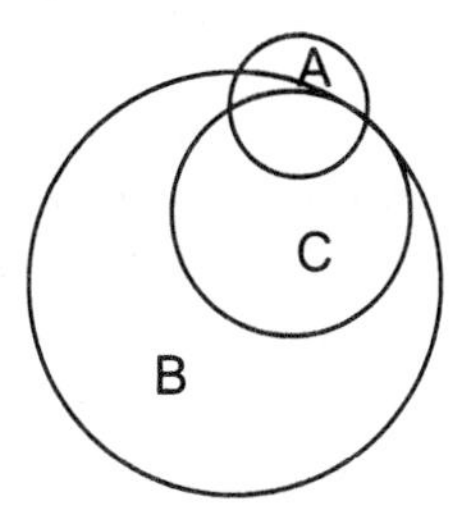

Figure 3. Final adjustment of all three definitional domains relative to each other.

who by contrast appear awake and alert. The elicitation of cognitive event-related potentials in some comatose and vegetative patients is not only a potentially useful prognostic sign but also an indirect indicator that mismatch between behavioral and subjective consciousness may be more common than is generally assumed.[36-40]

Taken together, all these considerations suggest that, in the Venn diagram, the proportion of B that is also C, though unknown and unknowable, is probably substantially less than 100%. Of course, if there is no inner consciousness to begin with (C-type definition), there will surely be no behavioral evidence for consciousness, so the proportion of C that is also B is necessarily 100%. Thus the final refinement of the Venn diagram, as depicted in Figure 3.

2.3. "C" for Consciousness – Cogitat, Ergo Est?

An example of the subjective-consciousness definition of VS is from the 1989 Position Statement of the American Academy of Neurology (AAN): "The persistent vegetative state is a form of eyes-open permanent unconsciousness in which the patient has periods of wakefulness and physiological sleep/wake cycles, but at no time is the patient aware of him- or herself or the environment."[41] It is a first-person definition from the perspective of the patient, asserting that the essence of VS is the absence of inner consciousness. The C-domain was also chosen as the definitional essence by the President's Commission,[42(pp.174-5)] the American Medical Association (AMA),[43] the American Neurological Association (ANA),[44] and the Multi-Society Task Force on PVS (MSTF).[45]

Implications for the other two domains are fairly straightforward. If there's an absence of consciousness "by definition," then of course there must be an absence of behavioral evidence of consciousness. As with B-type definitions, a relatively small proportion of cases will likely be due to purely or primarily cortical destruction or nonfunction. The main problem with C-type definitions is that, although philosophically interesting, emotionally poignant, and ethically relevant, they are clinically useless. We do not directly perceive the mental life of others, only their behavior. The inherent mismatch between clinical utility and ethical relevance is the crux of the definitional dilemma surrounding VS.

2.4. "Definitional dyslexia" – Jumbling the ABCs

The literature also features implicit definitions that seem to combine domains. It is often unclear whether one domain is meant to be the essence of the definition and the other(s) merely parenthetically mentioned as (supposedly) logical consequences, or the combination itself is intended as the essence of VS. The problem with such definitions is that, if you take them seriously, there may not be any actual cases of such "VS" at all. Some articles at the outset explicitly define VS in one domain, but later discuss it as though implicitly defined in a different domain. An important subcategory of this fallacy, which pervades the PVS literature, is when the behavioral domain is stated to be the definitional essence, but further along behavior is treated as an operational (diagnostic) criterion for a implicit consciousness definition. Such fallacies exemplify what I call "definitional dyslexia," since they confuse the "ABCs" of PVS.

Jennett, in his recent book on VS,[29] states on page 4 that, in the original deliberations over the name of their syndrome, he and Dr. Plum "wished to have one that did not presume a particular anatomical abnormality" and that "described behaviour ... as *the essence of the definition is observed behaviour*." (emphasis added) But then on page 11 he writes: "Much of the debate about diagnosis turns on what behaviours reflect cortical activity, and whether fragments of activity in the cortex necessarily indicates [sic] *awareness – lack of which is the crux of the diagnosis of the vegetative state*." (emphasis added) With the mere flick of a page, the definitional domain suddenly switches from B to C, and what determines whether a behavior is relevant as a diagnostic indicator of C is whether it is understood as A ("cortical").

A classic and influential example of such loose reasoning is the AAN's Position Statement on PVS.[41] As noted above, they defined VS in terms of lack of consciousness, but then they went on to say: "Neurologically, being awake but unaware is the result of a functioning brain stem and the total loss of cerebral cortical functioning." Is this an assertion that the consciousness definition implies diffuse cortical nonfunction, or the other way around, as though a tacit cortical definition implied lack of consciousness? Is the patient population at issue in this sentence implicitly defined in the A or C domain, or does the Academy mean to imply a perfect, one-to-one correspondence between the two domains?

A little further along they cite three independent bases for concluding that VS patients cannot experience pain or suffering. What began as a definition is now treated as a logical conclusion (from what definition or premise?). The first line of evidence is that "direct clinical experiences with these patients demonstrate that there is no behavioral indication of any awareness of pain or suffering." But who are "these patients"? Surely not those who *by definition* lack all awareness, for that would make the whole argumentation moot. Clearly the population referred to as "these patients" is implicitly defined behaviorally. As discussed above, the lack of "behavioral indication of any awareness of pain or suffering" does not *per se* imply a lack of pain or suffering. Moreover, who is to say in a given case that withdrawal, grimacing and crying are not "behavioral indication[s] of awareness of pain or suffering"?

The second line of evidence is: "In all [PVS] patients studied to date, post mortem examination reveals overwhelming bilateral damage to the cerebral hemispheres to a degree incompatible with consciousness or the capacity to experience pain or suffering."

This last qualifier ("to a degree incompatible with consciousness...") completely begs the question. Even if it were true (which it is not) that all post mortem examinations on patients given this diagnosis show "overwhelming bilateral damage to the hemispheres," to assert the damage to be "to a degree incompatible with consciousness" hardly constitutes *evidence* that it is incompatible with consciousness. It is merely a restatement of the cortical dogma, the truth of which is completely taken for granted.

The Academy's third line of evidence is that "recent data utilizing positron emission tomography indicates that the metabolic rate for glucose in the cerebral hemispheres is reduced in [PVS] patients to a degree incompatible with consciousness." This is essentially a functional, *in vivo* version of the claim about post mortem examination, and is just as question-begging.

The circular reasoning contained in these last two supposed "bases" can be summarized as follows. "(1) We know that cortical functioning is necessary for consciousness, because PVS patients, who lack cortical functioning, are unconscious. (2) We know that PVS patients are unconscious, because they lack cortical functioning." And of course the "PVS" within this logical circle is implicitly clinically (behaviorally) defined, whereas initially it had been subjectively defined. Such logical fallacies permeate the PVS literature with remarkable immunity.

2.5. Operational Diagnostic Criteria

If there is a lack of consensus on the very definition of VS, we should hardly be surprised to find inconsistencies among the proposed operational diagnostic criteria for it. All authorities concur that VS patients show no *evidence* of mental function, exhibit sleep-wake cycles, and have relatively intact brainstem reflexes and "vegetative" functions such as spontaneous breathing, cardiovascular regulation, and thermal stability. They typically have spastic quadriplegia and/or decerebrate rigidity. There is incontinence of bowel and bladder.

They can be motionless or hypermobile, mute or noisy.[46(p.6)] All authorities agree that sporadic movements of facial muscles, trunk and limbs are compatible with the diagnosis, provided the movements are not "voluntary," "purposeful," or "meaningful." But how can an observer know that some uncoordinated or stereotyped motor activity was not voluntary, purposeful or meaningful *from the patient's point of view*? Diagnostic guidelines also specify "no speech or signaling by eye movements."[28,45] How hard, by what means, and over how long a time, must examiners try to establish a system of communication, by eye movements or otherwise, before concluding that there is no signaling? Failure to facilitate or to notice subtle communicative gestures seems primarily responsible for the high rate of misdiagnosis of VS.[33]

Although Ingvar states there should be no emotional reaction,[1(pp.184,201)] most authorities allow both positive and negative expressions of affect. Brief smiling, laughing, frowning, grimacing, crying out and shedding tears are considered compatible with VS so long as they are "inconsistently related to any apparent stimuli."[44,45] How does the mere stereotypy of such responses guarantee that the severely disabled patient was not expressing pleasure, pain or discomfort? Moreover, why exclude the possibility that the responses might actually be consistently related to *inapparent* stimuli, such as gastroesophageal reflux, borborygmi, resolution of a cramp, an amusing thought, or simply feeling good or bad? If some patients are in fact in a kind of "super locked-in state," they would likely be very depressed at being considered and treated as vegetative with no way to communicate the contrary. It should hardly be surprising that they might moan, cry, or shed tears without apparent stimuli, because the stimulus is constant in their very situation.

Regarding claims that some patients "seem more relaxed when family members arrive and speak or when they stroke them, and become more agitated with they leave," Jennett plays the devil's advocate: "However, such behaviours are common in patients who, over a long period of follow-up, show no other signs of awareness or cognition, and most skilled observers accept them as part of the extensive repertoire of reflex responses shown by some vegetative patients"[29(p.16)] – as though it were self-evidently impossible for a patient to exhibit only one sign of awareness. By emphasizing that there are "no *other* signs of awareness," Jennett seems to implicitly acknowledge such behavior as a sign of awareness. At least it should be given the benefit of the doubt.

Yet it is dismissed. In the face of a cognitive-like behavior of uncertain nature, PVS authorities typically place the burden of proof on those who would interpret it as possibly indicating consciousness rather than on those who would dismiss it as a mere "reflex," perhaps on the basis that it is (presumably) not "cortical." This bias is so strong that even when the behavior *is* demonstrated to be "cortical," it is still dismissed on the grounds of involving not the cortex as a whole but mere islands of cortex[47,48] – including by one who claimed that the behavioral syndrome he helped to define "did not presume a particular anatomical abnormality."[29(pp.4,17)] Shouldn't the interpretation of a behavior (whether it indicates consciousness or not) be based on the qualities of the behavior, not on whether it is "cortical," or "cortical enough"?

Jennett and Plum wrote that VS patients have roving eye movements,[28] but Ingvar and colleagues stated that the eye movements are irregular and specifically *not* roving.[1(p.201)] To visual stimuli there should be "no evidence of sustained, reproducible, purposeful or voluntary behavioral responses."[45,49] But no practical definition is offered for any of those adjectives. They are merely a judgment call on the part of the clinician, so the operational diagnostic criteria are not really so "operational."

Both the AMA[43] and the MSTF[45] stated that VS patients do not blink to threat, but Jennett and Plum[28] accepted blinking to threat as compatible with the diagnosis. According to the latter pair and also to Ingvar,[1(p.201)] there should be no visual orientation or attention, but the AMA,[43] the IMEWP,[26] and the MSTF[45] stated that some VS patients turn their head or dart their eyes toward moving objects.

Moreover, the ANA said that "visual tracking can *occasionally* occur,"[44] and the MSTF said "sustained visual pursuit is lacking in *most* patients"[45] (emphasis added), implying that it is present in some. Jennett and Plum[28] and the IMEWP[26] also say that eyes can follow slowly moving objects, though inconsistently or briefly. What does "inconsistently" mean in operational terms? No more than four out of five trials, or seven out of ten, or ten out of twenty? What if today the patient tracks ten out of ten times, but tomorrow not at all? Would that count as "inconsistent," qualifying the patient as VS, or would it indicate fluctuation in and out of VS? What if the fluctuation were over hours instead of days? Many critical details are left open to the caprice of individual judgment. Besides, in very disabled patients, inconsistency does not necessarily imply lack of intention; a baby just learning to walk performs inconsistently but no less purposefully or meaningfully.

Regarding auditory stimuli, the AAN[49] and the MSTF[45] say "no evidence of sustained, reproducible, purposeful or voluntary behavioral responses to ... auditory ... stimuli," but the AMA[43] and the IMEWP[26] say that head turning or eyes darting toward noise is compatible with the diagnosis, without requiring that it be unsustained, inconsistent or irreproducible.

Most authorities state that grasping, groping, scratching, and rooting are compatible with the diagnosis, so long as they are stereotyped, reflex-like, and not purposeful or voluntary. But does the presence of a grasp reflex establish that all grasping movements are merely reflexive? Who can say that grasping, groping or scratching in a given instance is not voluntary or purposeful?

Authorities similarly state, "no evidence of sustained, reproducible, purposeful, or voluntary behavioral responses to ... noxious stimuli."[45,49] But compatible with the diagnosis are grunting, groaning, grimacing, crying-like behavior, movement of the hands toward the stimulus, avoidance movements, and stereotyped limb flexion. On what plausible ground can anyone confidently dismiss such behaviors as invariably *not* reflecting discomfort or pain? The illogic is the same as asserting that, just because pseudobulbar patients exhibit involuntary spontaneous affects, they experience no emotions at all.

The AAN[41] states "the capacity to chew and swallow in a normal manner is lost," implying that chewing or swallowing in a not quite normal manner is compatible with the diagnosis. Other authorities state explicitly that VS patients can exhibit "uncoordinated" chewing or swallowing of liquids or foods placed in the mouth.[1,26,28,42(p.175),44,45] Just because swallowing is difficult and poorly coordinated, does that make it necessarily not "meaningful," "purposeful," or "voluntary"?

In summary, apart from the many items of universal agreement, there is a disturbing amount of inconsistency and vagueness in the various proposed operational criteria for diagnosing VS.

3. TEMPORAL AXIS

We turn briefly to the temporal axis before concluding. The "P" originally stood for "persistent," intended by Jennett and Plum[28] simply to mean lasting some time (at least up to the present), although through common usage many came to understand it as meaning persisting indefinitely. In 1994 the MSTF introduced a new temporal qualifier, also beginning with the letter "P," so that henceforth the first term of the PVS acronym became ambiguously either "persistent" or "permanent." This innovation had both fortunate and unfortunate consequences. Prior to that, people generally had come to understand the acronym "PVS" simplistically and falsely as a stable diagnostic entity. The MSFT's distinction between "persistent" (until now) versus "permanent" (indefinitely into the future) was therefore a valuable contribution, emphasizing the difference between diagnosis and prognosis. But the fact that both of the distinguishing terms begin with "P" has generated more confusion than it eliminated. People still talk about "PVS" as though it were a singular, meaningful entity, although with the ambiguity of the "P" it means even less now than it did before. Even Jennett himself, one of the original co-coiners of "P" as "persistent," recently redefined it exclusively as "permanent."[29(p.5)]

Based on a meta-analysis of many studies, the MSTF proposed criteria for declaring that a VS had gone from "persistent" to "permanent," namely: for anoxic etiology, if it has persisted three months, and for traumatic etiology, if it has persisted twelve months – and this is so regardless of the age of the patient.[50] The Task Force was composed of prominent experts representing multiple professional associations, and its report was published amidst much fanfare by the *New England Journal of Medicine*, so it must be true! But the published studies that went into its meta-analysis are astonishingly incommensurate, both definitionally and methodologically, making it quite impossible to pool the data meaningfully. In the final analysis, the three- and twelve-month criteria for predicting "permanence" are one more example of official canonization of speculation as neurological dogma, analogous in the temporal dimension to the simplistic cortex-consciousness connection in the diagnostic dimension. Unfortunately space does not allow elaboration on this important issue.

4. CONCLUSION

Although PVS has been described as a "syndrome in search of a name,"[28,29(p.1)] it is more accurately described as an acronym in search of a name, and VS as a name in search of a definition. The vast literature on VS is inconsistent regarding the essence of its own topic, whether it lies in the anatomical (structural or functional), behavioral, or first-person mental/subjective domain. Supposed inferences or corollaries in the other two domains are based on premises taken for granted as established fact which are merely unproved (or unprovable) hypotheses. In particular, the key dogma linking cortical functioning and subjective consciousness is a product of circular reasoning. "Post-coma unresponsiveness" would have been an intellectually more honest and accurate term, but it is too late to attempt to change established vocabulary. If requirements for any scientific study include a clear definition of terms and the systematic ruling out of alternative hypotheses, then the scientific study of VS has yet to begin.

5. REFERENCES

1. Ingvar DH, Brun A, Johansson L, Samuelsson SM. Survival after severe cerebral anoxia with destruction of the cerebral cortex: the apallic syndrome. *Ann N Y Acad Sci* 1978;**315**:184-208; discussion 208-214.
2. Ingvar DH, Brun A. Das komplette apallische Syndrom. *Arch Psychiat Nervenkr* 1972;**215**:219-239.
3. Kretschmer E. Das apallische Syndrom. *Z gesamte Neurol Psychiat* 1940;**169**:576-579.
4. Brierley JB, Graham DI, Adams JH, Simpsom JA. Neocortical death after cardiac arrest. A clinical, neurophysiological, and neuropathological report of two cases. *Lancet* 1971;**2**:560-565.
5. Keane JR. Blinking to sudden illumination. A brain stem reflex present in neocortical death. *Arch Neurol* 1979;**36**:52-53.
6. Mizrahi EM, Pollack MA, Kellaway P. Neocortical death in infants: behavioral, neurologic, and electroencephalographic characteristics. *Pediatr Neurol* 1985;**1**:302-305.
7. Shewmon DA. "Brain death": a valid theme with invalid variations, blurred by semantic ambiguity. In: White RJ, Angstwurm H, Carrasco de Paula I, eds. *Working Group on the Determination of Brain Death*

and its Relationship to Human Death. 10-14 December, 1989. (Scripta Varia 83). Vatican City: Pontifical Academy of Sciences, 1992:23-51.
8. Kinney HC, Korein J, Panigrahy A, Dikkes P, Goode R. Neuropathological findings in the brain of Karen Ann Quinlan. The role of the thalamus in the persistent vegetative state. *N Engl J Med* 1994;**330:**1469-1475.
9. Schiff ND, Ribary U, Moreno DR, Beattie B, Kronberg E, Blasberg R, Giacino J, McCagg C, Fins JJ, Llinas R, Plum F. Residual cerebral activity and behavioral fragments can remain in the persistently vegetative brain. *Brain* 2002;**125:**1210-1234.
10. Shewmon DA, Holmes GL, Byrne PA. Consciousness in congenitally decorticate children: "developmental vegetative state" as self-fulfilling prophecy. *Dev Med Child Neurol* 1999;**41:**364-374.
11. Besson J-M, Guilbaud G, Peschanski M. *Thalamus and Pain.* Amsterdam: Elsevier Science Publishers, Excerpta Medica, 1987.
12. Bonica JJ. Anatomic and physiologic basis of nociception and pain. In: Bonica JJ, ed. *The Management of Pain.* 2 ed. Philadelphia: Lea & Febiger, 1990:28-94.
13. Bromm B, Desmedt JE. *Pain and the Brain. From Nociception to Cognition.* Hagerstown, MD: Lippincott-Raven, 1995.
14. Casey KL. Pain and central nervous system disease: a summary and overview. In: Casey KL, ed. *Pain and Central Nervous System Disease. The Central Pain Syndromes.* New York, NY: Raven Press, 1991:1-11.
15. Melzack R. Central pain syndromes and theories of pain. In: Casey KL, ed. *Pain and Central Nervous System Disease. The Central Pain Syndromes.* New York, NY: Raven Press, 1991:59-64.
16. Talbot JD, Marrett S, Evans AC, Meyer E, Bushnell MC, Duncan GH. Multiple representations of pain in human cerebral cortex. *Science* 1991;**251:**1355-1358.
17. Bromm B. Brain images of pain. *News Physiol Sci* 2001;**16:**244-249.
18. Hofbauer RK, Rainville P, Duncan GH, Bushnell MC. Cortical representation of the sensory dimension of pain. *Neurophysiol* 2001;**86:**402-411.
19. Peyron R, Frot M, Schneider F, Garcia-Larrea L, Mertens P, Barral FG, Sindou M, Laurent B, Mauguiere F. Role of operculoinsular cortices in human pain processing: converging evidence from PET, fMRI, dipole modeling, and intracerebral recordings of evoked potentials. *Neuroimage* 2002;**17:**1336-1346.
20. Rainville P. Brain mechanisms of pain affect and pain modulation. *Curr Opin Neurobiol* 2002;**12:**195-204.
21. Brodal A. *Neurological Anatomy in Relation to Clinical Medicine.* New York, NY: Oxford University Press, 1981.
22. Bouckoms AJ. Psychosurgery for pain. In: Wall PD, Melzack R, eds. *Textbook of Pain.* 2nd ed. Edinburgh: Churchill Livingstone, 1989:868-881.
23. Jannetta PJ, Gildenberg PL, Loeser JD, Sweet WH, Ojemann GA, Bonica JJ. Operations on the brain and brain stem for chronic pain. In: Bonica JJ, ed. *The Management of Pain.* 2 ed. Philadelphia: Lea & Febiger, 1990:2082-2103.
24. McQuillen MP. Can people who are unconscious or in the "vegetative state" perceive pain? *Issues Law Med* 1991;**6:**373-383.
25. Anand KJS, Hickey PR. Pain and its effects in the human neonate and fetus. *N Engl J Med* 1987;**317:**1321-1329.
26. Institute of Medical Ethics Working Party on the Ethics of Prolonging Life and Assisting Death. Withdrawal of life-support from patients in a persistent vegetative state. *Lancet* 1991;**337:**96-98.
27. Royal College of Physicians Working Group. The permanent vegetative state. *J R Coll Physicians Lond* 1996;**30:**119-121.
28. Jennett B, Plum F. Persistent vegetative state after brain damage: a syndrome in search of a name. *Lancet* 1972;**1:**734-737.
29. Jennett B. *The Vegetative State. Medical Facts, Ethical and Legal Dilemmas.* Cambridge: Cambridge University Press, 2002.
30. Borthwick C. The proof of the vegetable: a commentary on medical futility. *J Med Ethics* 1995;**21:**205-208.
31. Borthwick CJ. The permanent vegetative state: ethical crux, medical fiction? *Issues Law Med* 1996;**12:**167-185.
32. Howsepian AA. The 1994 Multi-Society Task Force consensus statement on the persistent vegetative state: a critical analysis. *Issues Law Med* 1996;**12:**3-29.
33. Andrews K, Murphy L, Munday R, Littlewood C. Misdiagnosis of the vegetative state: retrospective study in a rehabilitation unit. *BMJ* 1996;**313:**13-16.
34. La Puma J, Schiedermayer DL, Gulyas AE, Siegler M. Talking to comatose patients. *Arch Neurol* 1988;**45:**20-22.
35. Tosch P. Patients' recollections of their posttraumatic coma. *J Neurosci Nurs* 1988;**20:**223-228.
36. DeGiorgio CM, Rabinowicz AL, Gott PS. Predictive value of P300 event-related potentials compared with EEG and somatosensory evoked potentials in non-traumatic coma. *Acta Neurol Scand* 1993;**87:**423-427.

37. Glass I, Sazbon L, Groswasser Z. Mapping "cognitive" event-related potentials in prolonged postcoma unawareness state. *Clin Electroencephalogr* 1998;**29:**19-30.
38. Gott PS, Rabinowicz AL, DeGiorgio CM. P300 auditory event-related potentials in nontraumatic coma: association with Glasgow Coma Score and awakening. *Arch Neurol* 1991;**48:**1267-1270.
39. Kaga M, Inagaki M, Ozawa H. Discrimination of facial expression in vegetative state. Mismatch negativity of visual event-related potential study [abstract]. *Ann Neurol* 1996;**40:**316.
40. Yingling CD, Hosobuchi Y, Harrington M. P300 as a predictor of recovery from coma. *Lancet* 1990;**336:** 873.
41. American Academy of Neurology. Position of the American Academy of Neurology on certain aspects of the care and management of the persistent vegetative state patient. *Neurology* 1989;**39:**125-126.
42. President's Commission for the Study of Ethical Problems in Medicine and Biomedical and Behavioral Research. *Deciding to Forego Life-Sustaining Treatment.* Washington, DC: U.S. Government Printing Office, 1983.
43. American Medical Association Council on Scientific Affairs and Council on Ethical and Judicial Affairs. Persistent vegetative state and the decision to withdraw or withhold life support. *JAMA* 1990;**263:**426-430.
44. ANA Committee on Ethical Affairs. Persistent vegetative state: report of the American Neurological Association Committee on Ethical Affairs. *Ann Neurol* 1993;**33:**386-390.
45. Multi-Society Task Force on PVS. Medical aspects of the persistent vegetative state. (First of two parts). *N Engl J Med* 1994;**330:**1499-1508.
46. Plum F, Posner JB. *The Diagnosis of Stupor and Coma.* 3rd ed. Philadelphia: F. A. Davis Company, 1983.
47. Plum F, Shiff N, Ribary U, Llinas R. Coordinated expression in chronically unconscious persons. *Philos Trans R Soc Lond B Biol Sci* 1998;**353:**1929-1933.
48. Schiff N, Ribary U, Plum F, Llinas R. Words without mind. *J Cogn Neurosci* 1999;**11:**650-656.
49. American Academy of Neurology Quality Standards Subcommittee. Practice parameters: assessment and management of patients in the persistent vegetative state (summary statement). *Neurology* 1995;**45:**1015-1018.
50. Multi-Society Task Force on PVS. Medical aspects of the persistent vegetative state. (Second of two parts). *N Engl J Med* 1994;**330:**1572-1579 [erratum in *N Engl J Med* 1995;**333(2):**130].

BRAIN FUNCTION IN THE VEGETATIVE STATE

Steven Laureys, Marie-Elisabeth Faymonville, Xavier De Tiège, Philippe Peigneux, Jacques Berré, Gustave Moonen, Serge Goldman, Pierre Maquet*

1. INTRODUCTION

The vegetative state (VS) is a devastating medical condition characterized by preserved wakefulness contrasting with absent voluntary interaction with the environment (Figure 1). It can be diagnosed soon after a brain injury and can be partially or totally reversible, or it may progress to a persistent VS or death. It is important to distinguish between VS, persistent VS and permanent VS. Persistent VS is arbitrarily coined as a VS present one month after acute traumatic or non-traumatic brain injury or lasting at least one month in patients with degenerative or metabolic disorders or developmental malformations,[1] but does not imply irreversibility. Permanent VS implies the prediction that the patient will not recover. It was introduced by the American Multi-Society Task Force on PVS[1] in 1994 to denote irreversibility after three months following a non-traumatic brain injury and twelve months after traumatic injury. However, even after these long and arbitrary delays, some patients may exceptionally recover. Hence, the American Congress of Rehabilitation Medicine advocates abandoning the term "permanent" in favor of simply specifying the length of time patients have spent in VS.[2] The question which most concerns relatives and doctors caring for patients with vegetative state is whether a recovery is possible. The Task Force analyzed the prognosis of these patients and identified three factors that clearly influenced the chances of recovery: age, etiology, and time already spent in VS. The outcome is better after

* Steven Laureys, University of Liège, Cyclotron Research Center, Sart Tilman B30, 4000 Liège, Belgium and University of Liège, Department of Neurology, Sart Tilman B35, 4000 Liège, Belgium. Marie-Elisabeth Faymonville, University of Liège, Department of Anesthesiology, Sart Tilman B35, 4000 Liège, Belgium. Xavier De Tiège, University of Brussels, Erasme Hospital, PET Unit, 1070 Brussels, Belgium. Philippe Peigneux, University of Liège, Cyclotron Research Center, Sart Tilman B30, 4000 Liège, Belgium. Jacques Berré, University of Brussels, Erasme Hospital, Department of Intensive Care, 1070 Brussels, Belgium. Gustave Moonen, University of Liège, Department of Neurology, Sart Tilman B35, 4000 Liège, Belgium. Serge Goldman, University of Brussels, Erasme Hospital, PET Unit, 1070 Brussels, Belgium. Pierre Maquet University of Liège, Cyclotron Research Center, Sart Tilman B30, 4000 Liège, Belgium and University of Liège, Department of Neurology, Sart Tilman B35, 4000 Liège, Belgium.

Brain Death and Disorders of Consciousness, Edited by Machado and Shewmon
Kluwer Academic/Plenum Publishers, New York 2004

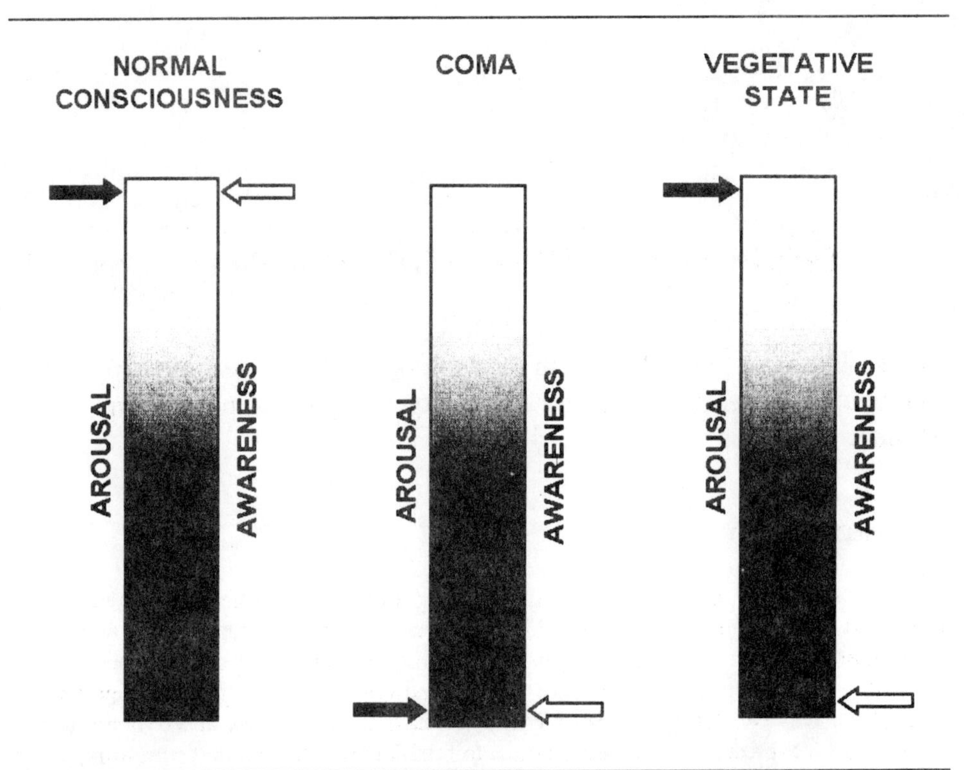

Figure 1. Graphical representation of the two components of consciousness (arousal and awareness) and their alterations in coma and the vegetative state.

traumatic than non-traumatic brain injury, better in children, and worse as time passes. Clinical, electroencephalographic (EEG), evoked potentials (EP), or structural imaging data do not permit an accurate prognostication in VS.[1]

The interest of functional imaging in VS is twofold. First, VS patients represent a clinical problem, in terms of diagnosis, prognosis, treatment and everyday management. Second, it offers a lesional approach to the study of human consciousness and adds to the international research effort to identify the neural correlate of consciousness. Indeed, these patients represent genuine cases of abolition of consciousness but, contrary to coma patients, with preserved arousal. Consciousness is thought to represent an emergent property of cortical and subcortical neural networks and their reciprocal projections. Its multifaceted aspects can be seen as expressions of various specialized areas of the cortex that are responsible for processing external and internal stimuli, short- and long-term storage, language comprehension and production, information integration and problem solving, and attention.[3]

2. GLOBAL IMPAIRMENT IN CEREBRAL METABOLISM

Positron Emission Tomography (PET) has shown a substantial reduction in global brain metabolism in patients in VS (Figure 2). In VS of various etiologies and durations, studies from our own[4-10] and other centers[11-15] have shown that cerebral metabolic rates for glucose (CMRGlu) are approximately 40 percent of normal values, whereas in patients in coma of hypoxic and traumatic origin, values are approximately 50 percent of normal.[16, 17] Compared to cerebral glucose metabolism, cerebral blood flow seems to show a larger interpatient variability in VS.[13] In long-standing post-hypoxic vegetative state, CMRGlu values decrease even further,[11,14] probably due to progressive Wallerian and transsynaptic degeneration. At present, there is no established correlation between CMRGlu depression and patient outcome.

A global depression of cerebral metabolism is not unique to vegetative state or coma. When different anesthetics are titrated to the point of unresponsiveness, the resulting reduction in CMRGlu is nearly as low as that observed in VS patients.[18-20] During propofol anesthesia, brain metabolism sometimes decreases to 30 percent of normal values. Another example of transient metabolic depression has been observed by our own

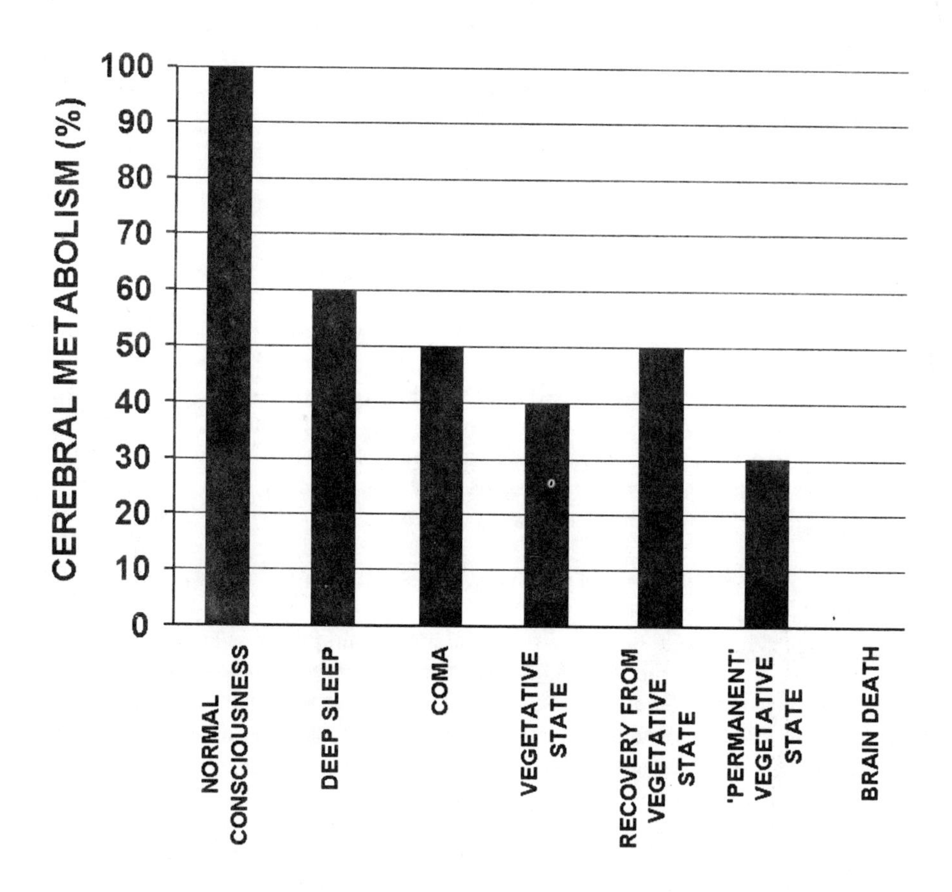

Figure 2. Cerebral metabolism in the different clinical entities (for references see text).

and other centers during slow-wave sleep.[21,22] In this daily physiological condition CMRGlu decreases to 60 percent of waking values (Figure 2).

More interestingly, we had the opportunity to scan a patient during VS and after recovery of consciousness.[10] To our surprise, global gray matter CMRGlu did not show a substantial increase after recovery. In this case, the recovery of consciousness seemed related to a modification of the regional distribution of brain function rather than to the global resumption of cerebral metabolism. Statistical Parametric Mapping (SPM)[23] analysis identified the most important decreases in metabolism, seen during VS but not after recovery, in the bilateral parietal associative cortices at the convexity and at the midline (precuneus and posterior cingulate).[10] To our knowledge there is only one other case published where PET scanning was performed during VS and after recovery of consciousness.[12] Again, global gray matter CMRGlu did not show a substantial increase after recovery (5.0 mg/100g.min versus 5.2 mg/100g.min). Although no SPM analysis was performed, region of interest (ROI) analysis showed the largest regional increase in parieto-occipital cortices. These data point to a critical role for the posterior associative cortices in the emergence of conscious experience.

It remains controversial whether the observed metabolic impairment in VS reflects functional and potentially reversible damage or irreversible structural neuronal loss. Rudolf and co-workers argue for the latter, using ^{11}C-flumazenil as a marker of neuronal integrity in evaluating acute anoxic VS patients.[24] We hypothesize that impairment in cortico-cortical and thalamo-cortical connectivity could explain part of the permanent, or in some fortunate cases transient, functional cortical impairment in VS.[25]

3. REGIONAL IMPAIRMENT IN CEREBRAL METABOLISM

3.1. Relatively Most Impaired Brain Areas

Using ROI analysis, previous PET studies have shown a reduction in overall cortical metabolism[11-14] with most profound reductions in the parieto-occipital and mesial frontal cortices.[12] By means of SPM analysis[23] we were able to identify the regional pattern of metabolic impairment common to our patients in VS.[9] The prefrontal, premotor and parietotemporal association cortices and the posterior cingulate/precuneus region showed the most severe functional impairment (Fig. 3B). This pattern is in agreement with postmortem findings where involvement of the association cortices is reported as a critical neuroanatomic substrate.[26] These associative cortices are known to be involved in various consciousness-related functions such as perception, attention, working memory, episodic memory, and language.

Interestingly, the medial parietal cortex (precuneus and posterior cingulate cortex) is one of the most active cerebral regions (together with the anterior cingulate and the prefrontal cortex) in conscious waking.[21,27,28] Moreover, it is systematically one of the least active regions in unconscious or minimally conscious states such as such as coma[17] (Fig. 3A), halothane anesthesia,[18] slow-wave sleep,[21] rapid eye movement sleep,[29] Wernicke-Korsakoff's and post-anoxic amnesia,[30] and hypnotic state.[31] This area is also the site of the earliest reductions in glucose metabolism in Alzheimer's disease.[32] These arguments suggest that the posterior parietal cortex might represent part of the neural network subserving conscious experience.

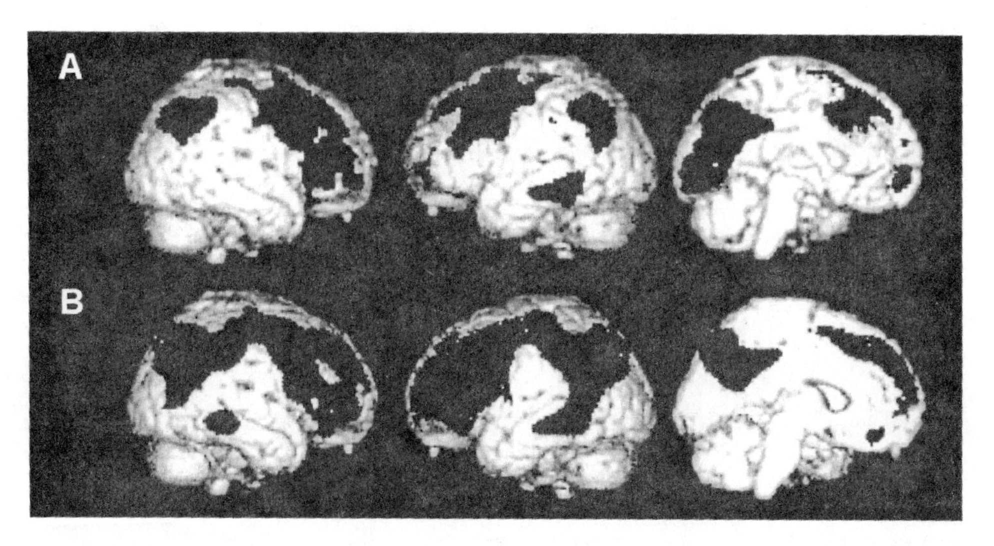

Figure 3. The common pattern of altered cerebral metabolism characterizing vegetative state patients. Using Statistical Parametric Mapping,[23] we identified areas where metabolism was relatively most impaired in comatose patients (A) and vegetative patients (B) compared to conscious controls (for methodological details see[9, 17]). A schematic representation of these areas is shown in black on a surface rendered normalized magnetic resonance image. Note that the functionally most impaired regions in both coma and the vegetative state are the associative cortices (frontal, parietotemporal and posterior cingulate/precuneus).

3.2. Relatively Spared Brain Areas

We observed another hallmark common to our patients in VS: the relative preservation of metabolism in the brainstem (encompassing the mesopontine reticular formation), basal forebrain, and posterior hypothalamus. This allows for the maintenance of vegetative functions in these patients: preserved sleep-wake cycles, autonomic and ventilatory control, and cranial nerve reflexes. This observation is in line with the post-mortem neuropathologic finding that these structures are relatively preserved in VS patients.[26]

4. FUNCTIONAL IMPAIRMENT IN CEREBRAL CONNECTIVITY

4.1. Cortico-Cortical Connectivity

Recently, functional imaging has offered an analytical tool to assess the functional connectivity between distant cerebral areas. Put simply, such a statistical analysis identifies brain regions that show condition-dependent differences in modulation with another (chosen) area. Using such a psychophysiological interaction analysis,[33] we were able to demonstrate that patients in vegetative state suffer from an altered cortico-cortical connectivity. Compared to control subjects, patients in VS showed an altered modulation between the left frontal cortices and the medial parietal cortex.[9] This impaired fronto-parietal connectivity in VS is in accordance with experiments in non-human primates demonstrating that the functional integrity of the prefrontal cortex and its interactions with modality specific posterior brain regions is critical for working memory.

4.2. Thalamo-Cortical Connectivity

Based on the putative role of high-frequency oscillatory thalamocortical circuitry underlying human consciousness in healthy volunteers,[34] our center has assessed the functional integrity of the thalamocortical connectivity in VS patients. Using the same analytical methodology[9] we identified brain areas that showed a different functional connectivity with both thalami in patients in VS compared to controls. We indeed observed an impaired functional relationship between the activity in the thalami and fronto-parietal associative cortices,[6] partially restoring normal function after recovery of consciousness.[25] The thalamus contains both specific thalamo-cortical relay nuclei and so-called nonspecific intralaminar nuclei. The former are the necessary relay for all sensory afferent stimuli (except some olfactory information). The latter have been implicated in the maintenance of thalamo-cortico-thalamic synchronous oscillations. Among these activities, 40 Hz oscillations seem to be deeply, although not exclusively, involved in conscious experience.[34,35] Thus, thalamic nuclei seem critical for the maintenance of human awareness.

5. CEREBRAL ACTIVATION AFTER EXTERNAL STIMULATION

In 1989, Momose and co-workers described a patient in VS whose CMRGlu increased after cervical spinal cord stimulation.[36] More recently, using magnetoencephalography[15,37,38] or $H_2{}^{15}O$-PET,[39,40] cerebral activation has been described during sensory stimulation in VS patients. De Jong and co-workers presented to a VS patient both a story told by his mother and non-word sound. They observed an activation [with one type of stimulus or both?] in anterior cingulate and temporal cortices which they interpreted as possibly reflecting the processing of emotional attributes of speech or sound.[39] Menon and co-workers presented to a VS patient photographs of familiar faces and meaningless pictures. The visual association areas showed significant activation when faces were compared to meaningless stimuli.[40] Our group has assessed the central processing of noxious somatosensory stimuli in the VS.[8] Changes in regional cerebral blood flow and event related potentials were measured during high intensity electrical stimulation of the median nerve in VS patients and compared to data obtained in healthy controls. Noxious stimulation activated midbrain, contralateral thalamus and primary somatosensory cortex in each and every VS patient, even in the absence of detectable cortical evoked potentials (Figure 4). However, a large network of hierarchically 'higher-order' multi-modal association areas failed to activate: the secondary somatosensory cortices, the insular regions, the posterior parietal and prefrontal areas and the anterior cingulate cortex (regions that are known to be involved in pain affect, attention and memory). Moreover, primary somatosensory cortex, the only cortical region that activated in vegetative patients, was no longer functionally connected (i.e., no longer communicated) with the rest of the brain (i.e., the 'higher order' brain regions thought to be necessary for conscious processing). Hence, somatosensory stimulation of VS patients, at intensities that elicited pain in controls, resulted in increased neuronal activity

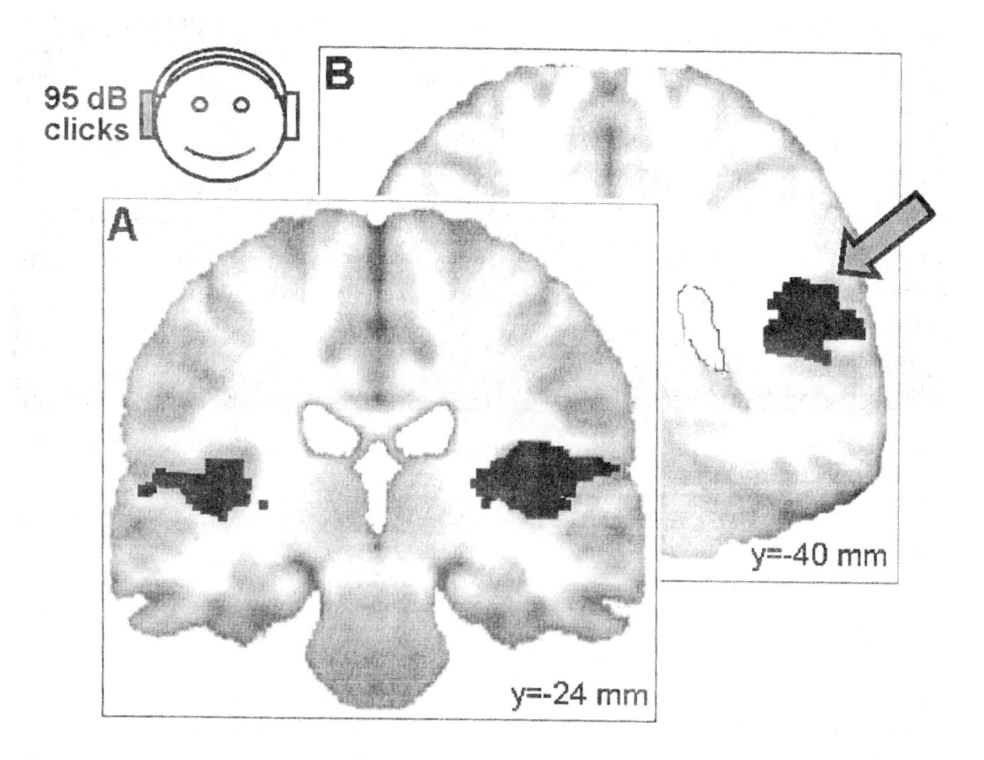

Figure 4. Areas of regional blood flow (rCBF) increase during auditory stimulation in VS patients projected on a coronal MRI section, 24 and 40 mm behind the anterior commisural line. The arrow points to the auditory association areas where rCBF showed significantly less activation in VS compared to controls. (Adapted from ref.[5])

in primary somatosensory cortex, even if resting brain metabolism was severely impaired. However, this activation of primary cortex seems to be isolated and dissociated from higher-order associative cortices.

Similarly, auditory stimulation activated bilateral primary auditory areas but, in contrast to controls, not the higher-order associative areas in the temporo-parietal junction (Figure 5). In VS patients, auditory association cortex was functionally disconnected from the posterior parietal association area, anterior cingulate cortex and hippocampus.[5] Thus, despite an altered resting metabolism, primary sensory cortices still activate during external stimulation, whereas hierarchically higher-order downstream multimodal association areas do not. In the absence of a generally accepted neural correlate of consciousness,[35] it is difficult to make definite judgments about conscious perception in VS patients. However, the cascade of functional disconnections along the sensory cortical pathways, from primary areas to multimodal and limbic areas, suggests that the observed activation of primary sensory cortex subsists as an island, dissociated from higher-order cortices that would be necessary to produce awareness.

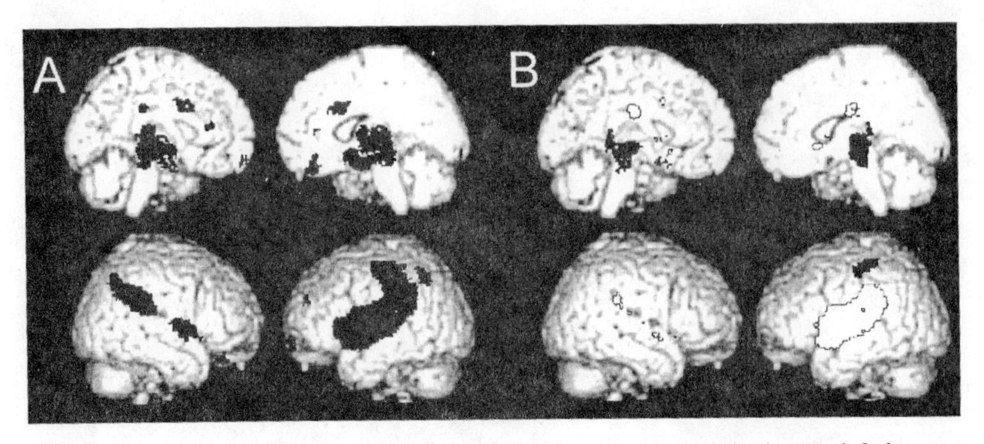

Figure 5. (A) Brain regions, shown in black, that activated during noxious stimulation in controls [subtraction stimulation-rest] projected on a 3-D spatially normalized brain MRI. (B) Brain regions that activated during stimulation in PVS patients, shown in black [subtraction stimulation-rest] and regions that activated less in patients than in controls [interaction (stimulation versus rest) x (patient versus control)], shown in white. (Adapted from ref.[8])

6. CONCLUSION

At present, the potential for recovery of awareness from the VS cannot be predicted reliably by any clinical or neurodiagnostic test. Functional imaging studies of residual brain function in VS provide an opportunity to understand the basic neural processes underlying human consciousness. Past studies from our own and other centers have used functional neuroimaging to study the residual brain function in such patients. These efforts identified a decrease in *global* metabolism of 40 percent. However, some patients who recovered from a VS, showed a modification of the *regional* distribution of brain function rather than a resumption of global metabolism. This leads to the hypothesis that some VS patients remain unconscious not because of a widespread neuronal loss, but due to the impaired activity in some critical brain areas and to an altered functional relationship between them. We were able to identify the common neural correlate of VS. The most severely affected brain regions were localized in the frontal and parietal associative cortices. On the contrary, brainstem, posterior hypothalamus, and basal forebrain were the most spared brain regions. By means of a psychophysiological interaction analysis[33] we subsequently demonstrated that patients in VS indeed suffer from an altered thalamo-cortical and cortico-cortical connectivity. Using cerebral activation paradigms, ongoing international research efforts will more closely correlate functional imaging with behavioral assessment, electrophysiological findings, and eventually, outcome.

7. ACKNOWLEDGMENTS

S. Laureys and P. Maquet are Research Associate and Research Director at the Fonds National de la Recherche Scientifique de Belgique (FNRS). This research was supported by FNRS, by the Reine Elisabeth Medical Foundation and by Research Grants from the University of Liège.

8. REFERENCES

1. The Multi-Society Task Force on PVS Medical aspects of the persistent vegetative state (1). *N Engl J Med* 1994;**330:**1499-1508.
2. American Congress of Rehabilitation Medicine Recommendations for use of uniform nomenclature pertinent to patients with severe alterations of consciousness. *Arch Phys Med Rehabil* 1995;**76:**205-209.
3. Laureys S, Majerus S, Moonen G. Assessing consciousness in critically ill patients. In: Vincent JL, ed. 2002 *Yearbook of Intensive Care and Emergency Medicine.* Heidelberg: Springer-Verlag, 2002:715-727.
4. Laureys S, et al. Brain function in the vegetative state. *Acta Neurol Belg* 2002;**102:**177-185.
5. Laureys S, et al. Auditory processing in the vegetative state. *Brain* 2000;**123:**1589-1601.
6. Laureys S, et al. In: Gjedde A, Hansen SB, Knudsen GM, Paulson OB, eds. *Physiological Imaging of the Brain with PET.* San Diego: Academic Press, 2000:329-334.
7. Laureys S, Faymonville ME, Moonen G, Luxen A, Maquet P. PET scanning and neuronal loss in acute vegetative state. *Lancet* 2000;**355:**1825-1826.
8. Laureys S, et al. Cortical processing of noxious somatosensory stimuli in the persistent vegetative state. *Neuroimage* 2002;**17:**732-741.
9. Laureys S, et al. Impaired effective cortical connectivity in vegetative state: preliminary investigation using PET. *Neuroimage* 1999;**9:**377-382.
10. Laureys S, Lemaire C, Maquet P, Phillips C, Franck G. Cerebral metabolism during vegetative state and after recovery to consciousness. *J Neurol Neurosurg Psychiatry* 1999;**67:**121.
11. Rudolf J, Ghaemi M, Haupt WP, Szelies B, Heiss WD. Cerebral glucose metabolism in acute and persistent vegetative state. *J Neurosurg Anesthesiol 1999;* **11:**17-24.
12. Volder A G. De et al. Brain glucose metabolism in postanoxic syndrome. Positron emission tomographic study. *Arch Neurol* **47:**197-204.
13. Levy DE et al. Differences in cerebral blood flow and glucose utilization in vegetative versus locked-in patients. *Ann Neurol* **22:**673-682.
14. Tommasino C, Grana C, Lucignani G, Torri G, Fazio F. Regional cerebral metabolism of glucose in comatose and vegetative state patients. *J Neurosurg Anesthesiol* 1995;**7:**109-116.
15. Schiff BD, et al. Residual cerebral activity and behavioural fragments can remain in the persistently vegetative brain. *Brain* 2002;**125:**1210-1234.
16. Tommasino C. Brain glucose metabolism in the comatose state and in post-comatose syndromes. *Minerva Anestesiol* 1994;**60:**523-525.
17. Laureys S, Berré J, Goldman S. Cerebral function in coma, vegetative state, minimally conscious state, locked-in syndrome and brain death. In: Vincent JL, ed. 2001 *Yearbook of Intensive Care and Emergency Medicine.* Berlin: Springer-Verlag, 2001:386-396.
18. Alkire MT, et al. Functional brain imaging during anesthesia in humans: effects of halothane on global and regional cerebral glucose metabolism. *Anesthesiology* 1999;**90:**701-709.
19. Alkire MT, Haier RF, Shah NK, Anderson CT. Positron emission tomography study of regional cerebral metabolism in humans during isoflurane anesthesia. *Anesthesiology* 1997;**86:**549-557.
20. Alkire MT, et al. Cerebral metabolism during propofol anesthesia in humans studied with positron emission tomography. *Anesthesiology* 1995;**82:**393-403.
21. Maquet P, et al. Functional neuroanatomy of human slow wave sleep. *J Neurosci* 1997;**17:**2807-2812.
22. Buchsbaum MS, et al. Regional cerebral glucose metabolic rate in human sleep assessed by positron emission tomography. *Life Sci* 1989;45:1349-1356.
23. Friston KJ. Analysing brain images: principles and overview. In: Frackowiak RSJ, Friston KJ, Frith CD, Dolan RJ and Mazziotta JC, eds. *Human Brain Function.* San Diego: Academic Press, 1997: 25-41.
24. Rudolf J, Sobesky J, Grond M, Heiss WD. Identification by positron emission tomography of neuronal loss in acute vegetative state. *Lancet* 2000;**355:**155.
25. Laureys S, et al. Restoration of thalamocortical connectivity after recovery from persistent vegetative state. *Lancet* 2000;**355:**1790-1791.
26. Kinney HC, Samuels MA. Neuropathology of the persitent vegetative state. A review. *J Neuropathol Exp Neurol* 199;**53:**548-558.
27. Andreasen NC, et al. Remembering the past: two facets of episodic memory explored with positron emission tomography. *Am J Psychiatry* 1995;**152:**1576-1585.

28. Gusnard DA, Raichle ME. Searching for a baseline: functional imaging and the resting human brain. *Nat Rev Neurosci* 2001;**2:**685-694.
29. Maquet P, et al. Functional neuroanatomy of human rapid-eye-movement sleep and dreaming. *Nature* 1996; **383:**163-166.
30. Aupee AM, et al. Voxel-based mapping of brain hypometabolism in permanent amnesia with PET. *Neuroimage* 2001;**13:**1164-1173.
31. Maquet P, et al. Functional neuroanatomy of hypnotic state. *Biol Psychiatry* 1999;**45:**327-333.
32. Minoshima S, et al. Metabolic reduction in the posterior cingulate cortex in very early Alzheimer's disease. *Ann Neurol* 1997;**42:**85-94.
33. Friston KJ, et al. Psychophysiological and modulatory interactions in neuroimaging. *Neuroimage* 1997; **6:**218-229.
34. Llinas R, Ribary U, Contreras D, Pedroarena C. The neuronal basis for consciousness. *Philos Trans R Soc Lond B Bio Sci* 1998;**353:**1841-1849.
35. Zeman A. Consciousness. *Brain* 2001;**124:**1263-1289.
36. Momose T, Matsui T, Kosaka N. Effect of cervical spinal cord stimulation (cSCS) on cerebral glucose metabolism and blood flow in a vegetative patient assessed by positron emission tomography (PET) and single photon emission computed tomography (SPECT). *Radiat Med* 1989;**7:**243-246.
37. Schiff ND, Ribary U, Plum F, Llinás R. Words without mind. *J Cogn Neurosci* 1999;**11:**650-656.
38. Schiff ND, Plum F. Cortical function in the persistent vegetative state. *Trends Cogn Sci* 1999;**3:**43-44.
39. Jong de B, Willemsen AT, Paans AM. Regional cerebral blood flow changes related to affective speech presentation in persistent vegetative state. *Clin Neurol Neurosurg* 1997;**99:**213-216.
40. Menon DK, et al. Cortical processing in persistent vegetative state. *Lancet* 1998;**352:**200.

GLOBAL NEURODYNAMICS AND DEEP BRAIN STIMULATION

Appreciating the perspectives of place and process

David I. Pincus*

1. INTRODUCTION

The study of coma and brain death necessarily involve the study of consciousness. A taxonomy of conscious states, and their correlation to the clinical conditions of persistent vegetative state, hyperkinetic mutism, delirium and akinetic mutism is currently lacking (see Watt and Pincus for a preliminary effort).[1] The neurobiological mechanisms and processes underlying conscious states are just now becoming better understood, and emerging technologies (electrical, magnetic, and chemical modulation) may help us to better understand the origins of consciousness, as well as enhance the consciousness of severely impaired or comatose patients.

It is now generally agreed that consciousness emerges in brains as the result of reentrant and widespread neuronal population behavior. There is not a consciousness button, center, processor, or other metaphorical location that turns-on wakefulness. Distributed models of brain function have overtaken locationist models for many years. Nonetheless, the appeal of a reduction to phrenological places continues to reassert itself with each new correlation of mind to a brain area that the latest scanner provides. However, in our drift towards a network doctrine, we must not forget about the more fixed parameters and more bounded structures. The risk is that global re-entrant networks, functioning according to the laws of mathematical chaos or stochastic resonance, will light up the sky in a fashion that overlooks the contributions of the specific landscape. It is perhaps human nature to irrevocably shift between the global and the discrete,† overshooting emphasis upon one in favor of the other, in an ongoing dialectical transition. Instead, can such a shifting occur because we do not have coherent answers to the mind/brain problem? I know that in my primary area, psychoanalysis,

* David I. Pincus, D.M.H., Psychiatry and Psychology, Case Western Reserve University, Cleveland, OH 44106, USA. Email: dpincus216@aol.com.

† I use the word discrete to refer to more localized, more bounded places in the brain. I do not mean to suggest that event the most conserved nuclei are isolated in any sense.

Brain Death and Disorders of Consciousness, Edited by Machado and Shewmon
Kluwer Academic/Plenum Publishers, New York 2004

models of mind are plagued with unacknowledged attachments to one variety of the mind-brain problem or another, often with very confusing results. Certainly, until we understand more, balance is what is desired. We can certainly assert that the living brain, along with the mind that emerges from it, surely result from a delicate balance and as yet poorly understood relationship between *place* and *process*.

This paper will deliberately focus upon examples of the global and the discrete, as both are relevant to our understanding disorders of consciousness, and therefore brain coma and brain death. I will begin with an a brief history of neurodynamics through the lens of the EEG, briefly review the history of research in the identification of brain stem nuclei and their functions, and then turn to the local stimulation of neurons, an even finer level of place, with the best example being deep brain stimulation (DBS). On the surface, neurodynamics and DBS are strange bedfellows (wide scale vs. local), but by viewing them side-by-side, I hope to heighten the reader's appreciation that both the large scale and the minute must be simultaneously considered. Neurodynamics runs the risk of deemphasizing the particular contributants; a local emphasis (and a stimulating technology that is very local, DBS) runs the risk of ushering in a new phrenology.

We shall begin broadly, go through a very narrow wormhole, and hopefully emerge in a widespread fashion at the other side.

2. TWO MOMENTS IN THE HISTORY OF NEURODYNAMICS: BERGER AND FREEMAN

Neurodynamics is the study of changes in neuronal organization and behavior over time. Dynamics of individual neurons are not considered here; instead the focus is upon the dynamics of groups of neurons acting in concert. Presumably it is the group behavior of neurons, in a large scale-population collaboration, that is the basis of a conscious brain and a living mind. The neurodynamic perspective is one that complements the other basic orientations, those of structure, connectivities, and neuromodulators. From our chapter (Watt and Pincus), we offer the following summary:

> *"One of the most puzzling and yet essential properties of consciousness is its seamless integration and fundamental unity. Many investigators have suggested that populations of neurons are coordinated via the generation of coherent patterns or 'oscillatory envelopes' that structure integrative communication between brain regions. Several investigators postulate that the synchronous firing behavior among these distributed populations could constitute the essential neurodynamic underpinnings for conscious states and their contents. Many if not most neurodynamic theories of consciousness are elaborations of basic neuroanatomical concepts that emphasize thalamocortical connectivities, and there is relatively little neurodynamic work looking closely at possible contributions of structures underneath the thalamus. These theories propose that essential features of functional integration are achieved thalamocortically, perhaps largely via the functioning and connectivities of the nonspecific thalamic systems (ILN/nRt)."*

It is agreed that Hans Berger deserves undisputed credit for the discovery of the human electroencephalograph, and it was Berger who was first to actually measure brain population activity at this level. Lord Adrian,[2] who was most responsible for introducing Berger's work to the English-speaking world, termed the alpha waves the 'Berger rhythm'. In the 1920's, when Berger made his initial discovery, the understanding of CNS physiology was very primitive. Berger looked to the large-scale brain activity rather than emphasizing the local, the latter being 'in vogue' because of better experimental paradigms and methods for isolation. As Gloor[3] stated, "By resolutely turning away from the then fashionable preoccupation with morphology, the minutiae of cerebral localization and an exclusive concern with basic electrophysiological mechanisms, he led the way to the investigation of the dynamics of brain function." (p.1)

Berger had had a near death telepathic experience from his childhood, and this led him into science and towards the physical representation of mental events. He nearly died in an accident during his military training, and received a telegram from his father that evening inquiring about his health. Berger's sister had insisted that the father send the telegram, because she was convinced that her brother had had a serious accident. Berger viewed this as a case of spontaneous telepathy, and while he never achieved much understanding of minds communicating across extensive physical space, the event did lead to his interest in the brain-mind and his pioneering research in the EEG.

In 1924, Berger applied electrical stimulation to the human scalp, focusing upon the scalp over the area of defect in a patient's skull. His interest in finding clinical changes from broad stimulation emphasized his belief that it was the broader, integrated levels of brain functioning that led to consciousness, coherent emotions, and thought. Later in the decade he had developed the EEG (which literally means 'electric brain writing'). Berger used simple tasks like the opening and closing of the eyes, measuring brain activity via scalp electrodes for the 'brain writing' and his eventual identification of the alpha blockade. Berger may be viewed as one bookend of the neurodyanamic perspective, and seventy years since have brought with them continued refinements in measurement and data analysis, so that we now have a better capacity to appreciate 'who' is doing the writing, 'where' it is coming from, and 'what' the text is saying. Hans Berger and the discovery of the EEG was an important development in the knowledge base for a neurobiology of consciousness. Appreciating this neurobiology of networking allows us to make more informed decisions about what constitutes brain death, developing a taxonomy of conscious (and unconscious) states, and considering stimulation strategies which might help to restore consciousness to impaired individuals.

In this new century, brain science is populated with researchers in neurodynamics, with clones of Hans Berger to be found all across the world. It is easy to predict that Virchow's neuron doctrine will be thoroughly complemented by a network doctrine. One such researcher at the other end of the bookend of neurodynamics who has utilized the EEG is Walter J. Freeman,[4,5] who for the past 40 years has studied the process of perception. He too has no less a personal basis for his interest in broad-scale networks: his father, also Walter Freeman, was an originator of the frontal lobotomy in America, the very local procedure for a very complex, non-local problem, that of human emotion. Beginning with the study of olfaction in rabbits, and taking advantage of Berger's EEG's, Freeman the son has developed a model of population behavior that is based upon the principles of chaos. He used a surgically implanted, tiny 64 lead epidural electrode array measuring four millimeters by four millimeters to study the dynamics of olfaction. Freeman, along with Bressler,[6] first coined the term 'gamma' to describe the macroscopic population effect. This population effect is made up of high frequency oscillations, and the oscillations occur at a frequency of 20-80 times per second, with the synchronization being aperiodic and chaotic, spreading across the entire gamma band and not tuned to a single frequency (as in 40 Hz). From a paper that links this neuronal population behavior with the psychoanalytic concept of transference[7], we discuss neuronal populations in the olfactory bulbs of rabbits:

> *"First, there is evidence that the whole population takes over in a way that limits the input from other sources, whether they are from the cortex or from the stimulus itself. Second, the macroscopic population effect is made up of high frequency oscillations and we see that the oscillation has the same waveform across the entire bulb. This repeating waveform rolls across the entire bulb, and the oscillations occur at a frequency of 20-80 times per second, which one can think of as a kind of brain purring. Since there is no pathway from other cortices or receptors that can drive the entire population to oscillate over this wide, high, gamma range, it must be concluded that this waveform is constructed by the bulb. Third, we saw that there was continual background activity when the meaningful odorant was not present. It was irregular and unpredictable, like the surface of a small pond when it is surrounded by fifty children with limitless piles of pebbles to toss into the water. Fourth, during an inhalation the waveform emerges over the entire bulb for a short period of oscillation that we describe as a burst. This burst forms a spatial pattern of amplitude modulations (AM) by the 64 micro-EEG's, but only if the odorant is familiar, that is, has been learned. We think of the burst as creating the meaning or relevance of the odorant for the creature. Fifth, each AM pattern is distinctive and individualized for each rabbit, like a unique signature, despite the fact that many rabbits sniff the same odorant. This suggests that the individualized pattern of amplitude modulation is unique for each rabbit, meaningfully constructed according to the individual history of each animal. (p.14-15)."*

Freeman's research has revealed that all sensory cortices, using these population effects, create meaning and consciousness on a far more global, integrative level of hierarchical organization. He is now observing these effects in humans using surface EEG's, and has found that the global aperiodic resetting occurs from six to nineteen centimeters in spread.[8] While Freeman utilizes chaos to model the brain's dynamics, others turn to models of complexity[9] or stochastic resonance.[10,11,12] At this point in our knowledge, we are far from reaching a good understanding of how this remarkable process occurs.

Berger initially, and now Freeman, have mined the EEG to measure and describe the cooperative neural organizations that are the correlates of conscious activity. Any thorough understanding of consciousness and the pathological conditions associated with an impairment of consciousness must include a neurodynamic appreciation in addition to the usual energic, hemodynamic, structural, neuromodulatory and other perspectives. A significant strength of distributed, neurodynamic models is that they help to solve the homuncular, phrenological simplifications of days gone by. Emergence replaces all-doing causes, switches and gates replace centers in the model builder's lexicon, and the hierarchically organized 'global workspace' of consciousness[13] comes into being. While 'emergence' and the dynamic implications of the network doctrine solve certain philosophical and epistemological problems in model building (homunculi, for example), an over-reliance upon these concepts neglects the very real bounded and discrete parameters which all of the successive layers of organization rely upon. A brain without oxygen will provide no population behavior for very long, and a small lesion to a very small area can lead to a sustained coma. In matters of consciousness, the neuroanatomic locale exists in a sophisticated relationship to the broad-scale processes. One can imagine, by way of an inverse analogy that the contents of consciousness rest upon a background sense of awareness. Focus upon one cannot obscure the contribution of the other.

We know that the man-on-the-street is a dualist by nature, separating body and mind, and self from the world-around in one's phenomenological distinctions. But we also know that naïve holisms, whether they come in the form of fundamentalist religions, philosophical essentialisms, or gestalt generalizations only focus upon one aspect of the complete picture: perhaps we must end up, in our model building, to think of figure and ground, and in reference to the discrete vs. the more global, that 'you can't have one without the other'. There are many reasons why the man-on-the-street, as well as the people in the lab, tend to focus on the discrete ***or*** the global, one to the exclusion of the other. Besides practical considerations like time and funding, unrecognized philosophical and psychological assumptions course their way through each human experience and every scientific model, including models of brains and minds.

My goal is to counterpose locationist and neurodynamic considerations for consciousness and coma, so that the dialectic between *place* and *process* is properly appreciated. Let us now turn to another history: the elucidation of the brain nuclei that appear necessary for consciousness.

3. REMOTE AND RECENT UNDERSTANDING OF BRAIN STEM NUCLEI AND THEIR CONNECTIVITIES IN THE DEVELOPMENT OF CONSCIOUS STATES

Cajal[14] and others at the turn of the 20th century determined that the white matter of the brainstem was organized into nuclear bundles. This finding seems to be in keeping with the observation that often times the phylogenetically older areas of anatomy/physiological function are ever more clustered into discrete, localizable packages. The early observations were based upon techniques that we now conclude were gross and superficial, the limitations imposed being by existing technology. Without modern markers, the early observers were left to make distinctions based upon color, cellular features and arrangements, and where fiber tracks appeared to begin or to taper off. For instance, referring to the unclear markers of earlier discoveries, Parvisi and Damasio[15] observe, "the core region of the brainstem was labeled as the reticular formation because neurons in that region were surrounded by interlacing fibers, which gave the region the appearance of a "reticulum" that is, a web."(p.140). However, despite the by now unsophisticated means of localization and identification, a sense of *place* for the origins of wakefulness was now possible.

The correlation of places to functions in the study of consciousness occurred as the result of a variety of experimental interventions: surgical isolation, electrical stimulation or chemical modulation. These manipulations were correlated with various behavioral and psychological measures and EEG findings. As noted above, Berger's observation of the process of wakefulness was the phenomenon of alpha blockade, which occurred when subjects would open their eyes. Alpha waves are typically about 10 Hz, and subjects with eyes closed had synchronized alpha activity in the occipital area. Eyes open created an end to alpha, and the dominant 'brain writing' was faster (beta, 20Hz), of smaller amplitude, and was desynchronized. Lord Adrian claimed that this shift in frequency, amplitude, and desynchronization was caused by the effects of afferent stimulation directly upon the cerebral cortex, a simple cortical stimulation model of consciousness. As we shall see, this model proved to be insufficient.

The notion of direct cortical stimulation could be questioned by at least four important observations. First, something akin to habituation occurred when repeated sensory stimulation was provided, and so a simple relationship between stimulation and cortical activation was unlikely. Second, when an activation pattern is induced in the cortex via stimulation within one sensory modality, the pattern is not constrained to the sensory cortex of the particular modality. More importantly, the corresponding area of the sensory cortex was not the first to be activated. Instead, all parts of the cortex generate this pattern simultaneously. Finally, the EEG pattern continued well after the afferent stimulus was removed, again calling into question the simple cortical stimulation model. All of these findings conflicted with Lord Adrian's theory. A simple, direct cortical stimulation model for consciousness was in trouble.

In the 1930's, Bremer[16] had isolated the cerebrum from the rest of the nervous system, creating the 'cerveau isole'. The resulting EEG looked like a normal brain under anaesthesia or natural sleep, leading Bremer to conclude that the mesencephalic transection created a deafferentation of the cortex, a view that would be in keeping with Lord Adrian's theory. However, afferents from olfaction and the visual system are still intact in the 'cerveau isole', and if Lord Adrian was correct, then there should be the appearance of alpha blockade in the olfactory and visual cortices. However, the anaesthetic appearing EEG was found across all sensory modalities, even though olfactory and visual afferents remain intact. Were the EEG changes that resulted from Bremer's cerveau isole caused by deafferentation, or because a more central activating system was put out of action?

Lord Adrian's notion of direct cortical stimulation was further assaulted by the observation of latency. Alpha blockade did not occur at the moment of afferent stimulation, which hinted that some mechanism beyond simple afferent stimulation to the cortex must be responsible for the alpha blockade. The end of Lord Adrian's theory came (at least from an evidential standpoint—Lord Adrian still argued it in 1947)[17] when Gerebtzoff[18] observed that lesioning the cortical area of the afferent system corresponding to vestibular stimulation still resulted in an arousal reaction in that area. Gerebtzoff suggested that there must be a gating system between the sensory afferents and the cortex, rather than direct stimulation of the cortex.

I have taken two researchers, Berger and Freeman, and utilized them to create an image of arbitrary bookends, so as to tell a partial story of the EEG and its role in the study of broad-scale neural dynamics. I have also told provided a brief history in the tradition of localization: the discovery of the 'reticulum', how it might function, and how it might be an important correlate and cause for conscious states. I will now introduce two other author 'bookends' (in this case each bookend is made up of two authors), again for the sake of illustration and again, appreciating that 'bookends' are held up by what has been before and what comes after. Publications by Moruzzi and Magoun in 1949 and Parvisi and Damasio in 2001 have each led to the elucidation of neuroanatomical structures crucial in understanding the origins of consciousness.

Moruzzi and Magoun[19] extended and integrated many findings into the notion of the reticular activating system (RAS). As with many scientific discoveries, several of the answers were already on the table, and it sometimes takes just one more piece of evidence or one imaginative thinker to put them together. Moruzzi and Magoun are generous in their appreciation of others' work and the good fortune they had to be able to assemble and extend it, but it was their vision and experimental finding that brought together the insight. Moruzzi and Magoun provided the definitive clarification: stimulation of the bulboreticular formation under barbiturate anesthesia did not lead to alpha blockade in any area of the cortex.

The lesioning and brainstem injury literature deepened the case of Morruzi and Magoun. In 1936 it was found that lesions to the basal diencephalon and anterior midbrain, which spared the afferent sensory pathways to the cortex, resulted in somnolence. Tumors in this same area, and lesions and encephalitis[20] were clinical conditions that had disastrous effects on consciousness, rounding out the support. But it was not until 1949 that various authors correlated the clinical symptomatology, EEG studies, and neuroanatomical location.

The sum total of the existing evidence led Marouzi and Magoon to conclude: "The conception of sleep as a functional deafferentation of the cerebrum is not opposed by this evidence if the term "deafferentation" is broadened to include interruption of the ascending influence of the brain stem reticular activating system, the contribution of which to wakefulness now seems more important than that conducted to the cortex over classical sensory paths." (p.471)

Direct activation to the sensory cortices was abandoned as a primary source of causality in the understanding of arousal, attention and consciousness. Brainstem activation was required. A place, if not a process, for the origins of consciousness seemed pretty well secured.

But a *place* for the origins of consciousness, in the sense of a mother lode, could not hold out for long, when that place is actually many places with many interactions, and its centrality is really not entirely that central after all. And so the finer and finer level of analysis yields smaller and smaller contributants (like particles in physics), and yet the finer level of analysis leads to a sense of connected, emergent process (as some also suggest in physics). In what follows, our other author-bookends juxtapose the concept of *place* with the complexity of *process*, and the notion of a network viewpoint of consciousness. I will only take the reader down into the dizzying details to the degree that is relevant for my point: a networked view emerges out of the complexity of very small places with remarkable connectivities and dynamics. When it comes to the brain, the conceptual limitations of *place* must always be complemented and integrated with that of *process*.

In a recent article by Parvizi and Damasio, the picture elaborated by Moruzzi and Magoun is revised with a different view of the centrality of purpose and the unity of structure within the RAS. The earlier work became a gospel like conviction of the nature of the relationship between the brainstem and consciousness. Parvisi and Damasio have updated this proposal. The homogeneity of the 'reticulum' gives way to significant specificity, in the delineation of some forty nuclei, the operation of several separable transmitter systems, and levels of structural connection not previously appreciated. This has even led some authors[21] to suggest that the term 'reticular formation' should be abandoned, as it does suggest a mistaken overall homogeneity.

Parvisi and Damasio, in their structural delineation, also wish to include the phenomena of subjective affective state and the functional context in which the brainstem does its activating. Their model becomes an anatomical-dynamic viewpoint. These authors seek to integrate the description of the nuclei in the reticular formation with the hypothalamic nuclei involved in homeostasis, with consciousness being a third spoke for systems devoted to life regulation. They offer a marriage of place to that of function and process. According to Parvisi and Damasio, if one looks at the afferents that feed into the brainstem nuclei and the efferents that depart, one can more convincingly put forward a model that describes a context for RAS functioning, a dynamic process view. The brainstem nuclei tie together various categories of state space for the organism. These categories range from pain to evaluations of the chemical profile of the bloodstream to vestibular/orientation evaluation to the state of the musculoskeletal frame to state input from rostral brain structures such as the amygdala, the cingulate gyrus, the insula, the prefrontal cortex, and the cingulate cortex. Activation of the cortex is viewed as but one function that results from brainstem nuclear activity. If one stretches the concept of homeostasis from strict definitions of physiological regulation and consider that the hypothalamic nuclei, for instance, are participants in a broader integration of life regulating functions, then one can view the nuclei of the reticular formation as participants in helping to regulate the overall state space of the organism. The activation of the cortex from this perspective helps to support mental processes (such as consciousness) that enrich homeostatic integration/regulation.

It is clear that as the research leads increasingly to more detailed analysis with ever more distinct places as contributants, the need to bring the parts into integrative envelopes becomes greater. Reductionism is crucial in scientific study, but at some point the reductive enterprise becomes obfuscating. We can see how Moruzzi and Magoun, and even more, Parvisi and Damasio, are in need of proposing how the parts contribute to an integrative state space for the organism.

Following the RAS upwards, midbrain targets include the intralaminar nuclei of the thalamus and the tegmental midbrain (mesencephalic reticular formation). In the 1960's and 1970's, a few attempts were made to stimulate these structures in humans.[22-24] Each of the patients who was stimulated existed in a persistent vegetative state (PVS), and it was thought that stimulating the thamalo-mesodiencephalic 'activating systems' would wake people up. This was based upon the premise that the RAS turns on the light in the metaphorical room of consciousness, and in order to see or know what is in the room, the lights must be on, and the switch must be in one place (this is in contrast to the current viewpoint, which emphasizes multiple structures and pathways, and very gradual gradations and qualities of illumination). The results were very disappointing: the stimulation did produce some pattern of physiological arousal, but all failed to achieve any semblance of responsive awareness. Why did they fail?

The interventional strategies of the 1960's and 1970's were based upon three prominent, and in retrospect, naïve assumptions. It was presumed that the reticular formation was a homogeneous mesh of neurons, and we now know that there are distinct nuclei within it. Second, it was thought that all paths led to Rome, in that projections from the reticular formation went to the intralaminar nuclei of the thalamus, which in turn are the source of the so-called diffuse thalamocortical projections, thought to be responsible for widespread attentiveness. And third, end targets were not considered adequately: patient selection involved PVS patients whose cortico-cortico networks may have been badly damaged or destroyed.

The diffuse projections were elaborated in a paper by Morison and Dempsey,[25] who in part derive their notion from Lorente De No,[26] who had suggested that there are thalamic cell groupings not normally thought of as projection nuclei which do in fact send fibers to the cortex, making a series of diffuse connections. Morison and Dempsey found two types of responses from their stimulation studies: "...the well known specific projection system with a more or less point to point arrangement and a secondary "non-specific" system with diffuse connections." (p.292). This more global and diffuse 'recruiting response' is the system that has figured, explicitly or implicitly, in the deep brain stimulation (DBS) strategies for patients with impairments of consciousness.

It is the notion of diffuse projection from a central *place* that informed the stimulating strategies. But a switch for a converging and then diverging electrical circuit is too simple. We now know that the nuclei in the reticular formation can influence the cortex by connecting with basal forebrain nuclei, and there are other projections that bypass the thalamus entirely and go to broad areas of the cortex. Additionally, there are projections from the reticular formation that travel to the reticular nucleus of the thalamus. The situation is more complicated by the supposition that other thalamic nuclei provide diffuse projection to the cortex[27] and/or that projections from the intralaminar nuclei are not quite as diffuse as imagined.[28] Furthermore, different channels ascending from the reticular formation utilize different neurotransmitters, creating a differential effect based upon the prevalence of a particular neurotransmitter in the resulting soup. Last, the idea that desynchronization alone (as observed in the EEG) is the result of reticular formation activation may prove to be naïve. Instead of 'activations' and 'switches', some authors, such as Llinas[29] speak of thalamocortical 'resonances' being needed for the emergence of consciousness. Alas, the situation is always more complex than we first imagined.

4. THE SMALLEST WORMHOLE YIELDS MATRIX CELLS AND BROAD-SCALE NETWORKS

We have one more narrowing in the wormhole, to some of the smaller structures in the brain. And yet the tiny locale is made up of special neurons--'matrix neurons', that are messengers and integrators to virtually all of the outlying communities.

Nicholas Schiff and his co-authors[30-32] have begun to develop a strategy for utilizing deep brain stimulation (DBS) for the treatment of patients in minimally conscious states (MCS).[33] These patients, as opposed to those with PVS, do demonstrate some degree of responsiveness and sensorimotor integration, even if it is fluctuating. It may be that previous stimulation strategies to the intralaminar nuclei of the thalamus were ineffective because of the choice of location or the type of stimulation utilized, though it is quite possible that the problem primarily involved patient selection: PVS patients may have extensive damage to the widespread cortical networks needed for integration, and even stimulation to the proper thalamic location would then have no where to go. MCS patients may be more likely candidates to benefit from stimulation strategies. Recent fMRI studies have shown that MCS patients do indeed have preserved these widespread distributed networks.[34]

For those not familiar with DBS, it involves the surgical placement of electrodes deep within the brain to treat intractable cases of Parkinson's disease,[35] dystonia, and even now some are reporting preliminary results with intractable obsessive-compulsive disorder.[36] As an example, the subthalamic nucleus is now the preferred site for treatment in Parkinson's, and one can observe dramatic and instantaneous effects from electrode placement that is in the right location and with the right stimulation parameters. With a host of scanning data and computer modeling, a thin, hair-like wire supporting four tiny electrodes (a total of 7 millimeters apart) is guided to the target site. The stimulation of each electrode, at differing parameters, is correlated with clinical evaluation. A very specific amplitude and frequency, at a very specific site, can lead to instantaneous and broad clinical effects: in the case of Parkinson's, tremors and unintentional movements, for example, disappear. Some are drawn to conclude that clinical improvement implies that the stimulated location is a 'center' for that functioning, creating a newfound phrenology. As we shall later see, there are other ways of understanding how the brain creates mind, and how DBS works as well.

Utilizing PET studies in traumatic brain injury, one readily finds reductions in cerebral metabolism remote to the site of injury, and these changes cannot be easily identified by alterations in cerebral blood flow. This phenomenon is termed 'diaschisis', and is a distal downregulation of neuronal populations[37] far from the site of injury. The result is a dysfacilitation of neurons, which results in their remaining in a hyperpolarized state. Contrarily, a lack of adequate stimulus may also lead to a hypersynchronization of neuronal populations, in which they 'lock on' in one fashion or another, an oversynchronized grip. These two findings suggest that healthy brain function involves a balance of optimum resonances in neuronal populations. Llinas has linked dysrhythmias of hypersynchronization to a variety of 'over-gripped' situations, including epilepsy and obsessive compulsive disorder. It is relevant to note that patients with TBI have been observed to evidence transient obsessive-compulsive disorder,[38] epilepsy,[39] and catatonia.[40] These clinical findings seem to support the notion that some optimum tonal resonances are needed in keeping proper functional balance in the brain. Exercise is good for our muscles and our brains, and if stimulating a specific site enhances consciousness, it does so through engendering optimal coordinated activity across broad-scale networks.

Dr. Schiff's proposal for using DBS for MCS patients refines Moruzzi and Magoun as well as my reading of Parvisi and Damasio. When we finally get to the fine, fine detail of the proposed site(s) of stimulation, we find that the proposed sites are rich in 'matrix cells'—cells that broadcast and integrate on a very wide level. Matrix cells represent a lovely convergence of the minute place and the networked process.

In proposing a DBS strategy for minimally conscious patients, Shiff takes us to a more clearly defined place--the intralaminar nuclei of the thalamus—with their now being a much finer map of this region. Remember that the intralaminar nuclei of the thalamus were the targets of some of the failed stimulation strategies of the last 30 years. Despite the failures, there are a number of findings that suggest that this group of structures is crucial for sustaining consciousness. One strong clinical argument is that injury to the paramedian mesodiencephalon is unique in its degree of devastation, creating widespread deficits in consciousness.[41]

Widespread afferents and efferents uniquely place the intralaminar nuclei (ILN) at a nexus point. The ILN has the greatest number of thalamic projections to the basal ganglia and the cortex. The *rostral* ILN includes the centralateral (CL), paracentralis (Pc), and central medial (CeM) nuclei, projecting to pre-frontal cortex, primary sensory cortices, and diffusely to the basal ganglia. Importantly, desynchronization of the EEG has been demonstrated with stimulation of the *rostral* ILN. The *caudal* ILN nuclei are composed of the centromedian-parafascularis complex (Cm-Pf), which project distinctly to basal ganglia and pre-motor and anterior parietal cortices.

Recall that previous stimulation strategies relied upon the notion of the nonspecific thalamic projections.[42] These neurons are found throughout the thalamus, have been determined to project to layer I of the cortex, and Jones in 1998 has categorized these neurons as either matrix or core, based upon differential calcium-binding protein expression. Matrix neurons are located throughout the ILN, through the rest of the thalamus, and even in some primary sensory relay nuclei. Recently, Jones[43] has taken the network concept and stated that it is the whole network of the matrix neurons, within the intralaminar nuclei and beyond, that lead to thalamocortical synchronization. Schiff, in an effort to shore up the distinctive nature of the ILN, notes the extensiveness of the clinical pathologies associated with ILN, as well as Munkle *et al.'s*[44,45] identification of the additional marker calcium binding calretinin unique to most ILN matrix neurons. Even more compelling is the clinical finding that Schiff and Pulver emphasize: in rehabilitative efforts such as cold water irrigation of the ear canal can lead to temporary recovery of significant cognitive functions. It is important to realize that the main outflow from the brainstem vestibular nuclei is to the rostral ILN.[46] Jones' broader network of matrix neurons gives no particular primacy to ILN neurons, a position that Schiff wishes to amend. For Schiff, all places are not created equally.

It is fascinating that the proposed site of ILN stimulation involves neurons known for their web-like connectivity--matrix neurons. Further, one can consider the debate between Jones and Schiff as representing the struggle to find the proper emphasis between p*lace* and *process* at the level of neuroanatomical organization. Jones appears to argue for an equipotentiality model for matrix neurons, Schiff gives primacy to rostral ILN matrix neurons.

5. STIMULATING SPECIFIC PLACES AND RESISTING A NEW PHRENOLOGY

We may be several years away from developing a rational strategy of DBS for minimally conscious patients. DBS treatment for enhancement of consciousness represents a noteworthy meeting ground for *place* and *process*. If successful, will DBS in the rostral ILN convince us that the center of consciousness is there? Or will we be flexible enough to appreciate that certain small structures are distinctly important to broad-scale, population dynamics? Those oriented towards *place* will be prone to overemphasize a site of stimulation, and correlate clinical findings with fMRI to bolster that conclusion. But we must remember that the workings of the mind, while grounded in brain structure and function, cannot simply be reduced to location. This is particularly true in stimulation and modulatory strategies that intervene at a very specific point in an enormously complex web of interconnectivity and dynamic integration. I will now mention two examples of how easily this complexity is confused.

The first example involves a case reported by Bejjani et al,[47] where a woman was treated for intractable Parkinson's disease with deep brain stimulation. She underwent precise bilateral insertion of the wire carrying the electrodes and then had testing for the optimal site of stimulation. There are four sites of activation (quadripolar electrodes) on a wire inserted into the brain, a total of seven millimeters apart in the usual setup. If the bottom electrode produces no clinical effect despite varying stimulation parameters, the next electrode 2 millimeters above will be tried, and so on. Effectiveness is determined by clinical evaluation of the patient intra-operatively, who has only local anesthesia to the scalp at the site of insertion. In this particular patient, stimulation at electrode 1 bilaterally led to a very good recovery from her Parkinson's. However, stimulation at electrode 0 on the left side only, 2mm below, led to an instantaneous and profound depression. This was in a patient who had no depressive history. The precise anatomical location was on the medial side of the left substantia nigra. These symptoms were abolished as soon as the stimulation parameters were changed or switched back to electrode 1. With the patient's permission, these results were reproduced at follow up, again with the same stimulation parameters, and again, with the same clinical effects.[‡]

These findings led some scientists to conclude that they had found the 'depression center' in the brain, a conclusion that exemplifies the all too often confusion between *place* and *process*. Other surgeons have reported euphoria and merriment with patients stimulated in the subthalamic nucleus, though none have suggested that they have discovered a comedy center, or that this woman's depression was turning off the happiness center rather than turning on the depression center. Bejjani, and most others who have commented on this phenomenon, are appropriately cautious in their conclusions, and merely suggest that left side basal ganglia appear to be involved in a network associated with affective regulation.

The second example involves the theories about how deep brain stimulation actually works, and how some theories are more phrenological than others. DBS involves continuous, regular high frequency stimulation to achieve its clinical effects. Surgical ablation of specific structures (for example the globus pallidus in the treatment of Parkinson's) has achieved clinical improvement for a number of conditions. Drawing upon this history, and the knowledge that neural tissue can be incapable of transmitting action potentials if stimulated at a high enough frequency (membranes are thought to remain in a state of depolarization blockade), DBS was thought of as a functional lesion, though with no actual ablation of tissue. The analogy of a functional lesion dominated the theoretical landscape until recently, though this view is gradually being replaced by a dynamic, process viewpoint.

To some extent, the lesion model for DBS, at least in Parkinson's, was based upon the supposition that relevant structures were over-active and not sufficiently inhibited as in normal physiological control. If actual lesioning worked, it surely inhibited output, and it was not a big step to conclude that DBS worked similarly. However, chemically induced Parkinson's in animal models has been found to result in no statistically significant changes in baseline or steady state neuronal activity in relevant structures. If background activity levels are not increased in the diseased state, what is causal in

[‡] Ali Rezai (personal communication) has suggested that the angle of insertion in this patient placed electrode 0 on the very medial side of the substantia nigra (SN), and that the affective changes might be due to stimulation of a neighboring structure just medial to the SN, the periaqueductal gray (PAG). Stimulation to the PAG has anecdotally been reported to produce dread, emotional changes etc. Jaak Panksepp[48,49,50] for many years has argued that the PAG is a prominent source of core affective consciousness, and that it is part of a system comprising the PAG, anterior cingulate, the dorsomedial thalamus, and insula, that are implicated in sadness.

understanding the clinical symptomatology, and how does DBS work? Attention was turned to phasic irregularity rather than to absolute levels of activity, a viewpoint that is more conducive to consideration of dynamic, interactional causality. Vitek,[51] Llinas, and others have stressed that all disease expression resulting from neuronal tissue is ultimately a problem with dynamics, whether it be at the level of individual neuron, small group effects, or at the global population level we discussed earlier. Montgomery has posited that it is the irregular activity of neurons that leads to the symptoms of Parkinson's disease. He concludes that DBS does not absolutely excite or inhibit, but rather may restore regularity of communication between neuronal populations, resulting in the smoothing out of symptoms and the reintegration of intention and movement. Montgomery created simulations that bore out this hypothesis: at lower frequencies and with irregular activity, sporadic and enhanced information transfer occurred. Imagine a Parkinsonian tremor, the initiation of movement not intended, or the difficulty in initiating intended movement. Montgomery suggests that the phenomenon of stochastic resonance may be the culprit: increasing noise (irregularity) into biological systems sometimes enhances information transfer, to the point of behavior loosened free of usual integrated will or intention, hence the symptoms. DBS, from this dynamic process perspective, reintroduces regularity into neuronal communication, restoring the proper level of what he terms information transfer.

6. GLOBAL POPULATION DYNAMICS: EMERGING FROM THE WORMHOLE

We have traveled from neurodynamics at the level of the scalp down to specific structures deep inside the brain that in turn are found to be composed of more specific and distinct substructures. And yet at the finest, most localized levels of analysis, multifactored, homeostatic state spaces are introduced by Damasio, matrix neurons by Jones, stochastic resonance by Montgomery and broad-scale thalamocortical and cortico-cortical resonances are emphasized by Freeman, Llinas, and Schiff and his colleagues. Neuroanatomical locale must be simulataneously considered with connectivities, and the population behavior of neuronal groups. The minute and the widespread become necessary bedfellows; *place* and *process* commingle.

We have yet to comprehend how the specific contributants emerge into collectivities that allow for the seamless integration and complexity we observe in conscious minds. As we have discussed, widespread neuronal cooperation can be modeled through nonlinear chaos, various gating, channeling and feedback mechanisms, or through stochastic resonance. In the case of stochastic resonance, irregularity of noise can lead to symptoms, as in Parkinson's disease, though proper regularity of noise can lead to widespread neuronal cooperation. Similarly, diaschisis (downregulation) can result from insufficient input, or, paradoxically, hypersynchronization. As Llinas has suggested, a variety of psychiatric and neurological diseases can be associated with the improper balance of thalamocortical and cortico-cortical resonances. While we do not yet understand the precise nature of particular contributants or just how the neuronal dynamics are constituted, we are making progress on these fronts. Given the failures of previous stimulations strategies, we have learned that in utilizing DBS to improve states of impaired consciousness, the site of stimulation and the targets of that stimulation must retain at least some portion of their structural integrity.

I had promised the reader that we would emerge from the wormhole and return to a more global perspective. To navigate the balance of place and process, and to integrate the global and the discrete, one must utilize the notion of a hierarchy, both in terms of neuronal organization and conceptual integration. Highly conserved nuclei may operate and be organized according to different principles than broad-scale networks, which have emerged at far later evolutionary epochs. As stated earlier, until we understand more, a balanced perspective that appreciates the complexity of the relationship between *place* and *process* is the one that will be our best guide. If we are to achieve a better understanding of consciousness as well as its disorders, and even develop neuromodulatory strategies that might improve clinical conditions, we are well advised to approach this vast territory with the deepest respect and the greatest breadth that we can muster.

7. REFERENCES

1. Watt DF, Pincus DI. The neural substrates of consciousness: implications for clinical psychiatry. In: Panksepp, ed. *Textbook of Biological Psychiatry.* New Jersey: J Wiley, 2004.
2 Adrian ED. The Basis of Sensation: the Action of the Sensory Organs. London: Christophers, 1928.
3. Gloor P. Hans Berger on the electroencephalogram of man. *Electroencephalogr Clin Neurophysiol* 1969; **suppl 28:**350.
4. Freeman WJ. *Mass Action in the Nervous System.* New York: Academic Press, 1975.
5. Freeman WJ. *How Brains Make Up Their Minds.* London: Weidenfeld and Nicolson, 1999.
6. Bressler SL, Freeman WJ. Frequency analysis of olfactory system EEG in cat, rabbit and rat. *Electroencephalogr Clin Neurophysiol* 1980;**50:**19-24.
7. Pincus DI, Freeman WJ, Modell A. (in preparation). Perception in the clinical hour: a proposed neurobiology of transference.
8. Freeman WJ, Burke BC, Holmes MD. Aperiodic phase re-setting in scalp EEG of beta-gamma oscillations by state transitions at alpha-theta rates. *Hum Brain Mapp* 2003;**19(4):**248-272.
9. Tononi G, Edelman GM. Consciousness and Complexity. *Science* 1998;**282:**1846-51.
10. Hidaka I, Nozaki D, Yamamoto Y. Functional stochastic resonance in the human brain: noise induced sensitization of baroreflex system. *Phys Rev Lett* 2000;**85(17):**3740-3743.
11. Russell D, Wilkens L, Moss F. Use of behavioral stochastic resonance by paddlefish for feeding, *Nature* 1999;**402:**219-223.
12. Montgomery EB, Baker KB. Mechanisms of deep brain stimulation and future technical developments. *Neurol Res* 2000;**22(3):**259-66.
13. Baars BJ. In the Theater of Consciousness: The Workspace of the Mind. Oxford: Oxford University Press, 1997.
14. Ramon y Cajal S. Estructura del ganglio habénula de los mamíferos. Anales de la Sociedad Española de Historia Natural 1894:Tomo 23.
15. Parvisi J, Damasio A. Consciousness and the brain stem. *Cognition* 2001;**79:**135-59.
16. Bremer F. Cerveau "isole" et physiologie du sommeil. *Comptes Rendus de la Societe Biologie* 1935; **118:**1235-1241.
17. Adrian ED. *The Physical Background of Perception.* Oxford: Clarendon Press, 1947.
18. Gerebtzoff MA. *Anatomical bases of the physiology of the cerebellum.* Dissertation. Leuwen: Catholic University of Leuwen, 1941.
19. Moruzzi G, Magoun HW. Brainstem reticular formation and activation of the EEG. *Electroencephalogr Neurophysiol* 1949;**1:**455-473.
20. Von Economo, C.F. *Encephalitis Lethargica.* Wien: W. Braumuller, 1917.
21. Blessing WW. Inadequate frameworks for understanding bodily homeostasis. *Trends Neurosci* 1997;**20(6):** 235-239.
22. Hassler R, et al. Behavioral and EEG arousal induced by stimulation of unspecific projection systems in a patient with post-traumatic apallic syndrome. *Electroencaphalogr Clin Neurophysiol* 1969;**27:**306-310, 689-690.

23. McLardy T, Ervin F, Mark, V. Attempted inset electrodes from traumatic coma: Neuropathological findings. *Trans Am Neurol Assoc* 1968;**93**:25-30.
24. Sturm V, et al. Chronic electrical stimulation of the thalamic unspecific activating system in a patient with coma due to midbrain and upper brain stem infarction. *Acta Neurochir (Wien)* 1979;**47**:235-244.
25. Morison RS, Dempsey EW. A study of thalamo-cortical relations. *Am J Physiol* 1941;**135**:281-292.
26. Lorente De No R. The cerebral cortex: Architecture, intracortical connections and motor projections. In: Fulton JF, ed. *Physiology of the Nervous System*. London: Oxford University Press. 1938:291-339.
27. Jones EG. Viewpoint: the core and matrix of thalamic organization. *Neuroscience* 1998;**85**:331-345.
28. Gronewegen H, Berendse H. The specificity of the 'non-specific' midline and intralaminar thalamic nuclei. *Trends Neurosci* 1994;**17**:52-66.
29. Llinas RR, Ribary U, Jeanmonod D, Kronberg E, Mitra PP. Thalamocortical dysrhythmia: a neurological and neuropsychiatric syndrome characterized by magnetoencephalography. *Proc Natl Acad Sci USA* 1999;**96**:15222-15227.
30. Schiff ND, Pulver M. Does vestibular stimulation activate thalamocortical mechanisms that reintegrate impaired cortical regions? *Proc R Soc Lond* 1999;**266**:421-423.
31. Schiff ND, Purpura KP, Victor JD. Gating of local network signals appear as stimulus-dependent activity envelopes in striate cortex. *J Neurophysiol* 1999;**82**:2182-2196.
32. Schiff ND, Rezai A, Plum F. A neuromodulation strategy for rational therapy of complex brain injury states. *Neurol Res* 2000;**22**:267-272.
33. Schiff ND, Purpura KP. Towards a neurophysiological foundation for cognitive neuromodulation through deep brain stimulation. *Thalamus and related systems* 2002;**2**:55-69.
34. Hirsch J, Kamal A, Rodriguez-Moreno D, Petrovich N, Giacino J, Plum F, Schiff N. fMRI reveals intact cognitive systems in two minimally conscious patients. In: *Proceedings of the Society for Neuroscience 30th Annual Meeting,* San Diego, California, 10-15 November, 2001:Abstract 529.
35. Limousin P, Krack P, Pollack P et al. Electrical stimulation of the subthalamic nucleus in advanced Parkinson's disease. *N Engl J Med* 1998;**339**:1105-1111.
36. Nuttin B, Cosyns P, Demeulemeester H, et al. Electrical stimulation in anterior limbs of internal capsule in patients with obsessive-compulsive disorder. *Lancet* 1999;**354**:1526.
37. Nguyen DK, Botez MI. Diaschisis and neurobehavior. *Can J Neurol Sci* 1998;**25**:5-12.
38. Berthier ML, Kulisevsky JJ, Gironell A, Lopez OL. Obsessive compulsive disorder and traumatic brain injury: behavioral, cognitive, and neuroimaging findings. *Neuropsychiatry Neuropsychol Behav Neurol* 2001;**14**:23-31.
39. Santhakumar V, Ratzliff AD, Jeng J, Toth Z, Soltesz I. Long-term hyperexcitability in the hippocampus after experimental head trauma. *Ann Neurol* 2001;**50**:708-717.
40. Wilcox JA, Nasrallah HA. Organic factors in catatonia. *Br J Psychiatry* 1986;**149**:782-784.
41. Szelies B, Herholz K, Pawlik G, Karbe H, Hebold I, Heiss WD. Widespread functional effects of discrete thalamic infarction. *Arch Neurol* 1991;**48**:178-182.
42. Jones EG. *The Thalamus*. New York: Plenum Press, 1985.
43. Jones EG. The thalamic matrix and thalamocortical synchrony. *Trends Neurosci* 2001;**24**:595-601.
44. Munkle MC, Waldvogel HJ, Faull RL. Calcium-binding protein immunoreactivity delineates the intralaminar nuclei of the thalamus in the human brain. *Neuroscience* 1999;**90**:485-491.
45. Munkle MC, Waldvogel HJ, Faull RL. The distribution of calbindin, calretinin and parvalbumin immunoreactivity in the human brain thalamus. *J Chem Neuroanat* 2000;**19**:155-173.
46. Shiroyama T, Kayahara T, Yasui Y, Nomura J, Nakano K. Projections of the vestibular nuclei to the thalamus in the rat: a Phaseolus vulgaris leucoagglutinin study. *J Comp Neurol* 1999;**407**:318-332.
47. Bejjani BP, Damier P, Arnulf I, et al. Transient acute depression induced by high-frequency deep-brain stimulation. *N Engl J Med* 1999;**340(19)**:1476-1479.
48. Panksepp J. *Affective Neuroscience*. New York: Oxford University Press, 1998.
49. Panksepp J. The periconscious substrates of consciousness: Affective states and the evolutionary origins of the SELF. *Journal of Consciousness Studies* 1998;**5**:566-582.
50. Panksepp J. Feeling the pain of social loss. *Science* 2003;**302**:237-239.
51. Vitek JL. Mechanisms of deep brain stimulation: excitation or inhibition. *Mov Disord* 2002;**17(Suppl.)**: S69-S72.

HONORING TREATMENT PREFERENCES NEAR THE END OF LIFE

The Oregon Physician Orders for Life-Sustaining Treatment (POLST) program

Terri A. Schmidt, Susan E. Hickman, and Susan W. Tolle*

1. INTRODUCTION

When people speak of end-of-life care in Oregon, the first thing that comes to mind for many is the legalization of physician-assisted suicide in 1997.[1] The Oregon Death with Dignity Act allows the attending physician of a terminally ill patient, under certain circumstances, to prescribe a lethal dose of medication that the patient then self-administers. The law has been used infrequently with a total of 129 people dying under the Act in the first five years of implementation. This translates to less than one in one thousand deaths in Oregon each year.[2] Over the past decade, health care providers in Oregon have been working quietly in other areas to improve the care of the dying. While much work is still needed, Oregon has also had some notable successes. One of these areas relates to the ability of health care providers to systematically document and communicate a patient's preferences to have or refuse life-sustaining treatments when the person is unable to speak for himself or herself.

One of the problems that patients and health care providers encounter with typical advance directives is that although they can be helpful, their usefulness is limited in the out of hospital setting. They are not physician orders and therefore cannot be honored by emergency medical technicians (EMTs) in many jurisdictions. In addition, forms in some states contain language that require one or two physicians to certify that the patient has a terminal condition and the proposed intervention will only prolong the dying process. This limitation prevents advance directives from being helpful in most out-of-hospital settings.

A decade ago it was not uncommon for the family of a dying patient to call 911 because of concern about the patient's discomfort due to severe shortness of breath or the sudden development of other distressing symptoms. The patient or family would call emergency services because they needed help with symptom management, not because

* Oregon Health & Science University, Portland, Oregon, 97239.

they wanted aggressive interventions. If the patient was too ill to speak for himself or herself, and the family lacked written orders confirming the patient's preference, paramedics would intubate the person and transport him or her to a hospital. As a result, many patients spent unwanted and uncomfortable days on a ventilator before dying. Other states continue to frequently report similar problems. In a recent New York Times article, the New York City Bellevue Hospital emergency department was described as a place that regularly treats patients with serious, advanced illnesses who are unable to express their wishes and have no advance care plan. Many of these patients receive aggressive treatment including intubation and tube feeding only to succumb to their illness after several days or longer of needless suffering because they were unable to express the wish to stop.[3]

Out-of-hospital do not resuscitate (DNR) orders have been developed as one solution to this problem. These documents provide instructions to Emergency Medical Services (EMS) regarding the use of life-sustaining treatments in the form of written medical orders. When advance directives are turned into medical orders, most EMTs will follow them. A recent study found that 89% of United States EMTs are willing to honor state-approved DNR orders.[4] The number of states in the United States authorizing these out-of-hospital DNR orders has increased from 11 states in 1992[5] to 42 states in 1999.[6] Connecticut, for example, has a statewide program using DNR bracelets for terminally ill patients who do not want resuscitation attempts. The program specifies that EMTs must honor a DNR order and provides immunity from liability for honoring the order.[7] These DNR orders still have limitations in that they only apply when the patient is pulseless and apneic. They do not provide direction to providers who are called to treat a patient who may be in extremis but is still breathing and has a pulse. Some patients in those circumstances would not want life-sustaining interventions while others would. Further, some patients may view bracelet programs as burdensome and they are infrequently used.

In Oregon, an alternative approach called the Physician Orders for Life-Sustaining Treatment (POLST) Program was created to address systems failures in honoring patient treatment preferences. The centerpiece of the Program is the POLST form, a two-sided bright pink medical order form designed to be filled out by a health care provider based on a patient's or surrogate's wishes for life-sustaining treatment (see figure 1). It can be used by anyone, including individuals who desire full, aggressive intervention under all circumstances, but is designed especially for individuals with serious or life-threatening illnesses. Because it is a signed physician order, it effectively converts patient preferences for life-sustaining treatments into medical orders. It is designed to transfer across treatment settings and should go with the patient every time he or she moves.[8,9,10]

2. SECTIONS OF THE POLST FORM

The front of the form includes four separate treatment circumstances. Section A (Resuscitation) applies if, and only if, the patient is pulseless and apneic. Under this section the patient and physician can choose either resuscitation or no resuscitation (DNR).

The original form should accompany the patient on transfer and remain with the patient where they reside.

The POLST form transforms patient treatment wishes into medical orders.

In any section left unmarked, the highest level of treatment must be provided.

Regardless of the level of medical intervention, comfort measures must always be provided.

A discussion about treatment preferences is required when completing the POLST form.

A physician/nurse practitioner must sign the POLST form, but it may be completed by a nurse, social worker, or other health care team member. This person is encouraged to sign the back of the POSLT (not shown) as the "preparer."

Copyright prevents the addition of organizational logos.

A brief summary of the patient's health status gives other providers a context for these orders.

SEND FORM WITH PATIENT/RESIDENT WHENEVER TRANSFERRED OR DISCHARGED

Physician Orders for Life-Sustaining Treatment (POLST)

This is a Physician Order Sheet. It is based on patient/resident medical condition and wishes. It summarizes any Advance Directive. Any section not completed indicates full treatment for that section. When the need occurs, first follow these orders, then contact physician.

Last Name of Patient/Resident

First Name/Middle Initial of Patient/Resident

Patient/Resident Date of Birth

Section A (Check One Box Only) — RESUSCITATION. **Patient/resident has no pulse and is not breathing.**
- ☐ Resuscitate
- ☐ Do Not Resuscitate (DNR)

When not in cardiopulmonary arrest, follow orders in Sections B, C and D.

Section B (Check One Box Only) — MEDICAL INTERVENTIONS. **Patient/resident has pulse and/or is breathing.**
- ☐ **Comfort Measures Only.** The patient/resident is treated with dignity, respect and kept clean, warm and dry. Reasonable measures are made to offer food and fluids by mouth, and attention is paid to hygiene. Medication, positioning, wound care and other measures are used to relieve pain and suffering. Oxygen, suction and manual treatment of airway obstruction may be used as needed for comfort. These measures are to be used where the patient/resident lives. The patient/resident is not to be hospitalized unless comfort measures fail.
- ☐ **Limited Additional Interventions.** Includes care above. May include cardiac monitor and oral/IV medications. Transfer to hospital if indicated, but no endotracheal intubation or long term life support measures. Usually no intensive care.
- ☐ **Full Treatment.** Includes care above plus endotracheal intubation and cardioversion.

Other Instructions: ______

Section C (Check One Box Only) — ANTIBIOTICS. **Comfort measures are always provided.**
- ☐ **No antibiotics**
- ☐ **Antibiotics**

Other Instructions: ______

Section D (Check One Box Only) — ARTIFICIALLY ADMINISTERED FLUIDS AND NUTRITION. **Comfort measures are always provided.**
- ☐ **No feeding tube/IV fluids**
- ☐ **Defined trial period of feeding tube/IV fluids**
- ☐ **Long term feeding tube/IV fluids**

Other Instructions: ______

Section E — Discussed with:
- ☐ Patient/Resident
- ☐ Parent of Minor
- ☐ Health Care Representative
- ☐ Court-Appointed Guardian
- ☐ Spouse
- ☐ Other:

Summarize Medical Condition

Physician/ Nurse Practitioner Name (print)	Physician/ NP Phone Number DAY EVE	Office Use Only
Physician/ NP Signature (mandatory)	Date	

SEND FORM WITH PATIENT/RESIDENT WHENEVER TRANSFERRED OR DISCHARGED

Figure 1. The Oregon POLST form.

Section B (Medical Interventions) is one of the unique aspects of the form. In this section, the patient and physician can choose one of three levels of medical intervention for situations in which the person is breathing and has a pulse. A person can choose "Comfort Measures Only," indicating that he or she does not want any intervention that deliberately prolongs the dying process, but should receive comfort measures such as food and fluids by mouth, hygiene, wound care, and measures to relieve pain and suffering. Under most circumstances, a person who desires "comfort measures only" does not want to be transported to a hospital. The next level of care is called "Limited Interventions." Persons who choose this category want many treatments including medications and monitoring but do not want endotracheal intubation or long-term life support measures such as a ventilator. Under most circumstances they should be transported to the hospital if indicated but do not want to be admitted to the intensive care unit. The third category is "Full Treatment." Patients choosing this category want all indicated medical interventions to sustain life including intensive care unit treatment.

Section C (Antibiotics) allows patients and physicians to indicate whether or not the patient wants to receive antibiotics to treat infections. It is noted here that comfort measures are always provided. It is assumed that there may be circumstances when antibiotics could be used for comfort, but that they would not be used to treat an infection and prolong life. An example often used is antibiotics for a urinary tract infection in a patient who has chosen "No Antibiotics." An elderly woman with moderate Alzheimer's Dementia may develop a urinary tract infection that makes her incontinent, causing burning and frequency when urinating. Her physician may choose to treat her with antibiotics to make her more comfortable even though her POLST form indicates "No Antibiotics," because the goal of this intervention is to increase comfort, not prolong life. On the other hand, if this same woman developed urosepsis, becoming confused and hypotensive, her physician may withhold antibiotics based on her wishes to withhold antibiotics but provide measures to assure her comfort in other ways.

Section D (Artificially Administered Fluids and Nutrition) contains physician orders regarding artificial nutrition and hydration based on the patient's preferences. It is assumed, as noted in comfort measures, that the patient would always be offered food and fluids by mouth unless the person lacked the ability to swallow and that other measures would be provided to assure comfort even if the patient refused treatment.

The final section on the front of the form, Section E, provides documentation of the basis for the orders reflected on the POLST form. When the patient has decision-making capacity, these orders are written based on a discussion with the patient about his or her treatment preferences. If the patient is unable to speak for himself or herself, the orders should be written based on conversations with the patient's surrogate decision maker. The form can be prepared by any health care professional based on a discussion with the patient/resident or his/her decision-maker, but must be signed by a physician or nurse practitioner in order to convert treatment preferences into medical orders. In order to function as a valid DNR order, the form must include three things: the patient's name, the DNR status in Section A, and the physician's or nurse practitioner's signature. All other sections are optional and, if left blank, treatments will be provided unless information to the contrary becomes available.

The back of the form is primarily administrative. It provides directions for how to update or modify the form. The POLST form should be reviewed any time the patient experiences a change in his or her health status, transfers from one care setting or level of care to another, or reports a change in treatment preferences. A section is available to document periodic review of the POLST form orders. In Oregon the form does not have an expiration date. Some institutions have their own internal policy requiring that DNR orders and the POLST form undergo periodic review. A wallet size version of the front side of the form is also available for ambulatory patients to carry with them outside their living setting. It is designed to be used in conjunction with the full size form.

Table 1. How the POLST form helps patients and providers

How Does POLST Help the Patient?

Documents	a patient's wishes for life-sustaining treatment in the form of a physician order
Streamlines	transfer of patient records between facilities
Clarifies	treatment intentions and minimizes confusion about patient preferences
Complements	advance directives when available
Assists	physicians, nurses, emergency personnel and health care facilities in promoting patient autonomy
Optimizes	comfort care of patients
Allows	periodic review and changing of orders as indicated by patient values and medical circumstances

How Does POLST Help Health Care Providers?

Ensures	readily available information for health care providers about the patient's treatment preferences
Provides	a system for communicating the physician's medical orders for the patient to other care facilities
Safe and effective	instrument that prevents unwarranted treatments and ensures that medically indicated treatments desired by the patient are provided
Offers	a practical way to assemble patient information on a one-page form
Allows	periodic review and changing of orders as indicated by patient values and medical circumstances

3. RESEARCH CONDUCTED ABOUT THE POLST PROGRAM

The Oregon POLST Program has been the focus of several studies to evaluate its effectiveness (Table1). The first study was initiated shortly after development and early implementation of the form. The purpose of the study was to determine if care providers correctly interpreted the POLST forms and would act according to patient preferences. Focus groups of long-term care providers and EMTs were asked to pick from a list of interventions those treatments they would provide to the patient based on a completed POLST form. This study found that with appropriate education, the form is generally a safe and effective instrument, preventing unwarranted treatments and insuring that medically indicated treatments desired by the patient are provided.[11] Subsequent research led to revisions that simplified the language and format of the form.[12]

A 1996 prospective study examined the use of the POLST form in eight of the first nursing facilities in Oregon to adopt the POLST Program. Patients in those facilities who had chosen DNR and comfort measures only were followed for a year. None of the 180 residents received unwanted CPR, ICU care, or ventilator support and only 2% were hospitalized to extend life. A majority of the residents who died during the study year had orders for narcotics to treat pain and only 5% died in an acute care hospital.[13] These findings suggest the POLST Program is effective in helping patients avoid unwanted life-sustaining treatments and hospitalization and may enhance attention to comfort measures. A later retrospective study evaluated the records for the last two weeks of life for enrollees in an Oregon Program of All-Inclusive Care for the Elderly (PACE) site, a program that cares for frail older adults who meet the criteria for nursing facility placement but are maintained at home. The study evaluated whether or not the care patients received matched the care that was requested on the POLST form. The authors found a high correlation between care received and care requested for resuscitation (91%), antibiotics (86%), IV fluids (84%), and feeding tubes (94%). The correlation was still good but lower for medical interventions in Section B of the form (46%). Specifically, 20% of patients received more invasive medical interventions than was indicated on the POLST form.[14] This led to revisions in the medical interventions section of the POLST form to make it easier to interpret and follow.

Research data also suggest that the POLST form is highly effective in improving communication and thus ensuring patients' treatment preferences are honored.[13,14] The POLST form is now the standard of care in most Oregon nursing homes and hospice programs and helps explain low rates of in-hospital deaths and reduced rates of intensive care in the final 6 months of life in Oregon.[15]

4. HISTORY OF THE POLST PROGRAM IN OREGON

At a 1991 statewide meeting of Oregon ethics committees, concern was voiced about the problem of respecting DNR orders when patients are transferred from nursing homes to hospitals. At the same time, EMTs were expressing similar concern about being required to provide interventions to patients when it seemed clear that they were not wanted. A multi-disciplinary task force was convened to address the problem of preventing unwanted transfers and intensive medical interventions. The group included representatives of the long-term care community, hospice, and emergency medical service providers. Together, they created a program, initially called the Medical Treatment Coversheet, and later changed to the Physician Orders for Life-Sustaining Treatment (POLST) Program, the centerpiece of which is the POLST form. Over the years, this group, the POLST Task Force, has overseen four revisions of the form to enhance its clarity for patients and providers alike.

In order to ensure that this form would be maximally effective, the POLST Taskforce worked with the state legislature in 1993 to remove restrictions in Oregon's advance directives laws. The Taskforce also worked with the Oregon Board of Medical Examiners in 1996 to redefine the Scope of Practice for EMTs First Responders, and their supervising physicians. EMTs/First Responders are now directed to respect patient wishes regarding life-sustaining treatments, by complying with life-sustaining treatment orders executed by a physician, such as those recorded on the POLST document.[16] In 2001, the EMT Scope of Practice was further modified to include respect for orders executed by nurse practitioners. It was decided not to seek a legislative mandate for the form because of concern that this might result in the inadvertent creation of barriers to use of the POLST Program by requiring a specific format, inhibiting the ability to make modifications as needed.

From the beginning, education and teamwork have been important parts of the implementation. In order for the POLST Program to be effective in an area, both the long-term care providers and the emergency providers must be familiar with it and willing to honor patient preferences as expressed by the form. The program began with implementation in a few nursing homes in select areas. It was successful because doctors in the area were willing to complete the forms, and care facilities and EMS providers were willing to honor the patient preferences. Education has been ongoing in Oregon for the last 10 years, including multiple inservices to EMTs, articles in newsletters, presentations at medical society meetings, agency inservices, and media coverage. From a small beginning, the POLST Program has become a common practice in Oregon. A recently completed survey (not yet published) found that a majority of Oregon nursing facilities (71%) now use the POLST Program for at least half of their residents. Records indicate that over 750,000 POLST forms have been distributed.

The Center for Ethics in Health Care at Oregon Health & Science University coordinates the Program. Educational materials and the form itself are provided to health care professionals at cost. The Center acts as a clearinghouse for questions about the POLST Program and provides speakers to meet educational needs as well as information on their website. In addition, the Center has developed educational materials both for patients and health care providers.

5. THE POLST PROGRAM AROUND THE UNITED STATES

Due in part to publications from The Robert Wood Johnson Foundation[17,18] and the 1997 Institute of Medicine report,[19] the POLST Program has caught the attention of communities across the United States. Other states including West Virginia,[20] Washington,[21] and Utah,[22] as well as parts of Wisconsin, Missouri,[18] New Mexico,[9] and Georgia[23] now use the POLST Program. Groups and individuals in 42 states and 6 countries have requested informational packets from the Center for Ethics in Health Care at Oregon Health & Science University.[24] It is anticipated that the POLST Program will continue to grow in the United States as additional research regarding its use is conducted and published.

6. CONCLUSION

The Oregon POLST Program offers a model of how to systematically elicit, document, and communicate patient treatment preferences regarding a range of life-sustaining interventions. Legislative and regulatory changes were made and an ongoing educational program was developed in order to ensure the successful implementation of this program. The resulting success of the POLST Program has captured the attention of individuals and coalitions interested in improving end-of-life care across the United States. As a result, individuals in Oregon are assured that their treatment preferences will be honored at the end of life, even when they are unable to speak for themselves.

7. REFERENCES

1. Oregon Death with Dignity Act, Oregon Revised Statute **127:**800-897.
2. Hedberg K, Hopkins D, Kohn M. Five years of legal assisted-suicide in Oregon. *N Engl J Med* 2003;**348:** 961-964.
3. Marco CA, Schears RM. Prehospital resuscitation practices: A survey of prehospital providers, *J Emerg Med* 2003;**24:**101-106.
4. Kleinfield NR. Elderly patients whose final wishes go unsaid put many doctors in a bind. *New York Times* 2003;**July 19:**A12.
5. Adams J. Prehospital do-not-resuscitate orders: A survey of policies in the United States. *Prehospital Disaster Med* 1993;**8:**317-322.
6. Sabatino C. Survey of state EMS-DNR laws and protocols. *J Law Med Ethics* 1999;**27:**297-315.
7. Leon MD, Wilson EM. Development of a statewide protocol for the prehospital identification of DNR patients in Connecticut including new DNR regulations. *Ann Emerg Med* 1999;**34:**263-274.
8. Hickman SE. Improving communication near the end of life. *Am Behav Sci* 2002;**46(2):**252-267.
9. Hickman SE, Newman J. *A decade of POLST in Oregon.* Portland: The Center for Ethics in Health Care, 2001.
10. Tolle SW, Tilden VP. Changing end-of-life planning: the Oregon experience. *J Palliat Med* 2002;**5(2):**311-317.
11. Dunn PM, Schmidt TA, Carley MM. Donius M. Weinstein MA, Dull VT. A method to communicate patient preferences about medically indicated life-sustaining treatment in the out-of-hospital setting. *J Am Geriatr Soc* 1996;**44(7):**785-791.
12. Dunn PM, Nelson CA, Tolle SW, Tilden VP. Communicating preferences for life-sustaining treatment using a physician order form. *J Gen Intern Med* 1997;**12(suppl):**102.
13. Tolle SW, Tilden VP, Nelson CA, Dunn PM. A prospective study of the efficacy of the physician order form for life-sustaining treatment. *J Am Geriatr Soc* 1998;**46(9):**1097-1102.
14. Lee MA, Brummel-Smith K, Meyer J, Drew N, London MR. Physician orders for life-sustaining treatment (POLST): outcomes in a PACE program. *J Am Geriatr Soc* 2002;**48(10):**1219-1225.
15. Tolle SW, Rosenfeld AG, Tilden VP, Park Y. Oregon's low in-hospital death rates: what determines where people die and satisfaction with decisions on place of death? *Ann Intern Med* 1999;**130(8):**681-685.
16. Oregon Administrative Rule: OAR 847-35-0030 [7].
17. Bain JW. Data-driven policymaking (an update): using statistics to shape agendas and measure progress. In: Christopher M, ed. *State Initiatives in End-of-Life Care No. 18.* Kansas City: Midwest Bioethics Center, 2003.
18. Spann J. Implementing end-of-life treatment preferences across clinical settings. In: Christopher M, ed. *State Initiatives in End-of-Life Care No. 3.* Kansas City: Midwest Bioethics Center, 1999.
19. Field MJ, Cassel CK. *Approaching death: improving care at the end of life.* Washington D.C.: National Academy Press, 1997.
20. Center for Health Ethics & Law (West Virginia, May 1, 2003); http://www.hsc.wvu.edu/chel/index.htm.
21. Washington Sate Department of Health, Washington, May 1, 2003; http://www.doh.wa.gov/hsqa/emtp/resuscitation.htm.
22. Utah Department of Health, Utah, May 1, 2003; http://hlunix.hl.state.ut.us/licensing/.
23. Georgia Collaborative Study Group to Improve End-of-Life Care. At a glance – Improving care at the end of life: a long term care initiative (Georgia Medical Care Foundation, Georgia, 2001).
24. Center for Ethics in Health Care May 1, 2003; Oregon: http://www.ohsu.edu/ethics/.

PALLIATIVE SEDATION IN TERMINALLY ILL PATIENTS

Paul C. Rousseau*

1. INTRODUCTION

Lamentably, the process of dying can be a time of considerable suffering and trepidation, engendering fear, loneliness, anxiety, and distress from intractable physical and existential anguish. Such concerns can be insufferable and abolish the hope of a tranquil and "dignified" death, and instead, foment a desire for physician-assisted death. Moreover, interminable suffering can portend a difficult and oftentimes irresolvable grief and bereavement process for surviving family members. To assuage the debilitating and callous symptoms of terminal illness, palliative sedation has emerged as a compassionate and empathic yet contentious intervention to manage refractory symptoms at the end of life. In the United States, palliative sedation was judicially sanctioned by the Supreme Court in its 1997 decision negating a constitutional right to physician-assisted suicide;[1-4] in their decision, Justices O'Connor and Souter both wrote concurrences supporting the use of medication to alleviate suffering in terminal illness, acknowledging that such treatment may bring about unconsciousness or hasten death.

Although not a frequent palliative intervention, the incidence of palliative sedation varies from 2% to 52%, with such variability attributable to diverse definitions of palliative sedation, the retrospective nature of many studies, and cultural and ethnic diversity.[2]

The definition of palliative sedation has been confusing and difficult to ascertain, however, two core definitions have arisen:

1. The intention of purposely inducing and maintaining a sedated state, but not deliberately causing death, in specific clinical situations complicated by refractory symptoms.[5,6]
2. A medical procedure used to palliative symptoms refractory to standard treatment by intentionally clouding consciousness.[7,8]

* Paul Rousseau, M.D., Associate Chief of Staff for Geriatrics/Extended Care, VA Medical Center, Phoenix, AZ 85012 USA.

The latter definition may broaden the scope of palliative sedation to incorporate length and depth, whether intermittent or continuous, and mild or deep.[5] Respite sedation is a form of sedation in which patients may be moderately-to-deeply sedated for 24-48 hours and then reawakened to assess the need for further sedation; this type of sedation has been specifically suggested for use in existential suffering.[1] With respite sedation, second guessing and reassessment by health care providers and family members may be accomplished, ensuring that further sedation is clearly and explicitly warranted in select cases.

The definition of a refractory symptom can be subjective, personal, and nonspecific, however, Cherny and Portenoy[6] define a refractory symptom as one that cannot be controlled adequately despite aggressive efforts to identify a tolerable therapy that does not compromise consciousness. In addition, they suggest that a symptom may be considered refractory when additional invasive and noninvasive interventions are incapable of providing adequate relief, or the therapy is associated with excessive and intolerable acute and/or chronic morbidity and is unlikely to provide relief within a reasonable time frame. Such a definition may be readily acceptable for sedation of intractable physical symptoms, but may be more problematic and ethically and emotionally charged when utilized for sedation of existential suffering.[1] Existential suffering encompasses various ill-defined psychological symptoms, including a sense of hopelessness, disappointment, loss of self-worth, remorse, meaninglessness, and disruption of personal identity, subjective symptoms for which there are no well-established clinical strategies or guidelines to evaluate, quantitate, or appropriately manage in terminally ill patients.[5] Moreover, sedation for existential suffering can be morally and emotionally challenging for health care providers and family members as the patient may be awake, cognitively intact, and socially interacting prior to initiation of sedation.[1,2] What's more, existential suffering may not be associated with significant physiological deterioration, further confounding many health care providers' clinical decision to use palliative sedation.[9] Differentiating inadequately treated maladies such as depression, delirium, anxiety, and familial discord from valid intractable existential distress may serve to further confound the medical decision to use palliative sedation.[1] However, as prudently noted by Rosen, a medical model that assesses psychological or spiritual aspects of patients' lives as a function of physical condition is shortsighted and may even be harmful;[10] we must remember that suffering is experienced by persons, not bodies. Accordingly, existential or psychological suffering can be just as refractory and engender just as much suffering as physical symptoms, consequently, palliative sedation should be considered a useful adjunct to mitigate the anguish of the disabling symptoms of refractory existential suffering, albeit in carefully selected patients.[1]

2. ETHICAL AND CLINICAL RATIONALE

The ethical rationale and approbation of palliative sedation arguably derives from the ethical principle of double effect, developed in the Middle Ages by Roman Catholic theologians[11,12] and applied to situations in which it is impossible to avoid all harmful actions.[12,13] The doctrine of double effect encompasses four tenets: 1) the nature of the act must be good or morally neutral; 2) the intended effect must be the good one, where the bad effect is merely foreseen and permitted; 3) death must not be the means to the good effect; and 4) the good effect must exceed or balance the bad effect (proportionality).[2,12,14]

While the appropriate use of palliative sedation indubitably incorporates the four tenets, some argue that deep and continuous palliative sedation inevitably causes death and that the concept of intent is ambiguous and elusive, particularly when a patient desires suicide.[13] In cases where patient intent is unknown or uncertain and concern arises regarding a surreptitious desire for hastened death, ethical and psychiatric consultations are obligatory in an effort to preclude the appearance and unintended consequence of physician-assisted death.[2] In addition, others contend that palliative sedation is nothing more than slow euthanasia,[15] with pharmaceuticals utilized to reduce suffering while ensuring a certain and hastened death. However, such apprehension is based on an erroneous interpretation and misunderstanding of the quintessence of double effect, and the actual practice of palliative sedation. Without doubt, not all palliative sedation is deep and continuous, the type of sedation that many opponents refer to in their argument that sedation is nothing more than physician-assisted death or euthanasia, nevertheless, deep and continuous sedation for refractory symptoms is a morally and ethically acceptable approach for the relief of refractory suffering at the end of life. Clearly, the ethical validity of palliative sedation remains polemic, divisive, and controversial, but such treatment must not be denied patients in whom all palliative therapies have been exhausted and devastating and distressing symptoms persist.

3. NUTRITION AND HYDRATION

The issue of terminating nutrition and hydration when instituting palliative sedation is contentious. Quill and associates argue that patients who are palliatively sedated die of dehydration, starvation, or some other intervening complication, not the underlying disease as so often promoted by practitioners of sedative pharmacotherapy.[11] However, many patients have already limited their oral intake or completely stopped eating or drinking secondary to anorexia, and for that reason, the absence of nutrition and hydration is not necessarily a conduit to a patient's death, but rather a "normal" physiologic response to the ravages of advanced disease.[1] Although withdrawing or withholding nutrition and hydration may appear to hasten death, in contemporary legal, ethical and moral dialogues, their provision is considered extraordinary care, and the ethical validity of discontinuing these interventions in patients selecting palliative sedation is based upon responsible ethical and legal judiciousness.[1] Controversy does arise when the continuance or initiation of artificial nutrition and hydration in a patient contemplating or undergoing palliative sedation is requested by the patient or surrogate decision-maker. In such cases, the futility of artificial nutrition and hydration should be discussed prior to initiating palliative sedation. However, the decision to proceed with palliative sedation in patients who elect to initiate or continue artificial nutrition and hydration should be individual and based upon honest and frank discussions, the clinician's ethical and moral values, knowledge of local regulatory policies, institutional requirements (if performed in a hospital, nursing home, or inpatient hospice unit), and consultation with an ethics committee.

4. GUIDELINES FOR PALLIATIVE SEDATION

Guidelines for palliative sedation have been suggested, and may serve as a basis for individual practitioners in developing criteria and policies. To assist in decision-making for the initiation of palliative sedation, the following guidelines are proposed:[1]

1. The patient must have a terminal illness, and should be in the advanced stages of the disease.
2. A do-not-resuscitate order must be in effect.
3. All palliative treatments must be exhausted, including treatments for depression, delirium, anxiety, and any other contributing maladies.
4. A psychological assessment by a skilled clinician should be considered, particularly in cases of existential suffering.
5. A spiritual assessment by a skilled clinician or clergy member should be considered, particularly in cases of existential suffering.
6. If nutritional support or intravenous or subcutaneous hydration is present, discussion should be initiated regarding the futility, benefit, and burdens of such therapy in view of impending palliative sedation.
7. Informed consent must be obtained from the patient or surrogate decision-maker.
8. Consideration be given to respite sedation, particularly in cases of existential suffering.
9. Consider the use of a scale to monitor depth of sedation.[16,17]

The choice of a medication for palliative sedation is dependent upon clinician preference, knowledge and comfort, local availability of medications, and the locale of sedation (at home or in an institution). The benzodiazepines and barbiturates have been favored agents, however, the neuroleptics chlorpromazine and haloperidol and the anesthetic agent propofol have also been used, with the most common routes of administration being intravenous, subcutaneous, and rectal. Also, since there is no definitive evidence that patients sedated to unconsciousness do not experience pain, and because abrupt discontinuation of opioid analgesics can precipitate withdrawal, opioid administration should continue while patients are sedated.[1]

Finally, once palliative sedation is initiated, the dose of the sedative agent should not be increased unless the level of sedation is not appropriate for alleviation of symptoms. To do so without a clinical indication, particularly with deep and continuous sedation, might imply the health care provider is intending to hasten death and would ostensibly cross the fine line between palliative sedation and physician-assisted suicide and/or euthanasia.[1,18,19]

5. REFERENCES

1. Rousseau PC. Existential suffering and palliative sedation: a brief commentary with a proposal for clinical guidelines. *Am J Hosp Palliat Care* 2001;**18:**226-228.
2. Rousseau PC. The ethical validity and clinical experience of palliative sedation. *Mayo Clin Proc* 2000;**75:**1064-1069.
3. *Vacco v. Quill 1997*; **No. 117 S.**: Ct. 2293.
4. *Washington v. Glucksberg* 1997; **No. 117 S.**: Ct. 2258.

5. Rousseau PC. Existential suffering and palliative sedation in terminal illness. *Prog Palliat Care* 2002;**10**:222-224.
6. Cherny NI, Portenoy RK. Sedation in the management of refractory symptoms: guidelines for evaluation and management. *J Palliat Care* 1994;**10**:31-38.
7. Morita T, Tsuneto S, Shima Y. Proposed definitions for terminal sedation. *Lancet* 2001;**358**:335-336.
8. Morita T, Tsunoda J, Inoue S, Chihara S. The decision-making process in sedation for symptom control in Japan. *Palliat Med* 1999;**13**:262-264.
9. Cherny NI. Sedation in response to refractory existential distress: walking the fine line. *J Pain Symptom Manage* 1998;**16**:404-406.
10. Rosen EJ. A case of "terminal sedation" in the family. *J Pain Symptom Manage* 1998;**16**:406-407.
11. Quill TE, Lo B, Brock DW. Palliative options of last resort: a comparison of voluntarily stopping eating and drinking, terminal sedation, physician-assisted suicide, and voluntary active euthanasia. *JAMA* 1997;**278**:2099-2104.
12. Quill TE, Dresser R, Brock DW. The rule of double effect—a critique of its role in end-of-life decision making. *N Engl J Med* 1997;**337**:1768-1771.
13. Garcia JLS. Double effect. In: Reich WT, ed. *Encyclopedia of Bioethics*. Vol 2. New York: Simon & Schuster Macmillan, 1995.
14. Beauchamp TL, Childress JF, eds. *Principles of Biomedical Ethics*. 3re ed. New York: Oxford University Press, 1989.
15. Billings J, Block SD. Slow euthanasia. *J Palliat Care* 1996;**12**:21-30.
16. Ramsay MA, Savege TM, Simpson BR, Goodwin R. Controlled sedation with alphaxalone-alphadolone. *Br Med J* 1974;**2**:656-659.
17. Rudkin GE, Osborne GA, Finn BP, Jarvis DA, Vickers D. Intra-operative patient-controlled sedation. *Anaesthesia* 1992;**47**:376-381.
18. Rousseau PC. Palliative sedation. *Am J Hosp Palliat Care 2002;***19**:295-297.
19. Alpers A, Lo B. The Supreme Court addresses physician-assisted suicide. Can its rulings improve palliative care? *Arch Fam Med* 1999;**8**:200-205.

INDEX

Abortion, 135
Abnormal electrocardiogram, 211
Abnormal echocardiography, 211
Absent brain functions
  signs of, 163
Absent systolic spikes, 165
Absent voluntary interaction, 229
Accidental carbon monoxide inhalation, 208
Accidental cyanide inhalation, 208
Acidosis, 169
Actuarial recipient survival rate, 210
Actuarial survival curves, 30
Acute disseminated encephalomyelitis, 115
Acute myocardial infarction, 124
Acute necrotizing pancreatitis, 210
Acute non-traumatic brain injury, 229
Acute pancerebral ischemia, 122
Acute poisoning, 207-212
Action potentials, 125
Acute traumatic brain injury
Adenosine, 198
Adenosine triphosphate, *see also* ATP, 197-204
ADP, 198-204
Adult brain, 80
Aerobic energy metabolism, 197
Afterlife, 118
Age, 30, 162
Agnosia, 220
Akinetic mutism, 239
Albumin infusions, 157
Alcohol dehydrogenase, 210
All brain function
  irreversible loss of, 100
Algor mortis, 141
Allograft organ transplantation, 207
Alpha activity, 191
Alpha2-agonist, 157
Alpha blockade, 242, 243
Alpha rhythm, 241
Alpha waves, 241
Alzheimer's disease, 232
AMP, 198-204
American Multi-Society Task Force on
  PVS, 229
Amphetamine derivatives, 211
Anaerobic energy metabolism, 197
Anaerobic glycolysis, 144, 198
Anaerobic metabolism, 144, 197
Anatomy, 216
Anatomical-dynamic viewpoint, 246
Anatomical substratum, 175
Ancillary tests, 182
Anima, 98
Animus, 98
Animal models, 123
Antibiotics, 257
Anticipatory activation, 125
Anti-entropy
  cessation of, 107
Antiparallel orientation, 198
Anencephaly, 1, 11, 19
Anencephalic neonates, 19
Anencephalic newborns, 135
Angiography, 18
Animal experiments, 197
Animalism, 53
Anoxia, 123, 141, 143, 193
Anoxic brain damage, 194
Anoxic blood
  weak circulation of, 110
Anoxic coma, 193
Anoxic insult, 195
Anoxic-ischemic brain injury, 194
Anoxic-ischemic coma, 195
Anti-hypertensive agents, 151
Apallia, 217
Apallic syndrome, 216

Apallisches Syndrome, 216
Aphasia, 124
Apnea, 18, 19, 162, 177, 190
Apnea test, *see also* apnea testing, 43, 161, 169-173
Apnea testing, *see also* apnea test, 169-173
    duration of, 170
    preconditions for, 169
    procedures for, 169
Apneic oxygenation, 172
Apoptosis, 144
Apraxia, 218, 220
Areflexia, 28
Arterial blood gas determinations, 170
Arterial blood pressure, 198
Arteriography, 172
Artificial $PaCO_2$ augmentation, 172
Artificially administered fluids and nutrition, 257
Artificial respiration, 120
Artificially ventilated patient, 134
Aristotelian-Thomistic substantial form, 91
Arousal, 4-11
Artifacts, 177
Ascending reticular activating system, 7-9
Asphyxia, 115
Association cortex, 20, 232-236
Associative cortices, 232-236
Asymptotic model, 92
Asystole, 31, 43, 169
ATP, *see also* adenosine triphosphate, 197-204
    α-ATP, 198-204
    β-ATP, 198-204
    γ-ATP, 198-204
ATP depletion, 144, 145
Atropine, 172
Atropine test, 172
Auditory brainstem responses, 178-183
Auditory cortex
Automatic defibillators, 143
Autonomic system deactivation, 28
Autonomous cord function, 37
Autoregulation, 154
Autoregulatory arteriole vasodilation, 155
Awareness, 4-11, 53, 218-226
    certain abilities of, 54
    certain qualities of, 54
    kernel of, 84
Awakened rational conscious life, 62

Babylonian Talmud, 134
Barbiturates, 208
Barbiturate overdose poisoning, 208
Basal forebrain, 233
Basilar artery occlusion, 179
Beating heart, 133
Behavior, 216
Benzodiazepines, 208
Berger rhythm, 241
Beta1-antagonist, 157
Bilateral temporal bone fractures, 179
Biochemical cascade, 143
Biological artifact, 52
Biological individual phase, 6
Biological organism, 24
Biological organisms, 54
Biological quantum coherence phenomenon, 128
Biological species Homo sapiens, 56
Biotechnology, 82
Bladder
    incontinence of, 223
Blood-brain barrier
    disruption of, 144
Blood flow, 18
    cessation of, 135
    lack of, 163
Blood gas levels, 170
Blood perfused hearts, 198
Blood pressure, 190
    transient rise in, 31
Bodily integration, 51
Body
    director of, 116
Body's central integrator, 23
Body-soul relationships
Body temperature, 170, 172
Border-zone, 108
Bowel
    incontinence of, 223
Brachial plexus, 180, 181
Bradycardia, 29-33
Bradyarrhythmias, 31
Brain, 1-11, 44, 48, 61
    anoxia of, 117
    all functioning of, 46
    cessation of its functions, 46
    characteristic functions of, 46
    extensive cell death in, 124
    global anoxia of, 124
    gray-matter mantle of, 216
    irreversible damage to, 117, 124
    irreversible destruction of, 141
    metabolic recovery of, 124
    neurophysiological processes of, 54
    surface of, 79
Brain anoxia, 20
Brain biopsy, 217

Brain circulation, 18
Brain's coordinating activity, 24
Brain-damaged patient, 163
Brain damaged person, 82
Brain dead, *see also* brain death, 10, 19, 43, 115
 classified as, 79
 diagnosed as, 79
Brain-dead bodies, 56
Brain dead donor, 197
Brain-dead individuals, 61, 62
Brain dead patients, 46, 48, 79, 163, 164
Brain dead pregnant mothers, 45
Brain death, 1-11, 15-20, 23-38, 43-48, 51-58, 61-76, 79-86, 89-112, 115-131, 133-137, 139-142, 161-166, 169-173, 175-182, 197-204,
 biological rationale of, 25
 concept of, 1, 15-20
 conceptual basis for, 51
 definitions of, 79
 determination of, 81
 diagnosis of, 45
 disintegration occurring in, 61
 loss of integration, 61
 loss of somatic integrative unity in, 34
 moment of, 162
 on irreversibility
 personhood-based "neo-cortical" position on, 79
 somatic integrative unity in, 24
Brain death criteria
 widespread acceptance of, 133
Brain death legislation, 135
Brain death-related hemodynamic instability, 197
Brain death standard, 83
Brain destruction, 19
Brain-disconnected bodies, 24
Brain edema, 151, 157
Brain events
 epiphenomena of, 65
 side-effects of, 65
 supervenient properties of, 65
Brain function, 46, 162
 cessation of, 47, 207
 distributed models of, 239
 lack of, 46
Brain functions, 162
 irreversible cessation of all, 47
Brain-functioning
 reversible cessation of, 46
Brain herniation, 28
Brain neurons, 164
Brain processes
 supervenient properties of, 61
Brain-injured animals, 48
Brain injuries, 80
Brain prostheses, 81-86
Brain-related criteria, 45
Brain repairs, 80
Brain tumor, 115
Brain swelling, 145
Brain stem, 4, 5, 9, 17, 19, 20, 24, 177, 233
 permanent functional death of, 24
 integrative functions of, 51
 death of, 176
 destruction of 135
 vertical plasticity of, 218
Brainstem activity, 47
 abolition of, 124
 loss of, 119
Brain-stem auditory evoked potentials, 136
Brain-stem death, 17, 166, 169
 concepts of, 176
Brain stem death doctrine, 166
Brain-stem formulation, 16, 24, 175
Brain-stem function
 absence of, 179
Brainstem medial lemniscus, 194
Brain's neuronal interactions, 80
Brain's neuronal structure, 46
Brainstem function, 51
Brainstem functions, 84
Brainstem's functions, 48
Brain stem functions
 loss of, 169
Brain-stem function
 direct information on, 177
 persistence of, 177
Brain stem nuclei, 243
Brain-stem reflexes, 18, 19, 179, 193
Brain stem reticular activating system, 245
Brain-stem structures, 193
 assessment of, 178
Breathing
 arrest of, 117
 cessation of, 98
British criteria, 45, 47
Broad-scale neural dynamics, 244
Bulbar level, 181
Bulboreticular formation, 244
Burst suppression, 194

Ca^{++} overload, 144
Cadaveric signs, 142
Cadavers, 48
Capacity for consciousness

irreversible loss of, 90
Carbon dioxide, *see also* $PaCO_2$, 170
arterial tension of, 170
Carbon monoxide, 208-212
Carbon monoxide poisoning, 208-212
Calcarine cortex, 178
Calcium antagonists, 144
Calcium uptake, 203
Calcium overload, 203
Cannabis, 208
Cardiopulmonary resuscitation, *see also* cerebral resuscitation, 117, 143-149, 189
CPR, *see also* cardiopulmonary resuscitation, 117, 143-149, 190
Cardiac arrest, 29, 115, 116, 143, 171, 177, 189
duration of, 118
survivors of, 118
Cardiac arrhythmia, 169
Cardiac arrhythmias, 29, 171
Cardiac autoresuscitation
loss of potential for, 90, 98, 99
Cardiac contractility
transient decrease of, 157
Cardiac transplant, 135
Cardio-circulatory criteria, 142
Cardio-circulatory functions
irreversible loss of, 141
Cardioprotective effect, 147
Cardiopulmonary death, 162
Cardio-respiratory functions, 133
Cardiovascular instability, 29
Carotid artery, 155
Catholic church, 135
Cause, 65
Cells, 116
Cell death, 123
initiators of, 145
Cellular detoxification, 144
Cellular ion homeostasis
energetic basis for. 144
Cellular life, 134
Cellular levels, 92
Cell systems, 129
Cellular damage, 81
Central nervous system, 18, 25
Central venous pressure, 155
continuous recordings of, 198
Cephalic reference, 178
Cerebellum, 19, 44
Cerebral anoxia, 119, 189
Cerebral autoregulatory mechanisms, 163
Cerebral blood flow, 18,123, 151, 234
Cerebral connectivity
functional impairment in, 233-236
Cerebral blood volume, 171
Cerebral circulation
arrest of, 177, 179
Cerebral circulatory arrest, 164
Cerebral cortex, 8, 9, 18, 19, 193
irreversible destruction of, 193
severe anoxic insult to, 194
Cerebral cortical functioning
total loss of, 223
Cerebral destruction, 20
Cerebral edema
consequences of, 210
Cerebral electrical activity, 191
Cerebral hemispheres, 19, 178
Cerebral metabolic rates for glucose, *see also* CMRGlu, 231-236
Cerebral metabolism
global impairment in, 230
Cerebral performance category, 145
Cerebral perfusion, 144
Cerebral perfusion pressures, 163
Cerebral perfusion pressure, 144, 151
Cerebral resuscitation, 143
Cerebral transtentorial herniation
syndromes of, 163
Cerebral oxygen uptake, 123
Cerebral tissues
replacement of, 83
Cerebrospinal fluid, 152
Cerebro-vascular accident, 124
Cerveau isole, 244
Cervical spinal cord
dorsal horn in, 181
Cervico-medullary junction, 25, 28, 181
Cervico-medullary junction infarction, 35
Cervico-medullary junction transaction, 36
Cervico-medullary level, 181
C-flumazenil, 232
Chaos
principles of, 241
Chief Rabbinate, 133-137
Children, 172
brain death in, 135
Character traits, 53
Chronic brain death, 161
Chronic ethanol abusers, 210
Chronic phase, 30
Chronic severe neurological dysfunction, 172
Circulation, 143
arrest of, 117
Circulationless body, 110
Civil code, 139-142
Clinical brain functions, 161

global loss of, 163
Clinical brain dysfunction
cause of, 162
Clinical criteria, 140
Clinical death, 115
definition of, 117
Clinical examination, 18
Clinical heart transplantation, 198-204
Clinical toxicology, 207
CMRGlu, *see also* Cerebral metabolic rates for glucose, 213-236
CO poisoned donors, 208-212
CO poisoning, *see also* carbon monoxide poisoning, 208-212
Cocaine, 208-212
Cocaine derivatives, 211
Cocaine exposure, 208
Cocaine use, 208
Cochlear ischemia, 179
Colloid osmotic pressure, 157
Coherent emotions, 241
Conceptual rationale, 26
Condition, 65
Cognition, 65
intentional act of, 65
Cognitive function, 16
Cognitive science, 82
Coherent patterns, 240
Coma, 18, 120, 162, 239-252
reversible causes of, 177, 178
Coma dépassé, 15, 175
Comatose patients, 177, 189, 194, 239-252
Comfort measures only, 257
Committee, 43
Complete cerebral ischemic model, 123
Complete electrocerebral inactivity development, 198
Computed tomography, 217
Computed tomography angiography, 165
Computerized prosthetic hippocampi, 80
Concept-criterion-tests, 91
Confirmatory tests, 18, 19
Confirmatory testing
use of, 162
Connectivities, 240
Conscious actuations, 64
Conscious experience, 63, 64
subject of, 62
Conscious inner awareness, 66
Conscious processes, 84
Conscious self, 64
Conscious state, 65
Conscious subject, 65
Consciously lived life, 74
Consciousness, 1-11, 20, 53, 54, 61, 115-131, 177, 194, 216-226, 229-236, 239-252
abolition of, 230
about the continuity of, 115
brain-stem plasticity for, 218
capacity for consciousness, 52
components of, 4
lack of, 23
level of, 70
irreversible loss of, 51, 54, 55
loss of, 51, 144
neurobiology of, 141
neurodynamics theories of, 240
permanent loss of, 100
potential for, 52, 57
recovery of, 100, 232
role of, 5-9
special state of 115
subjects of, 63
Consciousness-centered ethics, 85
Consciousness-related functions, 232
Contemporary technologies, 162
Contextually defined event, 107
Context-dependent notion, 91
Continuity, 92
Continuous positive airway pressure, 72
Continuous discrete personhood, 85
Contractile function
loss of 203
Contrast arteriography, 164
Controlled hypertension, 151-158
Core consciousness, 17
Cornea, 208-211
Corneal reflex, 124
Corneal reflexes, 191
Coronary artery angiography, 190
Corporeal characteristics, 55, 56
Cortex, 45, 48 177, 217
deafferentation of, 244
Cortical, 16
Cortical activation, 4
Cortical activity, 47, 177
Cortical electrical activity, 177
Cortical destruction, 216
Cortical function
loss of, 119
Cortical functions, 18
Cortical networks, 16
Cortical neural networks, 230
Cortical neurons, 18
Cortical responses, 194
Cortico-cerebral-thalamo-reticular complex, 4
Cortico-cortical connectivity, 233-236
Corpse, 52, 56

Cough, 190
Cranial vault, 163
Craniectomy, 157
Criteria
  National Institutes of, 45
Criterion, 175
Critical body functions, 16
Critical function, 24
Critical organ, 23
Critical system
  of the brain, 1-11
  of the organism, 4
Critical vital systems, 16
Cryomagnet
  magnetic field of,
Cryonicists, 81
Cryonic recovery, 81
Cryonic suspension, 81
CoQ10, 146
CT brain scan, 193
Cuban civil code, 139-142
Cuban law, 139-142
Cuban parliament, 140
Cyanide, 208-212
Cyanide poisoning
  severity of, 211
Cystein preoteases, 144
Cystein protease activation, 144
Cytochrome c, 145
Cytoskeleton, 144
Cytotoxic brain edema, 144

Dead, *see also* death, 15, 45, 46, 51, 161
Dead brain, 82
Dead donor rule, 139, 108-111, 90-112
Dead enough
  future operational definition of, 83
Decapitated human bodies, 56
Deceased relatives, 120
  encounter with, 122
Decerebrate rigidity, 28
Decompressive craniectomy, 152
Decorticate rigidity, 28
Deafness, 179
Death, *see also* Brain death, 1-11, 46, 61, 89-112
  afraid of, 120
  biological, 23
  biology, chemistry and physics, 2-5
  brain criteria, 16-18
  brain-oriented formulations of, 175
  cause of, 143
  certification of, 139-142
  consciousness-related formulation of, 52-55
  criteria of, 53
  criteria for, 133
  criterion of, 15, 25, 46, 47
  concept of, 53, 90-112
  context-dependent-event notion of, 98
  current definition of, 85
  definition of, 15, 20, 52, 54, 175, 90-112
  death of, 79, 80
  determination of, 142
  diagnosis of, 177
  exact timing of, 100
  fear of, 118
  higher-brain formulation of, 54
  kinds of, 103
  moment of, 98, 134
  neurocentric definition of, 6
  neurological criterion of, 52, 53
  neurological criterion for determining, 53
  new criterion for, 43
  normative concept of, 91
  on-going redefinition of, 79
  ontological concept of, 91
  psychological 23
  semiotics of, 89-112
  signs of, 140
  sociological, 23
  somatic dysfunctions in, 25
  times of, 162
  timing of, 133
  traditional concepts of, 96
  traditional criterion of, 136
  whole-brain criterion of, 164
  whole-brain neurological criterion of, 52, 53
Decreased cerebral perfusion, 118
Death concept, 81
Death-concept axiom, 106
Death-related events, 90-112
Death-related phenomena, 91-112
Death-scenarios, 90
Decerebrate rigidity, 223
Decapitation, 100
Deep brain stmulation, *see also* DBS, 239-252
DBS, *see also* deep brain stmulation, 239-252
Deep tendon reflexes, 190
Deeply comatose, 179
Definitional dyslexia, 222
Degenerative disorders, 229
Déja vu, 121

Delayed cerebral damage, 143
Delayed neuronal damage, 144
Delayed neuronal injury, 144
Delayed posthypoxic brain injury, 143
Deliberate methanol ingestion, 208
Delirium, 249
Delta activity, 123, 192
Dendrites, 64
Dendritic trees, 126
Destroyed brains, 83
Destruction, 46
Desynchronization, 243
Developmental malformations, 229
Developmental vegetative state, 218
Deep sleep, 81
Detectable cortical evoked potentials, 234
Diagnostic criteria, 175, 177
 sets of, 175
Diagnostic protocols, 81
Diaschisis, 248
Diastole, 163
Diastolic pressure, 163, 171
Diencephalon
 vertical plasticity of, 218
Diffuse cerebral edema, 163
Diffuse delta, 194
Diffuse neuronal damage, 161
Diffusion-weighted imaging, 165
Dignity, 70, 71
 loss of, 61
 sense of, 75
Digital computerized EEG, 191
Dihydroergotamine, 157
Direct multisystem damage, 28
Diabetes insipidus, 23-38
Disability
 cause of, 143
Discontinuity, 92
Discontinuous personalities
 legal status of, 80
Distributed identity, 85
Do not resuscitate order, *see also* DNR, 82, 265
Dobutamine, 157
Dogma, 136
Dopamine, 171
Dopamine infusion, 198
DNA, 5,6, 12
DNA fragmentation, 144
DNR, 82
Drug intoxication, 18
Drug overdose, 194
Drugs abusers, 211
Dying, 54
 process of, 263-266
 technological deconstruction of, 79
Dying/decaying process, 105
Dying process, 83, 210, 257
Dynamic body/mind relationships, 67
Dynamic radionuclide brain scanning, 165
Dystonia, 248

Early infant brain
 plasticity of, 218
EEG, *see also* electroencephalogram, 6, 8, 9, 45, 119, 123, 171, 176, 177-183, 189-195, 230, 240-252,
Effective circulation, 110
Effective heartbeat
 cessation of, 110
Eigenstates, 93, 94
EKG, 92-102
EKG artifact, 191
EKG electrical activity, 96
End-of-life care, 255-262
End-tidal capnometry, 172
Endogenous catecholamines, 197
Effect, 65
Electric brain writing, 241
Electrical activity
 cessation of, 79
Electrocardiogram,
 QRS complex of, 110
Electroretinogram, 178-183
Electrocerebral silence, 179, 194
Electrochemical gradient
 collapse of, 145
Electroencephalogram, *see also* EEG, 18, 123
Electroencephalograms, *see also* EEG, 79
Electrophysiological study, 195
Emergency medical technicians, 255
Emergency medical services, 256
Emitted signal, 198
Emotions, 116
Encephalo-somatic communication, 35
Endogenous opposition to entropy
 loss of, 105
Endorphins
 release of, 117
Endotracheal intubation, 257
Energy consumption, 145
Engendering fear, 263
Entire brain, 25
 complete destruction of 47
 irreversible cessation of all functions of , 24, 45
 total destruction of, 46

Entropy, 96-98, 105
Environment
    discriminative awareness of, 217
Environment modern ICUs, 96
Epidural balloon, 198
Epileptic discharges, 217
Epinephrine, 171, 190
Episodic memory, 232
Essential nature, 84
Ethanol, 208
Ethical implications, 20
Ethical rationale, 264
Etiology, 162
Event related potentials, 234
Evoked potentials, 18, 176-183, 230
    brain-stem auditory 18
    diagnosis of brain death
    multimodality, 18
    somatosensory, 18
    visual, 18
Excessive hypercarbia, 169, 171
Excitation-contraction coupling system
    subcellular level of, 203
Excitatory amino acids
    massive release of, 144
Explantation procedure, 203
Explorable brain-stem structures, 177
Extra-uterine feti, 86
Eyes-open permanent unconsciousness, 220
Eye opening, 190

Family's death-bed scene, 98
Far-field potentials, 180
Fatal poisoning
Fast activity, 123
Favorable outcome
Fear-death experiences, 115
Fearful affect, 217
Feline myocardial energy metabolism, 197
Fetal brain
    plasticity of, 218
Fetus, 3, 6, 10, 83
    killing of a, 135
Flaccidity, 28
Flat EEG, 124, 179
Flat line EEG, 124
Flexible criteria, 45
Flexion response, 191
Fiberoptic catheter, 151, 155
Forebrain hemispheric function, 178
Forensic cases, 162
Forensic circumstances, 141
Forma corporis, *68*
Formate, 210

40 Hz oscillations, 234
Four-vessel angiography, 18
Fourier transformation, 198
Free induction decay, 198
Frequencies
    spectrum of, 198
Frontal lobe
    a thin sliver of, 217
Fronto-parietal associative cortices, 233
Full treatment, 257
Function, 46
    irreversible lack of, 48
Functional magnetic resonance imaging,
    239-253
Functional reversible damage, 232
Functions, 46
Functional-anatomical definition, 218
Functionalism, 54
Functioning
    cessation of, 46
    permanent cessation of, 51
Functioning midbrains, 79
Functioning brainstems, 79
Future technology, 162

Gag, 190
Gag reflex, 124
Gamma band, 242
Gamma oscillations, 8
Gastrointestinal motility, 31
    recovery of, 33
Gemara, 133
Gene therapies, 80
Generalized epileptiforme activity, 194
Generalized voltage suppression, 194
Generalized tonic-clonic seizure, 190
Genetic code, 64
Glasgow Coma Score, 189
Glasgow outcome scale, 146, 155
Glial liquefactive necrosis, 164
Gliotic tissue
    a thin layer, 217
Global apraxia, 220
Global ischemia, 145
Global brain metabolism
    substantial reduction in, 230
Global neurodynamics, 239-252
Glossopharyngeal nerve, 34
Glossopharyngeal nerve function, 33
Glutamate
    massive release of, 144
Glutamate receptors, 144
Glutamate release, 144
Goggles, 178

Graf
toxic origin of, 208
Grafts, 207
potential toxic effect on, 208
Graft survival rate, 210
Grimacing, 222
Group identity, 85
Gyrus angularis, 126

Halakha, 133-137
Halakhic significance, 134
Halothane anesthesia, 232
Harvard Committee, 47, 48
Harvard criteria, 45
Harvard Medical School,
Ad Hoc Committee of, 51, 175
energy metabolism of, 198
Head injured patients, 151
Head trauma, 207
Heartbeat
signs of a, 134
Heartbeat-driven thorax excursions, 172
Heart, 197-204
contractile capacity of, 198
warm ischemia of, 203
Hearts, 209
Heart/lung retrieval, 110
Heart rate, 32
Heart transplant, 129
Head trauma, 19
Hemiplegia, 124
Hemispherectomy, 218
Hemispheric function
electrophysiologic assessment of, 178
Hemodynamic decompensation, 31
Hemodynamic deterioration, 203
Hemodynamic instability
inotropic treatment, 198
treatment of, 198
Hemodynamically unstable brain dead donor, 197
Heart/breathing-arrested person, 83
Heart transplant surgeons, 207
Heart transplantation, 135, 197
High-frequency oscillatory thalamocortical circuitry, 233-236
High intracranial pressure, 151
High spinal cord transection, 24-38
High spinal cord injury, 25
Higher brain formulation, 16, 17, 20, 26, 175
Higher-dimensional space, 127
Higher-order associative areas, 235
Hippocampus, 80
Holographic life review, 120,121
Holoprosencephaly, 217
Homicidal assault,
brain dead victims of, 46
Hopf bifurcation, 93, 94
Human awareness
maintenance of, 234
Human body, 67
Human bodies, 57
Human face, 67
Humanness, 17
Humanoid, 52
Humanoid bodies, 57
Human consciousness, 19
substantial nature of, 64
substantial subject of, 64
Human biological life, 53
Human brain-dead donors, 197
Human being, 15, 52, 55
traditional Judeo-Christian concept of, 55
Human beings, 57
functional specifications of, 55
qualitative specifications of, 55
Human being phase, 6
Human death, *see also* death, 23, 24
determination of, 161
Human characteristics, 52
Human dignity, 70-76
first source of, 71
Human donor issues, 197
Human life
infinite value on, 134
Human life cycle, 4-7
Human-machine cyborgs, 86
Human organism, 56
Human organism as a whole, 52
Human person
dignity of, 70
substantial being of, 72
Human persons
actual death of, 61
Human personal life, 53
Human soul, 69
dignity of, 67
nature of, 67
Human spirit, 69
Humanistic values, 136
Huntington's disease, 145
Hydranencephaly, 217
Hydranencephalic children, 218
Hydration, 265
Hydrogen ion concentration, 170
Hydrostatic capillary pressure, 157
Hypercarbia, 119

Hyperkinetic mutism, 239
Hyperoxia, 123
Hypertension, 27-29
Hyperventilation, 170
Hypnotic state, 232
Hypotension, 27-29, 43, 169, 171
Hypotensive episodes, 155
Hypothermia, 29
Hyperemia, 154
Hypothermic human donor heart, 198
Hypercarbia, 169
Hyperthermia, 30, 36, 145, 194
Hyperventilation, 119, 154
Hyperpolarization, 126
Hypersynchronization, 239-252
Hypothermic patients, 81
Hypothermically suspended, 81
Hypothalamic-anterior pituitary function, 30
Hypothalamic-pituitary function, 37
Hypothalamic-posterior pituitary function, 32
Hypothalamic thermoregulatory centers, 36
Hypothalamus, 7, 8, 34
Hypovolemic shock, 29
Hypoxemic body, 110
Hypoxia, 169, 170, 171, 211
Hypoxic encephalopathy, 149
Hypoxic ischemia
    severity of, 144

Identity, 65
Identity cloning, 85
Identity-critical information
    loss of, 86
Identity malleability, 85
Illicit drugs, 208
Inanimate corpse, 52
Incommunicable existence, 70
Individual death, 20
Individual existence
    reality of, 70
Individual living person himself, 70
Infants
    death in, 135
Inflammatory mediators, 144
Information integration, 230
Information technology, 82
Increased venous capacitance, 29
Inner conscious experiences, 220
Inotropic agent, 155
Inotropic support, 210-212
Insulin, 209
Intracellular acidification, 144
Intravenous radionuclide angiography, 165
Instrumental criteria, 140
Intellectual nature, 70
Interconnectedness, 127
Interstitial fluid volume, 151
Intracellular sodium influx, 144
Intracellular Ca^{++} influx, 144
Intracranial blood flow
    reduction of, 164
Intractable obsessive-compulsive disorder, 248
Intraparenchymal bleeding, 157
Intraparenchymal haemorrhages, 201
Isoelectric EEG, 18, 216
Immaterial subject, 65
Implanted electrodes, 80
Intelligent chimeras, 86
Integrated organisms, 56
Integrative-unity rationale, 61
Integrative unity, 35, 105
Intensive care environment, 142
Intentions, 53
Intentionally clouding consciousness, 263
Intensive care unit, 151-158
Internal carotid, 165
Internal defibrillators, 123
Intra-cerebral haemorrhage, 115
Intracortical facilitation, 126
Intracortical inhibition, 126
Intracranial blood flow
    total absence of, 163
Intracranial circulation
    complete absence of, 162
Intracranial hemodynamics, 163
Intracranial hypertension
    pathophysiology of, 177
Intracranial pressure, 19, 151, 163, 171
Intracranial volume, 163
Intraventricular catheter, 151
Integrating critical system, 177
Intercurrent infections, 31
Interventional resuscitation
    loss of potential for, 99
Impaired cardiac contractility, 29
Impaired cerebral perfusion, 119
Intra-arterial blood gas analysis, 172
Irreplaceable individuality, 65
Irreversible coma, 43, 51, 100
Irreversible apnea, 26, 28
Irreversible structural neuronal loss, 232
Irreversibility, 46, 161-166
Irreversible brain-stem nonfunction, 136
Ischemia, 141
Ischemic brain damage, 80
Isoelectric EEG, 191, 217
Iso-electricity, 123

Isolated donor heart, 197
Isolated neurons, 47
Interconnectedness, 121, 122

Jewish attitude
    scriptural basis for, 135
Jewish law
    modern interpretation of, 133
Jewish religious authorities, 135
Jewish religious interpretation, 133
Jewish religious jurisprudence, 133
Jewish religious perspective, 133
Judaism, 133-137
    concepts in, 133
    principles in, 133
Jugular bulb venous oxygen saturation, 152

Ketamine, 119
Kidney, 208-212
Kidneys, 209
Kidney group, 208
Kidney transplantations, 208

Lactate, 144
Lactic acidosis, 211
Language, 232
Language cortex
    a few words without, 217
Language comprehension, 230
Language production, 217
Lateralized intracranial pressure cones
    development of, 163
Law, 139-142
Lawyers, 46, 139-142
Lead, 208
Left middle fossa, 218
Left ventricle, 197
Legal personhood, 83
Lemniscal pathways, 178
Leukocyte-derived mediators
    inflammatory response with, 144
Lidocaine, 190
Life, 1, 43, 116
    another form of, 122
    end of,
    consciousness-based definition of, 81
    definition of, 2
    giving back of, 98
    loss of, 46
    quality of, 70
    sanctity of, 134
    spirit of, 134
    thermodynamical concept of, 90
Life-death boundary, 108
Life after death, 122
Life support, 79, 142
Life support efforts, 45
Life sustaining treatment, see also POLST
    Oregon physician orders for, 255-262
Lifeless body, 122
Lifeless corpse, 56
Light-emitting diodes, 178
Living human being, 11
Liver, 208-211
Liver transplantation, 135, 208-212
Limbic lesional surgery, 218
Limbic system, 6, 7
Limited interventions, 257
Linguistic respect, 94-96
Livers, 209
Liver transplantations, 208
Living systems
    entropy, 2-5
    dynamics of, 93, 94
Live human, 43
Lived body, 66
Living being, 52
Living birth, 139
Living organism, 52
Living organisms,
    unicellular, 2, 10
    multicellular, 2, 10
Living patient, 46
Living person, 52
Living state, 5, 6
Livor mortis, 141
Local hypercapnic vasodilation, 171
Localized anti-entropy, 90
Localized electrical stimulation, 126
Locationist models, 239
Locked-in state, 80
Long-term life support measures, 257
Long-term storage, 230
Long-term survival, 148
Loneliness, 263
Lower brainstem
    destruction of, 136
LSD, 119
Lowly spinal cord, 23
Lucid remembered experiences, 119
Lungs, 209
Lung transplantation, 170

Macroscopic systems, 93, 94
Magnetic resonance, 165
Magnetic resonance angiography, 165
Magnetic resonance imaging, 125

Magnetic resonance spectroscopy,
Magnetoencephalography, 125, 234
Man
  spiritual soul of, 69
Man-on-the-street, 242
Mannitol, 154
Material entity, 64
Massive brain stem hemorrhage, 166
Massive intracranial bleeding, 207
Massive sympathetic discharge, 36
Matrix metalloproteinases, 144
Maximal stimulation
  threshold for, 172
Mean arterial pressure, 151-158, 163, 171
Mean blood flow velocity, 155
Medication
  lethal dose of, 255-262
Mesencephalon, 182, 201
Meaningful motor activity, 220
Mechanical device, 84
Medial parietal cortex, 232
Median somatosensory evoked potentials, 189-195
Medical debate, 18, 20
Medical Interventions, 257
Medico-philosophical debate, 16, 19, 20
Medulla, 44
Medulla oblongata, 181
Membrane potential, 125, 144
Memory, 53, 54
Memories, 54, 80-86, 116-131
Meninges, 217
Mental function, 52, 57
  no evidence of, 223
Mental processes
  nanorobotic replication of, 85
Mesencephalic transaction, 244
Mesial-basal temporal lobe tissue, 218
Mesopontine reticular formation, 233
Metabolic acidosis, 210
  correction of, 210
Metabolic cascade
  intensity of, 144
Metabolic disorders, 229
Metabolic encephalopathy, 163
Metabolic factors
  presence of, 177
Metabotropic glutamate receptors, 144
Methanol, 208-212
Methanol poisoned donors, 210-212
Methanol poisoning
  target organ of, 210
Methanol toxicity
  transmission of, 210
Methaqualone, 209
Microcirculation, 157
Microvascular perfusion, 144
Mitochondrial cytochrome oxidases, 211
Midbrain, 44
  central transtentorial herniation of, 163
  uncal transtentorial herniation of, 163
Middle cerebral artery blood flow, 123
Midline fissure, 217
Mind, 61, 62, 68
  contemporary functionalist theories of, 54
  immaterial nature, 66
  models of, 239
  philosophy of, 54
  spiritual nature, 66
Mind-brain relation, 128
Mind/brain relationships, 66
Minnesota criteria, 45, 47
Minimally conscious states
  least active regions in, 232
Minimally conscious patients, 239-252
Mishna, 133
Missing person, 82
  deadness of, 83
Missing person's death, 82
Mitochondria, 144, 145
Mitochondrial dysfunction, 145
Mitochondrial permeability, 144
Mitochondrial permeability transition pores, 144
Moist thin paper membrane, 173
Molecular levels, 92
Monoxide poisoning, 207-212
Morbidity rates, 151
Mortality rates, 151
Modern medical technology, 133
Motor cortex, 80, 217
Motor control regions, 80
Motor response, 190, 193
MRI scan, *see also* magnetic resonance imaging, 217
Multimodality evoked potentials, 152, 178-183
Multiple criteria, 44
Multisystem damage, 26
Multisystem failure, 19
Muscle contractions
  loss of, 141
Muscle artifact, 191
Myocardial infarction, 117
Myocardial ischemia, 198
Myocardial energy status, 197
Myocardial energy metabolism, 197-204
Myocardial energy preserving cardioplegic

solution, 203
Myocardial injury, 197
Myocardium
integrity of, 197
Myoclonic movements, 190
Myoclonus, 189, 194, 195

Nanomedicine, 81
Nano-neuro network, 82
Nanoneurological repair, 81
Nanorobots
complete intracranial network of, 82
Nanotechnology, 81-86
Nasopharyngeal derivation, 181
Nanotechnological neuro-prosthetics, 80
National commission, 139-142
Near death, *see also* near-death experiences, 189
Near-death experiences, 115-131
Near death telepathic experience 241
NDE, *see also* ear-death experiences, 115-131
frequency of, 118
Near-drowning, 115
Nerve growth factors, 80
Neocortical death, 175, 176, 216
Neo-cortical death, *see also* death, 79
Neo-cortical standard, 79
Neshama, 134
Neshima, 134
Network doctrine, 242
Neural-computer prostheses, 80
Neural network software, 80
Neural stem cells, 80
Neurodynamics, 240-252
Neurologic prognosis, 195
Neurodegenerative disorders, 145
Neuromodulators, 239
Neuronal damage
amount of, 144
markers of, 144
Neuronal endonucleases, 144
Neuronal firing, 80
Neuronal function, 144
cessation of, 164
Neuronal injury, 144
Neuronal liquefactive necrosis, 164
Neurogenesis, 80-86
Neurogenic pulmonary edema, 28
Neurologic lesion
induction phase of, 26
Neurological criterion, 51
Neuronal membrane depolarization, 144
Neuronal networks, 119, 126
electromagnetic fields of, 128
Neurons, 64
isolated pools of, 176
Neuroremediation, 79
Neuroremediation technologies, 79
Neuromuscular blockade, 164
Neuroprotection, 143-149
Neuroprotective agents, 144
Neuroprotective potential, 143
Neuroprotective strategy, 143
Neurprotective treatment, 145
Neurophysiological processes, 118
Neurophysiological tests, 18
Neurotrophics, 80
Neurotrophic drugs, 80
Neutrons, 198
New Jersey's brain death law, 83
New York State legislature, 135
Nitric oxide, *see also* NO, 144
Na-K-pump function
energetic basis for, 144
NO, *see also* Nitric oxide, 144
NO synthetase
ischemic upregulation of, 144
Noncephalic derivation, 178
Noncephalic reference, 178
Non-critical function, 24
Non-destructive technique, 198
Non-heart-beating organ donors, 98, 99
Non-heart-beating donor controversy, 83
Nonlinear dynamics, 93, 94
Non-locality, 121, 125
Non-local interconnectedness, 127
Non-measurable phase-space, 128
Non-sensory perception, 119
Nonspecific intralaminar nuclei, 234, 240
Non-traumatic injury, 82, 220
Norepinephrine, 154, 171
Normal blood flow
a fraction of, 144
Normothermia, 147
Normoventilation, 170
Noxious somatosensory stimuli
central processing of, 234
Noxious stimuli, 31
Noxious stimulation, 190
Nuclei
resonance frequency signal of, 198
Nucleus cuneatus, 180
Nutrition, 265
N9, 180
N11, 180
N18, 180
N20 cortical response, 189
N20/N13 amplitude ratio, 147

Occipital area, 178
Oculocephalic maneuver, 190
Oculocephalic movements, 191
Oculocephalic reflexes, 193
Old Testament, 135
Optic fundi, 190
Optic nerve, 178
Oregon death with dignity, 255
Organic integration
    loss of, 51
Organs
    collection of, 35
Organ donation, 19, 116, 133-137
Organ donors, 32
Organ harvesting, 19, 47
Organic function, 16
Organic integration, 52
    loss of, 51
    irreversible loss of, 52
Organ procurement, 210
Organ procurement and transplantation
    network, 207
Organ transplants, 139
Organ transplantation, 115, 133-137
Organism, 53
Organism as a whole, 24, 35, 36, 37, 51, 52
    cessation of, 90
Organism's biological unity, 24
Organism phase, 5
Outcome, 146
    prediction of, 189-195
Organic levels, 92
Ototoxic drug administration, 179
Oscillatory envelopes, 240
Out-of-body experience, 115-131
Out-of-hospital do not resuscitate (DNR)
    orders, 256
Oxidative phosphorylation, 198-204
Oxygen, *see also* PaO_2
    partial arterial tension of 170
Oxygen consumption, 144
Oxygen free radical detoxification, 145

Pain
    affective component of, 218
Pain sensation
    neuroanatomical pathways of, 218
Painful stimulation, 191
$PaCO_2$, *see also* carbon dioxide, 170-173
PaO_2, *see also* oxygen, 170
Palliative intervention, 263
Palliative sedation, 263
Pallium, 216
Pancreas, 208-211
Pancreases, 209
Pancreatic enzymes
    increase of, 210
Pancreatic injury, 210
Pancreatic transplantation, 210
Panhypopituitarism, 35
Paracetamol poisoning, 208-212
Parallel processing, 82
Parietal cortex, 182
Parietal associative cortices, 232
Parietotemporal association cortex, 232
Parkinson's disease, 145, 248
Pattern reversal, 178
Pathophysiological approach, 18, 19, 20
Perception, 232
Permanent vegetative state, *see also* PVS,
    51, 54, 73, 79, 215-226, 239-252
Permanent VS, *see also*, permanent
    vegetative state, 229-236, 100
Persistent vegetative state, *see also* PVS,
    229-236
Persistent VS, *see also* persistent vegetative
    state, 229-236
Peroxynitrite, 144
Person, 53, 55, 57, 80-86
    absolute uniqueness of, 72
    character of, 64
    conscious actualization of, 73
    death of, 46
    dignity of, 62, 71
    destruction of, 62
    essence of, 67
    functionalist view of, 53, 56
    non-substitutability of, 72
    notion of, 65
    ontology of, 64, 65
    qualitative view of, 53
    substantive concept of, 55
Person phase, 6,7
Persons, 57
    species view of, 56
Person's life, 135
Personal dignity
    origin of, 72
Personal identity, 53, 54
    contemporary functionalist theories
    of, 54
    loss of, 55
    memory criterion of, 57
Personal soul, 65
Personal subject, 64
Personal uniqueness
Personality, 82-86
Personality traits, 80
Personhood

continuity of, 84
loss of, 26
functionalist theory of, 55
nature of, 67
Personhood-centered ethics, 85
PET, *see also* positron emission tomography, 231-246
Pharmacologic neuroprotection, 146
Phase-speed, 128-131
Phase-space, 128-131
Phenylephrine, 154
Phenylpropanolamine, 208
Phenomenological approach, 18, 19
Philosophical debate, 15, 16, 20
Phosphocreatine, 197
Phosphorus spectroscopy, 198
Phosphorus MR spectrum, 198-204
Phosphodiesters, 198
Phosphomonoesters, 198
Phosphorus nuclei, 198
Phosphorus magnetic resonance spectroscopy, *see also* ^{31}P MRS, 197-204
Phrenology, 239-252
Physiological activity, 43
Physiological decapitation, 100
Physical examination, 190
Physiological kernel, 24
Piloerection, 31
Pituitary, 34
Placebo group, 147
Poikilothermia, 29
Poison control center, 208
Poisoned patients, 208
POLST, 255-262
Polyradiculopathies, 181
Pons, 181, 198
ponto-mesencephalic level, 181
Positron emission tomography (PET) scanning, 125
Postnatally acquired apallia, 218
Post-anoxic coma, 20, 179
Posterior associative cortices, 232
Posterior hypothalamus, 233
Posthuman persons, 85
Post mortem pregnancy, 52
Potentially reversible damage, 232
Parasympathetic tone, 32
Personality, 54
Permanent vegetative state
Multi-Society Task Force on, 82
Permanent visual impairment, 210
Persistent vegetative state, *see also* PVS, 1, 10, 11
Plantar response, 190
Plantar responses, 28
Pneuma, 98
Poisoned donors, 207-212
Pontine damage, 179
Population behavior
a model of, 241
Potential organ donors, 208
Potential organ recipient, 134
Positron emission tomography, *see also* PET, 231-236
Post-anoxic amnesia, 232
Posterior cingulate cortex, 232
Posthypoxic brain injury, 144
Post transplantation heart performance. 197
Postmortem coldness, 141
Postmortem lividity, 141
Postmortem rigidity, 141
Pre-blastula stage, *see also* pre-embryo stage, 4, 5
Precuneus cingulate cortex, 232
Pre-embryo stage, *see also* pre-blastula stage, 4, 5
Prefrontal association cortex, 232
functional integrity of, 233
Pregnant mother, 135
Pregnant woman, 134
Premotor association cortex
Preserved arousal, 230
Preserved wakefulness, 229
Preoxygenation, 172
Primary asystole, 143
Primary auditory areas, 235
Primary brain stem catastrophe, 166
Primary consciousness, 17
Primary cortex, 20
Primary sensory cortices, 20
Primary somatosensory cortex, 234
Prior person, 83
President's Commission, 45
guidelines of, 45
Problem solving, 230
Propofol anesthesia, 231
Prolonged cardiac arrest, 211
Process
definition of, 2,3
Process-vs.-event debate, 90-112
Programmed cell death, 144
Progressive pressure gradient-induced ischemia, 163
Proteases, 144
Prostacyclin, 157
Prostheses, 81
Prosthetic replacements, 80
Protons, 198
Pulse, 135